CRAGSIDER

PAST ...... *J. Batey & Son Ltd.'s advertisement of c.1905.Probably from a calendar. (Company archives)*

*...... **and PRESENT** Lawson-Batey's Wearsider on trials in 1980. (Walter Fussey & Son)*

150 Years of The Maltese Cross 1840 - 1990

THE STORY OF
• TYNE, BLYTH and WEAR •
TUG COMPANIES

John H. Proud

Tyne & Wear Tugs Ltd.

Front endpaper:
Lawson-Batey's Cragsider *on charter to Tyne Tugs Ltd. in early 1977.*
(Turners (Photography) Ltd.).

Back endpaper:
France Fenwick's Hendon *(2) assisting the Ellerman liner* City of Port Elizabeth.
(Company archives).

By the same author:-

1985 Seahorses of the Tees
The story of Tees Towing Co.Ltd.

1988 North Eastern Record, Vol.I
A Survey of the Appearance of the
North Eastern Railway and the Hull &
Barnsley Railway (in collaboration with others)

First published in Great Britain in 1993 by Tyne & Wear Tugs Ltd.,
27 Market Place, South Shields, Tyne and Wear NE33 1JF

ISBN 0 9522721 0 5

Typeset by Room for Design,
43 High Street, Great Ayton, Middlesbrough, Cleveland TS9 6NH

Printed by Studio Print,
Admiral Chaloner House, Belmangate, Guisborough, Cleveland TS14 7AD

• CONTENTS •

Foreword, by Douglas F. Hamilton		7
Acknowledgements		8
Introduction		10
Notes and Abbreviations		14
Chapter One	The Background	15
Chapter Two	The Lawson Companies c.1840 - 1920	25
Chapter Three	The Batey Companies c.1840 - 1920	37
Chapter Four	Lawson-Batey Tugs Ltd. 1920 - 1990	47
Chapter Five	The Blyth Companies 1887 - 1988	89
Chapter Six	The Crosthwaite Companies c.1854 - 1950	103
Chapter Seven	The Ridley Companies c.1848 - 1989	107
Chapter Eight	France, Fenwick Tyne & Wear Co.Ltd. 1918 - 1977	125
Chapter Nine	Sunderland Towage Co.Ltd. 1900 - 1918	171
Chapter Ten	Anchor Steam Tug Co.Ltd. c.1867 - 1920	177
Chapter Eleven	The Coulson Companies c.1866 - 1925	181
Chapter Twelve	The Brown Companies 1888 - 1930	185
Chapter Thirteen	The Irwing Companies c.1867 - 1934	189
Chapter Fourteen	The Redhead Companies c.1855 - 1929	195
Chapter Fifteen	The Dry Companies c.1857 - 1929	201
Chapter Sixteen	Redhead & Dry's Tugs Ltd. 1929 - 1944	207
Chapter Seventeen	The Lambton Companies 1858 - 1945	211
Chapter Eighteen	The Cook Companies c.1865 - 1947	215
Chapter Nineteen	Clyde Shipping Co.Ltd. 1815 - 1990	220
Chapter Twenty	The Charter Companies 1959 - 1990	221
Chapter Twenty-one	Postscript 1991 - 1992	241

General Appendices:

1. Funnel and Flag colours 243
2. Tug whistle signals 247
3. Origin of tug names 248
4. Numbers of tugs owned 1880 - 1990 249
5. The North East Coast Tugboatmen's Association 250
6. Tyne Tugowners' Association 251
7. Purchasing power of sterling 254

Index of Tug Names, Tug Builders and Tug Engine Builders 255

This book is dedicated to
my wife, Win,
whose understanding
made the writing of this book possible.

• FOREWORD •

by DOUGLAS F. HAMILTON

People find tugs intriguing. Each week we receive letters at our office which ask questions about tugs - tug companies, tug names, tug builders. Which tugmaster? What engine? Which funnel marking? What colour? Which flag? And when; always when!

My initial reaction to John Proud's suggestion that he might write a history of tugs and tug companies on the Tyne was therefore positive, for two reasons. Firstly, I was sure that an authoritative reference book on the subject was needed, and secondly, having read "Seahorses of the Tees" I had no doubts about John's passion for tugs or of his ability to research and write an interesting, well-documented book.

What I had not realised was the enormity of his task, although this soon became clear as his early researches began to reveal the vast number of tugs and companies which were involved, and his investigation spread from the Tyne to cover the Wearside and Blyth areas. I was then able to admire not just John's patient research and close attention to detail, but also his ability to see the wood through the trees, so that a cohesive pattern emerged from the labyrinth of tug family details.

Information was gleaned from all sorts of sources with a quiet persistence; interviews were conducted with patience and a polite good nature. But above all, on reading the drafts the author's good humour and love of tugs shone through the text to produce a readable story and an accurate account of local towage over 150 years.

The North East of England is steeped in marine tradition and folklore. Most families living alongside the three rivers, the Blyth, Wear and Tyne, have members or relatives with some experience of building ships, repairing ships or sailing in them. This book looks at developments in those marine industries over the years from a tug operator's perspective. I believe it will be a worthy addition to the historical records of the region, and a very good read.

D.F. HAMILTON.
Managing Director, Tyne & Wear Tugs Ltd.
Chairman, The British Tugowners' Association.

• ACKNOWLEDGEMENTS •

My first acknowledgement must be to the chairmen and directors of Clyde Shipping Tugs Ltd., Lawson-Batey Tugs Ltd. and Tyne & Wear Tugs Ltd. for their encouragement and assistance given to me in the writing and publishing of this history. I am very grateful for the access to their company minute books, other records and photographs which they allowed. In particular Douglas Hamilton, the managing director of Tyne & Wear Tugs Ltd., has been a great help in reading and criticising the original manuscript notwithstanding his many other commitments during the more than three years it has taken to research and write this book. He has also been kind enough to write the Foreword.

Without exception, the employees of the present day companies with whom I have been in contact have always made me feel very welcome and have answered my many questions without hesitation. Especially I must thank Jean Bell who has typed, and retyped, the text so well and without ever once complaining about the nuisance which I must have been.

Retired employees of the companies have been very helpful, notably Gordon Bell, Bill Rutherford, Hector Stewart (over ninety years of age) and Boswell Whitaker.

Many individuals, societies and organisations have helped in the supply of information. I am most grateful to them. My thanks go to the staff of the Local History Room of South Tyneside Central Library at South Shields for their information and access to their ship photograph collection. Similarly the staff of the Newcastle Custom House were extremely helpful in guiding me through the current Newcastle, North Shields, South Shields, Blyth, Sunderland and West Hartlepool ship registers. I very much appreciate the permission which was given by their London headquarters for me to have access to these registers. Cardiff Custom House also provided useful information by post. Lloyds Registers at Middlesbrough Central Reference Library have yielded much useful information regarding tugs greater than 100 gross tons register. Montrose Reference Library provided information on John Batey's activities at that port in the 19th century.

The older Custom House Registers lodged with the Tyne & Wear Archives Service at Newcastle have been consulted. The efficient service provided by that organisation has been appreciated. Also I am grateful to Adrian Osler, Keeper of Maritime History at Newcastle Museum of Science and Engineering for permission to examine the Coulson Cairns collection of tug photographs.

John and Sheila Vickers were very helpful in reading and correcting my faulty grammar in the early drafts of the text thus avoiding a large number of amendments at a later stage. My World Ship Society colleague Harold Appleyard gave continuous help in digging out pieces of obscure information from his many records always quickly and without complaint. His assistance was very much appreciated.

I am most grateful to Phil Thomas who provided details and photographs from his extensive tug index; Bernard Kent, Les Turner and other members of the North East Tug Society for photographs and other information; Frank Smith, Bookseller, Newcastle for access to Mercantile Navy Lists; the Misses Softley for documents and information about their tug skipper father; Paul Hood and J.W. Heron for information about the Port of Blyth; Mike Hawkins for obtaining information from Cardiff Archives Department and to Ann Los for consulting Hull Archives Department on my behalf.

Those helping with photographic information included Chris Reynolds, Tom Cavanagh, Ken Groundwater, George Scott, Ken Turrell, G.A. Lewis, John Hill, Dr. 'Harry' Harrington, Paul Hood, R.F. Wray, R. Clasper, W.T. Parker, and J. Wilkinson. Dennis Harris was most helpful in reproducing photographic material suitable for publication in this book.

Mrs. Joan Batey kindly provided information about her late husband John Francis Batey and the Batey family in general.

J.L. Loughran provided information regarding funnel colours of some of the tug companies, Peter and Nigel Hogg of Smith, Hogg & Co., Hartlepool allowed me access to their Mercantile Navy Lists and other archival material, and Reg Sowler assisted in obtaining information from Whitby Museum.

The Port of Tyne Authority and the Port of Sunderland Authority kindly provided statistical information regarding trade in their areas of jurisdiction. The German Bugsier and Danish Em.Z. Svitzer tug companies provided information regarding tugs which had previously been owned by them. Bill Robson, currently secretary of the North East Coast Tugboatmen's Association, helped with information, including a video, about his union and its history. Many members of the Teesside and Tyneside Branches of the World Ship Society have helped in numerous ways, each perhaps only in a relatively small way but adding up to a considerable amount of useful information.

There must surely be some people and/or organisations who have not been mentioned above who should have been. My apologies for any such omissions. I would like to thank most sincerely all those who have helped in any way, included above or not, for the massive amount of information which they have provided. Without their help the book would have been less than comprehensive.

Credits for the photographs are given along with the appropriate captions where the photographer or the owner of the copyright is known. Where none is given then the origin of the photograph is not known. Wherever possible an attempt has been made to obtain permission to reproduce photographs and to give due credit. If this has not been done in any particular instance then it is due to the fact that information as to the original photographer is not available to me. It is, therefore, inevitable, particularly with the older photographs, that some credits may have been omitted unintentionally. If this is so then I offer my apologies.

Finally, my thanks must go to my wife, Win, for her strong and continuous support and encouragement during the compilation of this book. Without this the book would never have been written. Her practical help in reading and criticising the manuscript has been of great value, and her immense tolerance of the mental absences and incomplete household maintenance that are the domestic price of authorship have been very much appreciated.

John H. Proud
Middlesbrough 1993.

• INTRODUCTION •

Following the publication of my history of Tees Towing Co.Ltd. in 1985 I was asked frequently 'what are you going to write next?' In answer to the question, apart from saying that I needed a rest after three years of research and writing, my thoughts began to turn to the idea of writing a history of the tug companies based on the River Tyne. The more I thought about the idea the more I wondered why it had not been done before, knowing the great interest which still exists in the Tyne tug companies today. I also felt that an interloper from Teesside was not perhaps the best person to undertake the task.

Thus things rested for some four years until 1989, by which time I had realised that nobody else was contemplating taking on the project, when I approached Tyne & Wear Tugs Ltd. to ask for their support and access to their records and photographs. This co-operation was soon forthcoming and I was away once again on the laborious voluntary task of producing another tug company history only this time many companies and over three times as many tugs were to be involved.

It was not long before I realised why the task had not been attempted before. The story was indeed a complex one due to the interdependence of some of the companies, to the number of companies taken over by others and to the fact that the companies of not only the Tyne but of Blyth and Sunderland also would have to be included if the story was to be fully and properly told. Further complications were that company minute books were not available before 1920 and some not at all and details of the tugs, a great many of which were under 100 gross tons, did not appear in Lloyds Register and had to be sought in Custom House Registers and Mercantile Navy Lists. Some of these latter volumes were not readily available within reasonable distance of home.

No less than 269 tugs were involved and confusion had to be avoided when it was found that in eighteen cases there were two tugs of the same name, in five cases three of the same name, and in one case each four and even five of the same name. Not only that, some of these multiple namings were in existence at the same time. Also sixty-five of the tugs had been owned by two of the companies in this history, thirteen by three of the companies, three by four of the companies and one by five of the companies. It was obvious that a readily understood logic would have to be followed in deciding the layout of the book.

Therefore, each chapter covers either a single company or a related group of companies and has been written so that it is complete in itself thus allowing the reader, if he or she so wishes, to pick out the history of one particular company in which they may be particularly interested. This has resulted in a small amount of repetition between some chapters where take-overs of companies occurred and it is hoped that the reader of the book as a whole will understand the reason for this. Each chapter is complete with its own appropriate fleet list.

This book does not cover the history of all Tyne, Blyth and Wear tug companies or tugowners on those rivers. What it does include is all the companies which by liquidation, take-over, absorption or acquisition have formed part of the 'family tree' which descends to the present-day single company of Lawson-Batey Tugs Ltd. over the one hundred and fifty year period c.1840 - 1990. This 'family tree' and a further diagram showing the interdependence between certain of the companies since 1920 are provided as Appendices A and B to this introduction. It is suggested that the reader should make regular reference to these charts to enhance his or her understanding of the sometimes complex organisation of the companies involved.

The final chapter refers briefly to developments which have taken place during 1991 and 1992.

Where vessel names, printed in italics, are followed by figures in brackets thus (1), (2) etc., this indicates that the vessel is the first, second, etc. of that name in order of acquisition in this history. Figures thus, (118/1974), after a vessel name refer respectively to the gross tonnage and the year of build.

Where values of vessels have been recorded they are in the units of the time to which they apply with decimal equivalents given in brackets immediately following. A graph of the relative purchasing power of the pound sterling from c.1890 to the present time is given in the General Appendix 7 so that, if required, contemporary prices can be compared with today's values.

Items of a general nature which may affect more than one tug company have been included in chapters three or four.

Some aspects of the extensive research which has been necessary to produce this history require comment. A growing number of museums and archive departments will only look up information for a postal inquirer if a prohibitively heavy fee is paid. One such is known at £12.80 per hour of staff time! This charge, in particular when it is open-ended, means that a private individual is unlikely to be able to afford information from anywhere beyond normal travelling distances in his home area. Whilst understanding the need for financial economies in Local Authority departments there is a danger that, if this practice becomes widespread, valuable archives will be inaccessible to many people. For this reason alone I take the opportunity to publicise the matter here.

Many people are mentioned in this book, but even so it is not possible to mention all those one would wish. This is one of the hardest choices to make and I would apologise in advance to those who may feel that they have been overlooked.

Whilst every effort has been made to make this book as accurate and complete as possible it is inevitable that, because of the very nature of the research and because of the dearth of early company records, there will be errors and omissions. I would be very grateful to learn from readers of any such shortcomings.

APPENDIX A TO INTRODUCTION

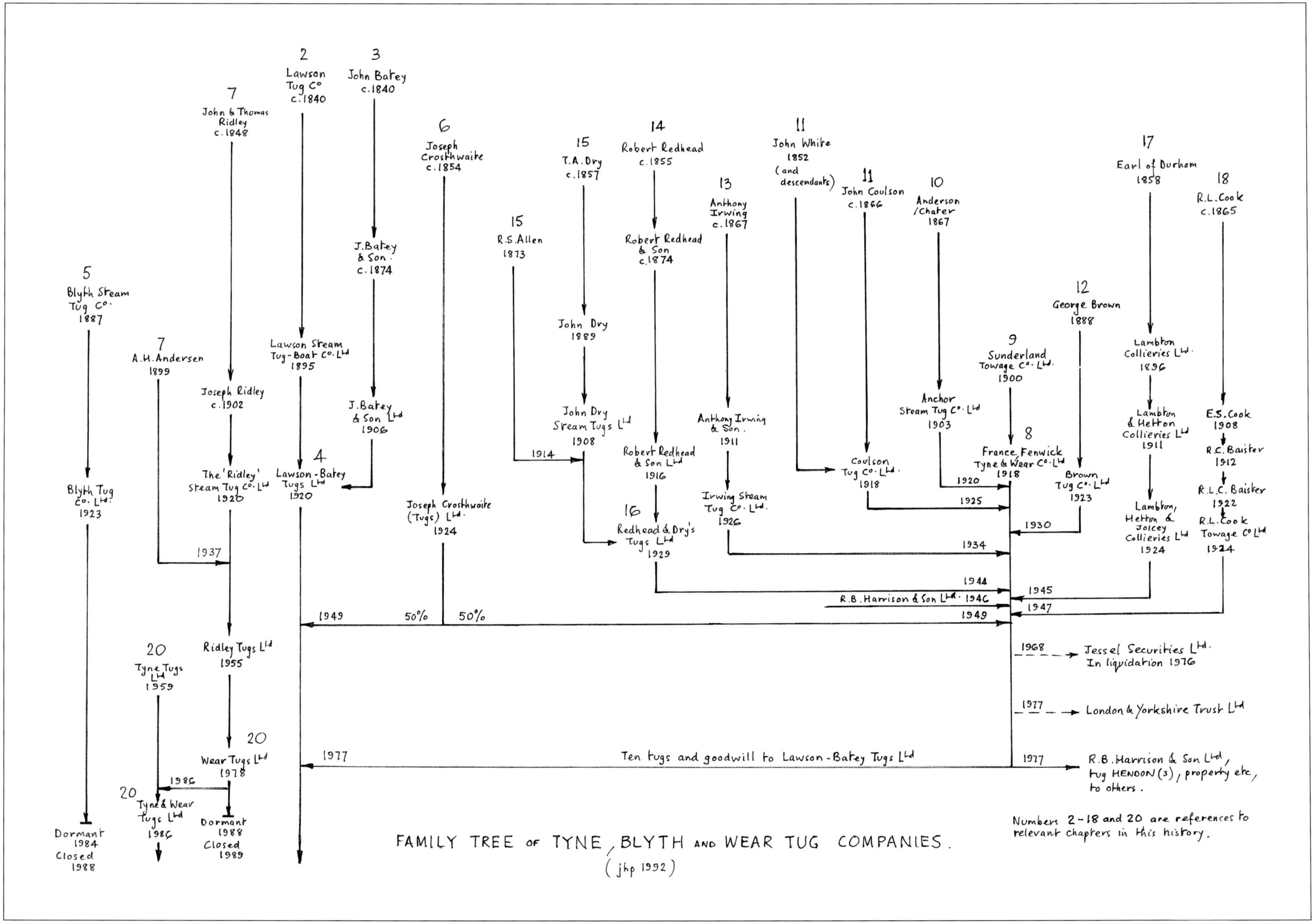

APPENDIX B TO INTRODUCTION

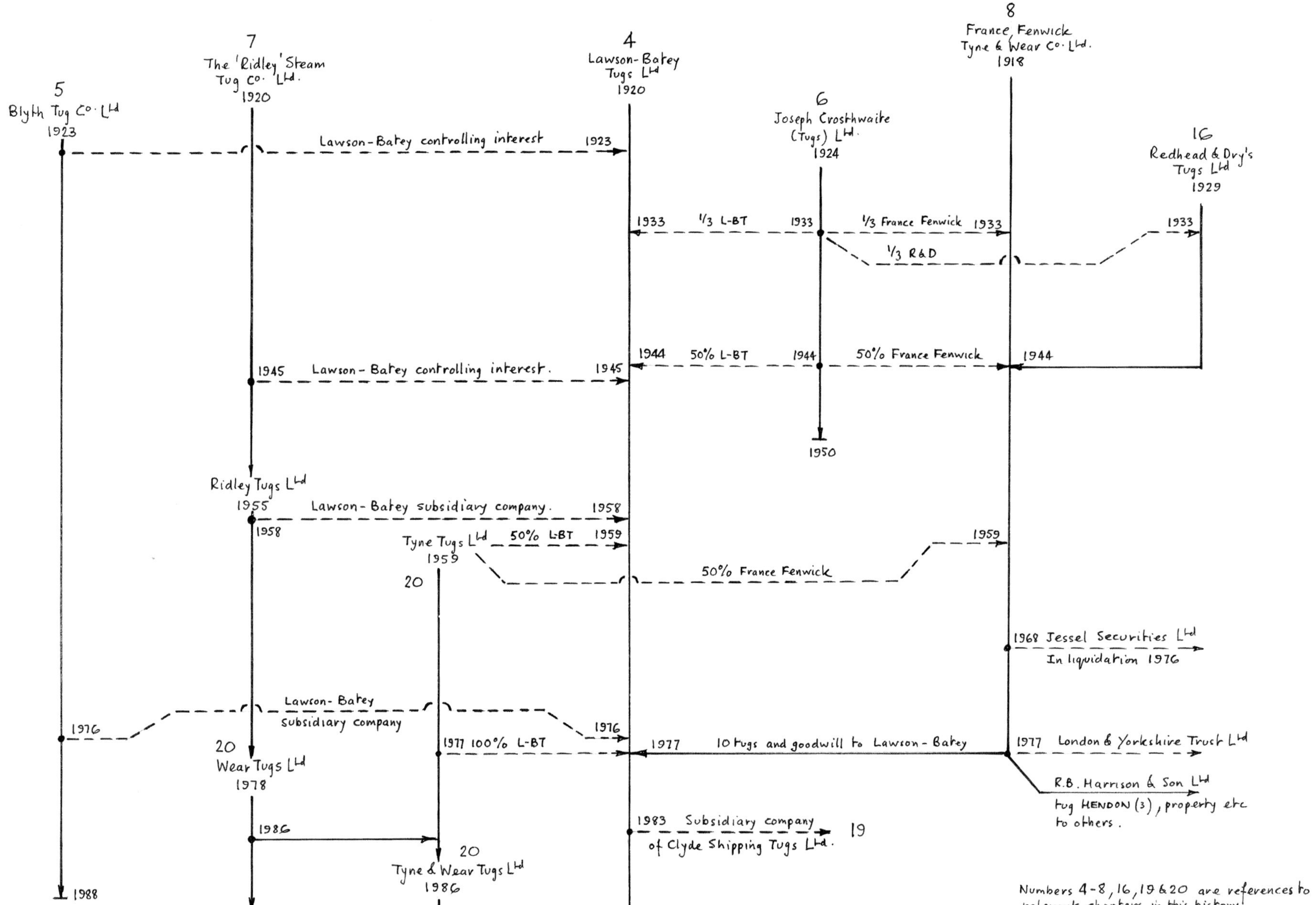

INTER-DEPENDENT INTERESTS
OF TUG COMPANIES
(jhp 1992)

• NOTES AND ABBREVIATIONS •

1) General abbreviations in the text

IHP	Indicated horsepower.
BHP	Brake horsepower.
BP	Bollard pull in tons.
RPM	Revolutions per minute.

2) Key to symbols used in tug ownership chronology charts

●	Built
	Owned (or managed) by company which is the subject of the particular chart
	Owned by other companies
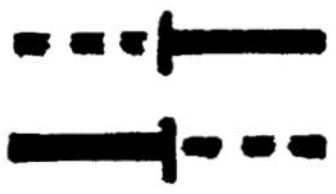	Change of ownership
	Broken up
	Foundered
P	Preserved

3) Explanation of the arrangement of Fleet Lists

In the fleet lists which are given as appendices to Chapters Two to Eighteen the following terminology applies:-

The vessel histories are numbered sequentially in each fleet list, the numbers being preceded by key letters identifying the company concerned. A list of these key letters is given in the first page of the index.

The numbers (1), (2) etc. in brackets after a vessel's name indicate that she is, chronologically in order of acquisition, the first, second, etc. of that name in this history but not necessarily in any one particular fleet.

The dates following the name are the years of service in that particular fleet in which the vessel is listed.

On the line below the name is given the Official Number (ON) in the British Registry followed by the gross tonnage and dimensions. These dimensions are registered length x breadth x depth in feet and tenths of a foot for vessels acquired before mid-1954, overall length x extreme breadth x summer draught in feet and inches acquired mid-1954 to 1974 and metres for vessels acquired from 1975. To the right of the dimensions is given the material of construction of the hull and the method of propulsion.

On the next line is given the name of the builder (B) followed, where known, by the yard number (YN). Below this is given the type and size of the engines (E), their power output and builder. The following abbreviations are used for the engine details:-

Steam	C	–	compound-expansion engine.	
	T	–	triple-expansion engine.	
	OB	–	overhead beam engine.	
	L	–	side lever engine.	
	D	–	diagonal engine.	
Motor	SA	–	single acting	prefixed by 2 for two-stroke or 4 for four-stroke
	DA	–	double acting	
All	Cy	–	cylinders - prefixed by 1, 2, 3, etc. for the number of cylinders	
	NHP	–	nominal	horsepower
	IHP	–	indicated	
	BHP	–	brake	
	BP	–	bollard pull in tons	
	K	–	knots	
	CP	–	controllable pitch propeller	

Next follows the port(s) of registry (PR) with the relevant dates and finally the detailed chronological history of the vessel.

• CHAPTER ONE •

The Background

A contemporary engraving of Perseverance *towing a brig in the River Tyne c.1820.*

Before embarking on the story of the many individual tug companies which have operated in the Tyne, Blyth and Wear areas of north-east England it is necessary to give the reader at least an outline of:-

a) the contribution made by this locality to the birth of the steam tug for ship towage.
b) the development of tugs from that time forward.
c) the geography of the area of the three rivers.

In other words, to put into context the story which follows.

Nevertheless, no attempt has been made to recount the full story under any of these headings, all of which are covered adequately in many other publications. It should also be remembered that it is not possible to be dogmatic about claiming 'the first of this' or 'the last of that' for events which took place so long ago.

The Birth and Development of Tugs and Ship Towage

There are many claims as to which area produced the first practical steamboat and at this distance in time it is difficult to determine which should have the honour. This is no doubt due to the likelihood that similar experimentation was being pursued in different places at home and abroad at the same time in the early years of the 19th century. However, it is known that the River Tyne was to the forefront in the introduction of steam propulsion to vessels of various types.

In February 1814 a steam-driven paddle vessel was built at Gateshead for the carriage of passengers between Newcastle and Shields at the mouth of the river. She has variously been referred to as *Tyne Steamboat* or *Tyne Packet* but whether these were actual names or just a popular reference to what she was is not known. The vessel was first put to work in June 1814 but the venture was not a success, such that in 1817, after a period of idleness, she was sold to Joseph Price, who, very aptly, named her *Perseverance.*

In July 1818 Price had the idea of using his steamboat to tow sailing vessels from Newcastle to the sea. His first attempt came when he arranged to tow Mr. Robson's sailing vessel *Friends' Adventure* which was outward bound from Newcastle. The journey down river and two miles beyond the bar at Shields - a total of thirteen miles - was accomplished in the very respectable time of 2 hours 10 minutes. Thus the concept of tugboats was born and thereafter their use became popular when it was realised that sailing colliers, for example, bound to London from the Tyne could make thirteen round trips per year instead of eight, due to the time which was saved in negotiating the difficult river of that time. Only three years later, c.1821, did ship towage begin to be taken up in other ports and rivers of the British Isles.

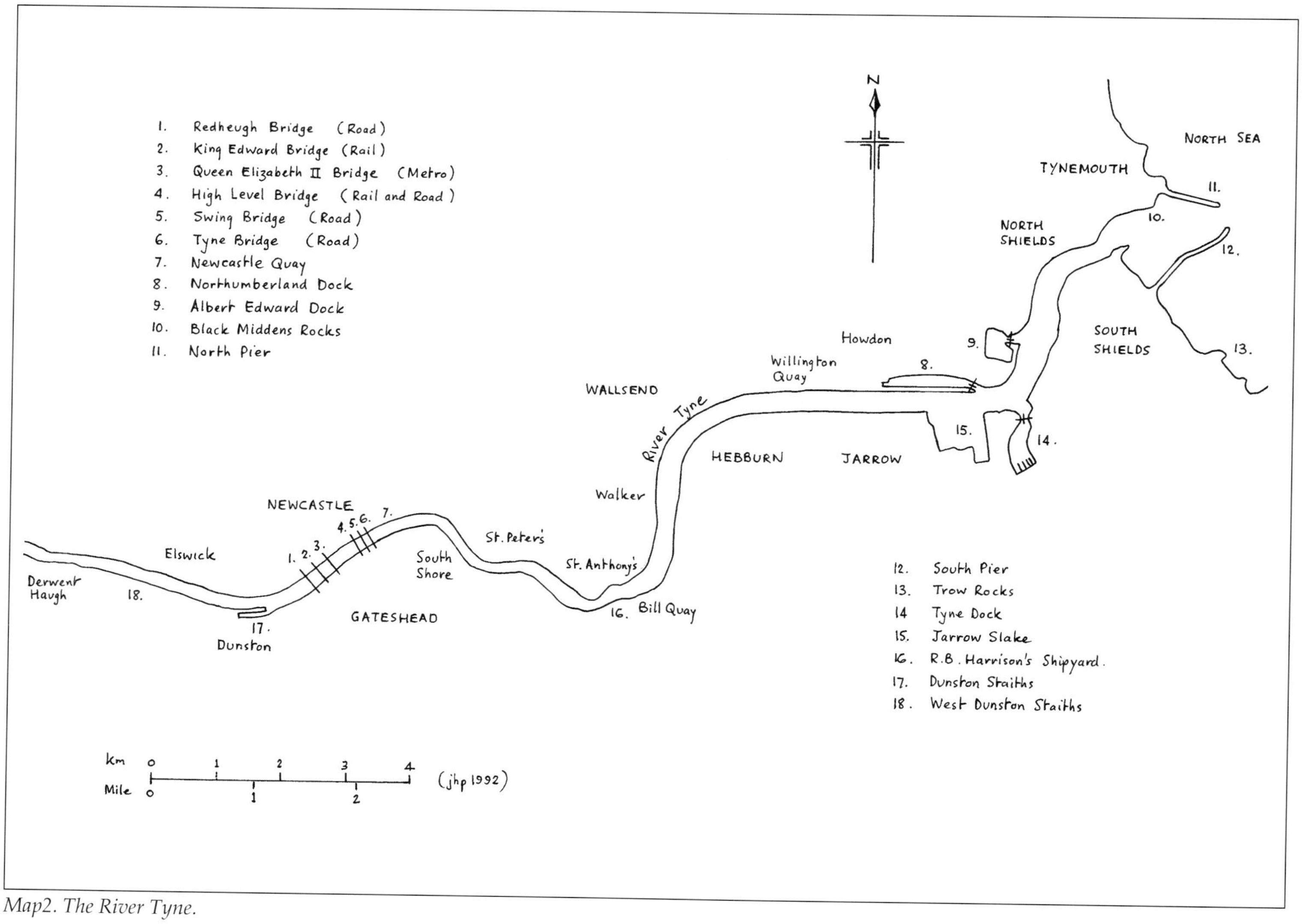

Map2. The River Tyne.

Newcastle-upon-Tyne has always been the trade and commercial hub of the towns bordering the lower reaches of the river Tyne. Although some eleven miles from the sea it was, until modern times, the location of the lowest crossing point for road traffic. Newcastle had been created a town and county by the year 1400 and become a city in 1882. The Custom House for the river and the surrounding area was located in Newcastle until 1848, when Shields (North Shields) was designated a separate port with its own Custom House. Later, in 1865, South Shields was hived off as a port in its own right. Thus the three Custom Houses of Newcastle, North Shields and South Shields remained until recent times, when they and others in the area were combined at Newcastle once again.

The bridges over the river at Newcastle are a major feature of the landscape. Originally there was an ancient stone bridge (said to be the fifth on the site) at river bank level which was removed in 1874-5 and replaced by a swing bridge opened in 1876. One hundred yards upstream of these low level bridges there is the High Level Bridge which is double-decked and carries rail track above a roadway. It was designed by Robert Stephenson and opened by HM Queen Victoria in 1849. A further 900 yards upstream stood the Redheugh road bridge, originally opened in 1871 and reconstructed in 1901. The huge amount of north/south railway traffic through Newcastle necessitated a second railway bridge across the river. This bridge, located 700 yards upstream of the earlier railway/road bridge, was opened by HM King Edward VII in 1906 and is named after him. There was also need for a further road bridge particularly to relieve congestion over the low-level swing bridge. Not until 1928 was the huge Tyne Bridge opened by HM King George V, this being located 130 yards downstream of the swing bridge. In 1981 a further bridge was added to the collection to carry the new Metro Rapid Transit System over the river between Newcastle and Gateshead. The bridge, located between the two earlier railway bridges, was opened by HM Queen Elizabeth II and bears her name. Finally the Redheugh road bridge of 1871 was replaced by a completely new structure which was opened by HRH The Prince of Wales in 1983. Thus there are today six fine bridges within a distance of less than three-quarters of a mile - Tyne (road), Swing (road), High Level (rail and road), Queen Elizabeth II (Metro), King Edward (rail) and Redheugh (road).

An engraving showing early steamboats on the River Tyne and the stone bridge at Newcastle.

The stone bridge was obviously a barrier to ships of any reasonable size getting further upstream than Newcastle. This problem was relieved from 1876 when the Swing bridge was brought into use. The remainder were all relatively high bridges, with over 80 feet clearance at high water, but masters and pilots of vessels had to be wary at high tide times to make sure that they dodged the bridge piers which had a minimum width of 96 feet. They did not always succeed!

Until the middle of the 19th century the town of Newcastle, through its civic authorities, had been in control of the river much to the chagrin of the people of Gateshead, Tynemouth, Shields and elsewhere. The river was in a poor state, being very narrow in places, containing many shoals and rocky outcrops and being fordable near the bar and at Newcastle at low tides. Because of this discontent a bill was laid before parliament which after due consideration, received the Royal Assent in 1850 as the River Tyne Improvement Act. Arising from this act a body named the Tyne Improvement Commission (TIC) was established, comprising, as well as people with maritime and river interests, representatives of the major towns along the banks of the navigable portion of the river so that many bodies had a say in the management, maintenance and development of the river. Their jurisdiction was from the mouth of the river (later the pier ends) to the Hedwin Streams, a distance of some nineteen miles.

One of the first tasks of the TIC was to begin the construction of piers at the river entrance. Started in 1855-6 both were completed in 1895. The length of the north pier was 2,950 feet and that of the south 5,150 feet with a clear space of 1,180 feet between the ends, on each of which stood a lighthouse. Work was also in progress to widen and deepen the river up to Newcastle bridge. By 1864 there was 17 feet of water at low tide at the bar, and the river was being used by 18,410 ships per year, of which 85% were of less that 300 net tons and 12% were steam powered.

Two enclosed docks had been completed, namely Northumberland Dock on the north side of the river opened by His Grace the Duke of Northumberland for the TIC in 1857, and on the south side, Tyne Dock, constructed for the North Eastern Railway Co. (NER) and opened in 1859 by Lord Ravensworth. Both of these docks contained a multiplicity of staiths for loading coal into colliers. Coal was of prime importance to the prosperity of the river and the surrounding area and the huge amount shipped out of the river is illustrated in the graph which is given as Appendix C to this chapter. The work available for the tugs was, of course, significantly affected by the variations in the amount of coal and coke traffic, which by 1860 had reached some four million tons per year. Nevertheless, the elements (as always) played a part in the maritime scene in that same year when Shields harbour was completely blocked by ice.

Improvements to the river and its surroundings continued apace. Some of those notable events were:-

1874 Whitehill Point Nos. 1 and 2 TIC staiths opened.

1880 60% of vessels trading to the Tyne were steam-powered.

1884 Albert Edward Dock on the north side of the river was opened by TRH The Prince and Princess of Wales.

1888 13,645 vessels cleared the river of which 3,570 were under 300 net tons.

1891 Whitehill Point Nos. 3 and 4 TIC staiths opened.

1893 NER Dunston staiths opened. They remained in use until 1980.

1897 A 300-feet section of the north pier swept away in a storm. Re-construction was completed by 1909.

1903 Dunston staiths were extended by construction of inside berths.

1904 Whitehill Point No. 5 TIC staith opened.

1920s Twenty ships per day were passing through the Newcastle Swing Bridge.

1922 Jarrow Slake fuel-oil berth opened.

1923 London and North Eastern Railway Co.'s (LNER) West Dunston staiths opened having been completed in 1914.

1928 Tyne Commission Riverside Quay opened at Albert Edward Dock. Extended 1937.

1932 TIC Howdon staith opened.

1936 Jarrow staiths opened by HRH The Duchess of York.

1937 Tyne Dock acquired by TIC from LNER.

1942 Sutherland Quay (Riverside) at Tyne Dock opened.

1953 Tyne Dock Iron Ore Quay came into operation. Closed 1974.

1966 Car Ferry Terminal at Whitehill Point opened.

1968 Port of Tyne Authority constituted. Tyne Improvement Commission dissolved.

1985 Tyne Coal Terminal opened by HM Queen Elizabeth the Queen Mother on land reclaimed from Jarrow Slake at Tyne Dock. All other coal shipping points were closed.

As well as the number of ships trading to the river the number of vessels launched from the large number of shipyards had a considerable influence on the amount of work available for the tug companies. The graphs, Appendices D and E to this chapter, illustrate the change of these factors over the years. It will be noted that these diagrams, together with that related to coal shipments, show both dramatic rises and falls. This volatile pattern, as will be seen, is mirrored in the fortunes of the tug companies whose histories are related in this volume.

The River Blyth and the Port of Blyth (Maps 1 and 3)

The town and port of Blyth is situated in Northumberland at the mouth of the River Blyth which enters the sea approximately nine miles north of the Tyne. For the last mile or so before flowing into the sea the River Blyth turns south and is separated from the ocean by only a narrow neck of land and a long man-made pier. It is this portion of the river which forms the port of Blyth, with the town of the same name standing immediately to the west of the harbour.

Like the Tyne, Blyth has been involved predominantly with the shipment of coal. In fact, in 1961 it was described as 'the largest coal port in Europe' when 6.9 million tons/year were leaving the harbour. As elsewhere, coal shipments have varied dramatically over the years and this is illustrated in the graph provided as Appendix F to this chapter. Coal had been exported from Blyth since mediaeval times albeit probably from small quays further upriver than from the area of the modern port.

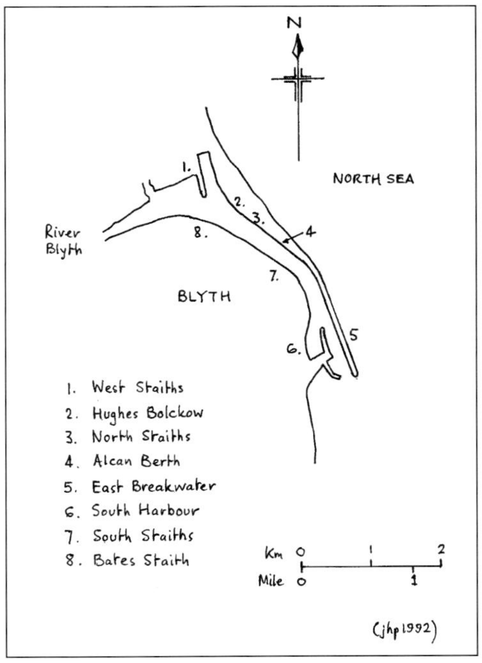

Map 3. Blyth.

Appendix G illustrates the number of vessels launched at Blyth over the years.

Notable events concerning the port were:-

1765	First breakwater (North Dyke) built of loose stones.
1788	First coal staith opened (south side).
1819	Experiment with a steamboat used as a tug. This vessel belonged to the Tyne; perhaps she was *Perseverance*.
1822	Stonework east breakwater built.
1848	Up to this time, Blyth had been part of the port of Newcastle and now became part of the port of Shields.
1849	Blyth & Tyne Railway Co. opened south side staiths.
1854	Blyth Harbour & Dock Co. formed.
1856	West pier commenced.
1868	Cowpen north staiths opened.
1882	Blyth Harbour Commission created as a Public Trust.
1885	New west breakwater built.
1886	East breakwater extended by 300 feet.
1888	Additional staiths opened on south side of harbour by NER.
1896	NER north staiths opened.
1897	Blyth constituted a port in its own right.
1908	East breakwater extended a further 900 feet.
1911	Hughes Bolckow shipbreaking yard opened.
1915	Work on NER west staiths began but not completed by the LNER until 1926. First shipment 1928.
1934	Bates staith opened.
1964	South staiths closed.
1966	Shipbuilding ceased.
1968	North side Cowpen staiths closed after 100 years of operation.
1970	Alcan (UK) Ltd. berth in use.
1972	Remainder of north staiths closed.
1980	Hughes Bolckow shipbreaking yard closed.

The River Wear and the Port of Sunderland (Maps 1 and 4).

The River Wear is situated about eight miles south of the River Tyne and the city and port of Sunderland covers both north and south banks of the river near its mouth. Though now part of the County of Tyne and Wear since 1974 Sunderland was, before that, in County Durham. The city, a town until 1992, has with justification often been referred to as 'the largest shipbuilding town in the world' - even as long ago as the mid-19th century. Nevertheless the claim surprises many, when they see the narrow and twisting layout of the River Wear, with shipyards wedged into odd corners here and there in what appear to be impossible sites. Like the Tyne at Newcastle, the Wear runs through a steep-sided valley which leaves only small areas for the development of riverside enterprises. In the past numerous buoy berths for the colliers awaiting loading added to the congestion.

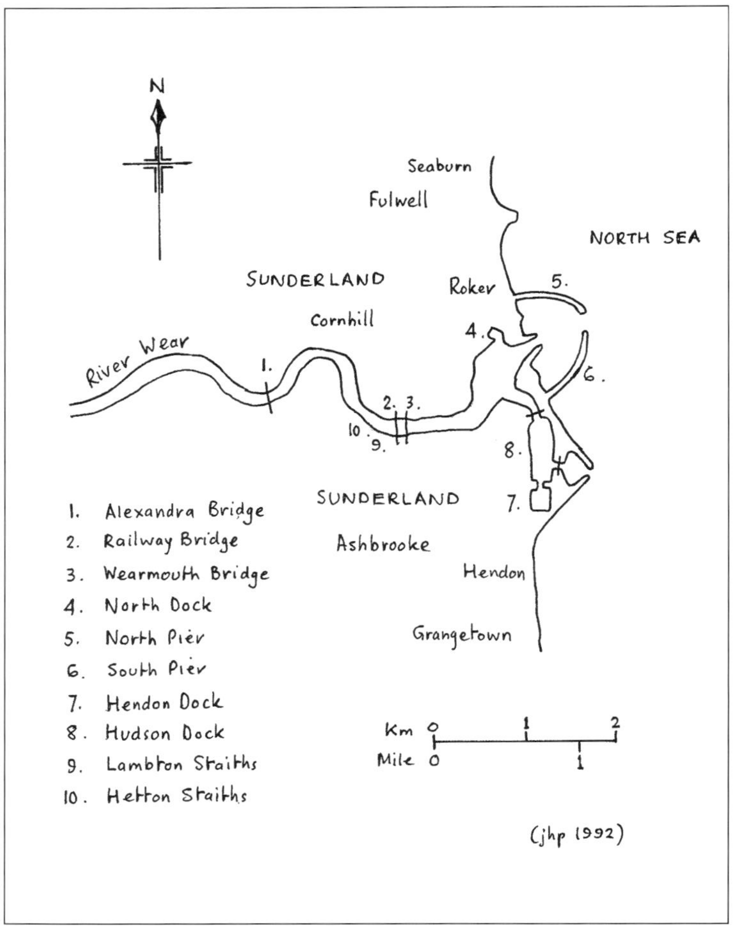

Map 4. Sunderland and the River Wear.

As on the Tyne and at Blyth, Sunderland's major maritime activity was the shipment of coal; first recorded in 1396. Even before that there was reference to shipbuilding in 1346

Entrance to the River Wear.

and maritime commerce in 1154, when the ancient name for the Wear was Vedra.

The work available for the tug companies was very much related to fluctuations in the level of coal shipments, to the number of ships trading to the port and to the number of ships launched. The graphs (Appendices H, I & J to this chapter) illustrate these changes.

Notable events on the River Wear were:-

1726 Inner south pier completed.
1796 Original iron Wearmouth bridge completed.
1802 Inner north pier completed.
1837 North Dock opened by Wearmouth Dock Co. Later owned by railway company.
1850 Enclosed Hudson Dock opened on south side of river. Extended 1855.
1859 River Wear Commission incorporated. Jurisdiction was 9 miles upstream from the mouth of the river. Later became conservators and owners of docks and quays.
1859 Wearmouth bridge re-built by Robert Stephenson.
1868 Hendon Dock opened.
1879 Railway bridge opened. This was a hog-backed iron girder bridge and was located a very short distance upstream of Wearmouth bridge.
1885 Building of north outer pier commenced. Completed in 1903. Opened by Earl of Durham.
1893 Building of south outer pier commenced. Worked stopped 1907. Completed in 1914.
1909 Queen Alexandra bridge completed. Double deck with rail over road. One-and-a-quarter miles upstream of 1879 railway bridge and two-and-three quarter miles from pier heads.
1922 North Dock acquired by River Wear Commission.
1928 Twelve major shipyards on the river.
1929 New Wearmouth bridge opened by HRH The Duke of York. Replaced the 1859 iron bridge.
1934 Corporation Quay opened.
1960 Seven major shipyards on the river.
1972 The jurisdiction of the port and river was taken over by the Borough of Sunderland and named the Port of Sunderland Authority.
1979 The port is now jointly administered by the Borough of Sunderland and Tyne and Wear County Council.
1986 Coal shipments ceased after almost 600 years. Last cargo was 6,578 tonnes for the Thames in *Rora Head* on 31st October 1986.

APPENDICES TO CHAPTER ONE

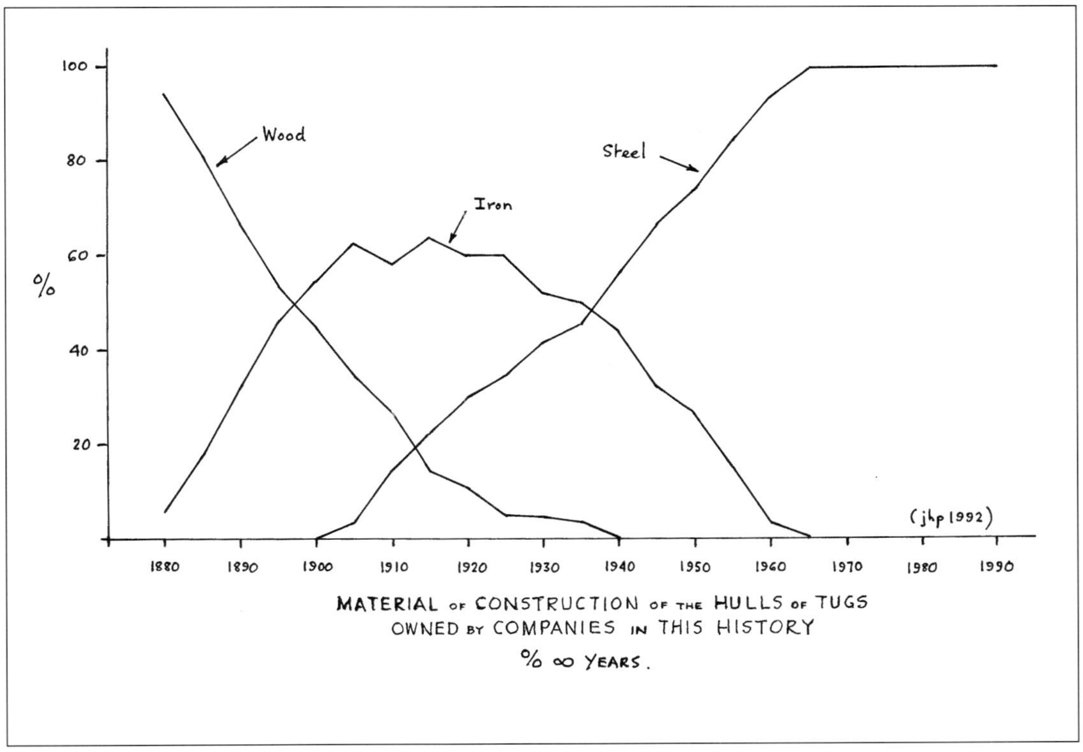

Appendix A to Chapter One

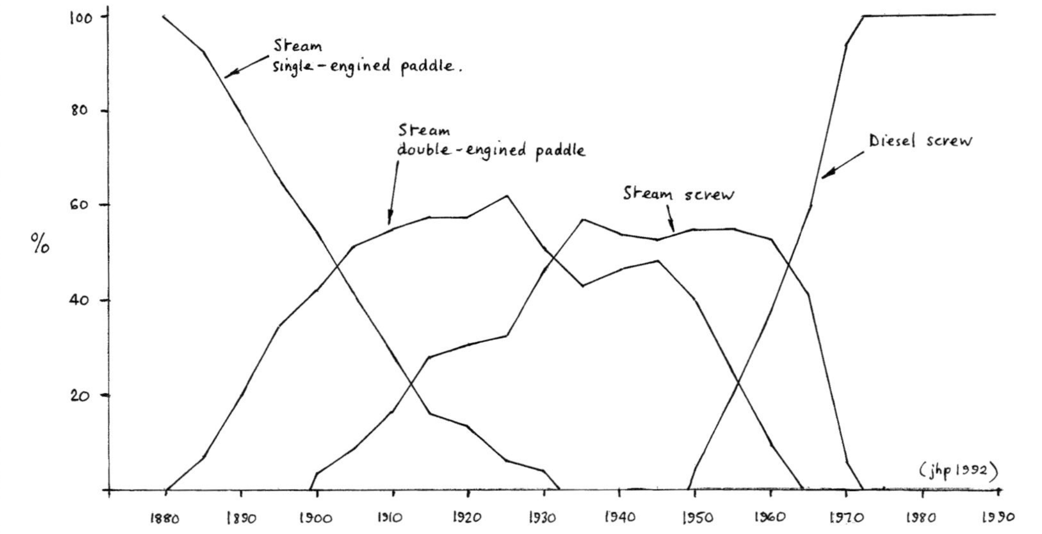

METHOD OF PROPULSION OF TUGS OWNED BY COMPANIES IN THIS HISTORY

% ∞ YEARS

Appendix B to Chapter One

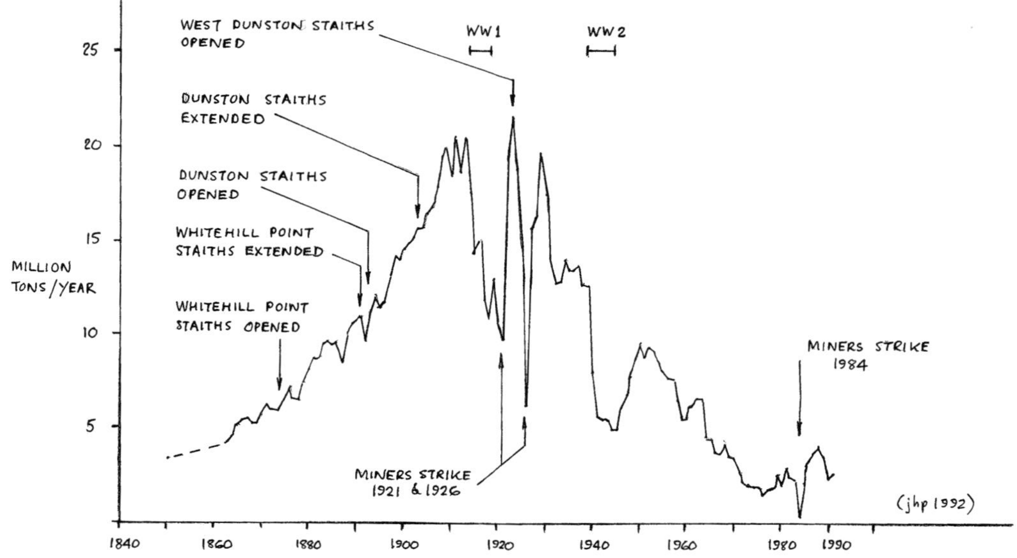

COAL AND COKE SHIPMENTS FROM THE TYNE

Appendix C to Chapter One

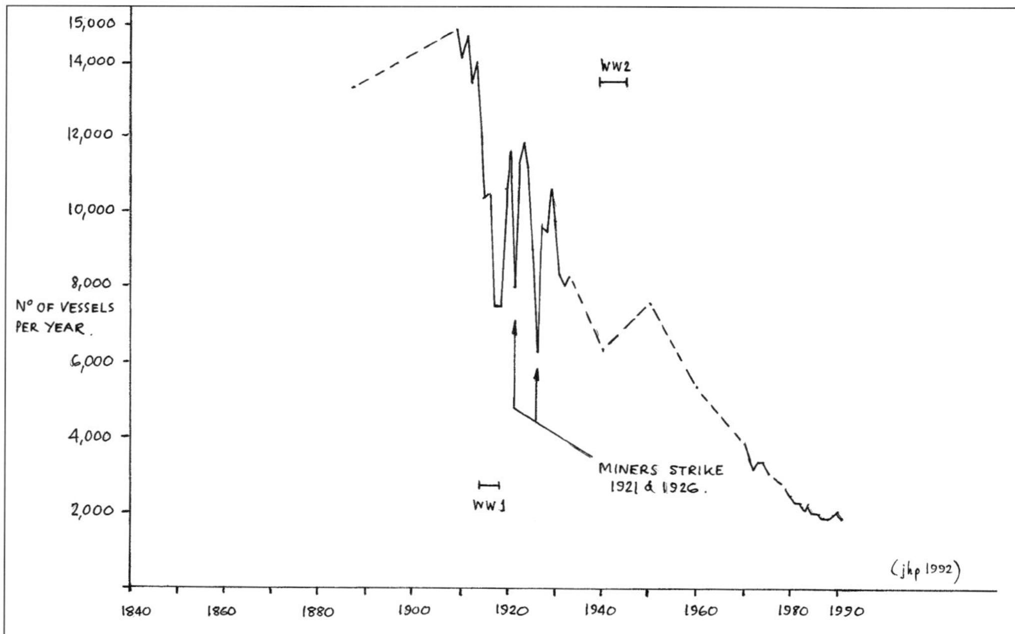

NUMBER OF VESSELS TRADING TO THE TYNE

Appendix D to Chapter One

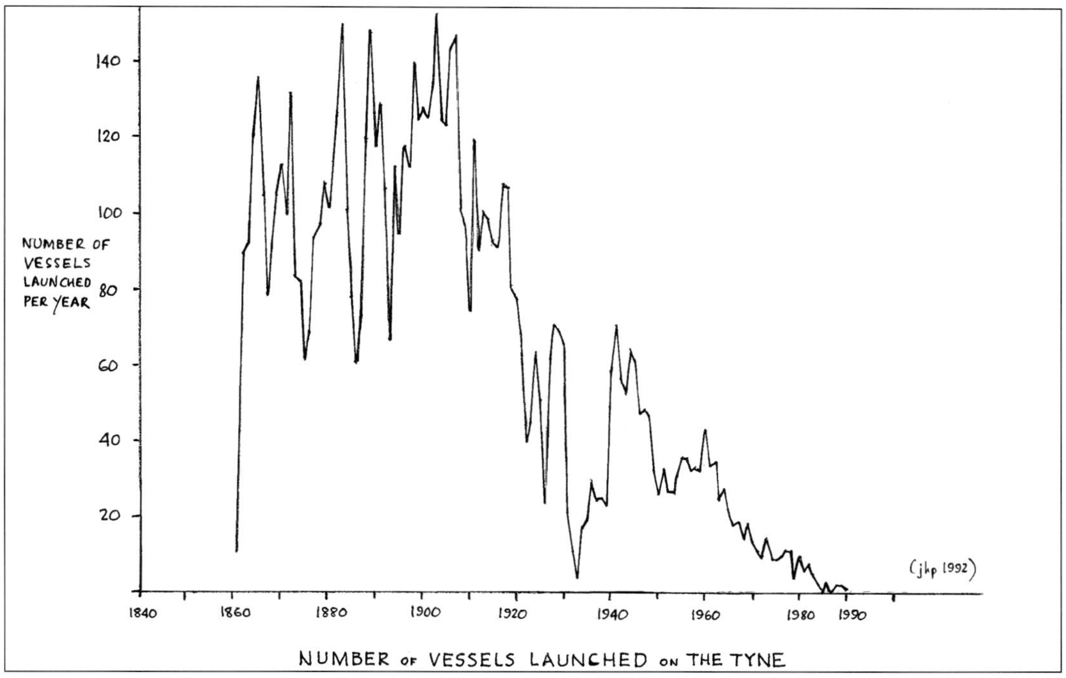

Appendix E to Chapter One

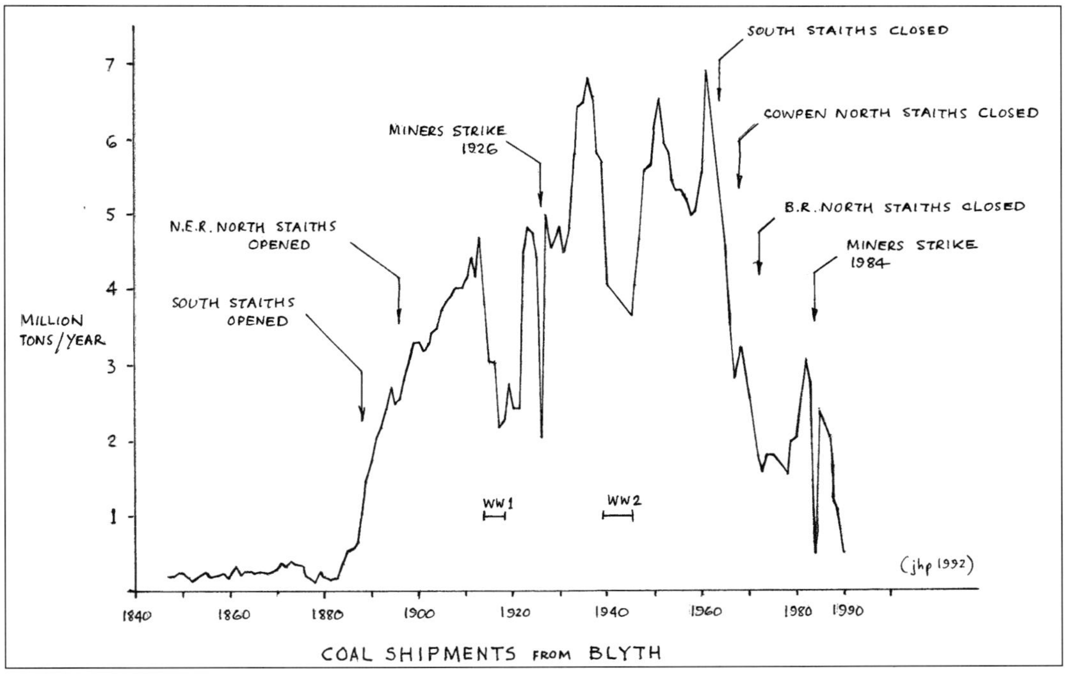

Appendix F to Chapter One

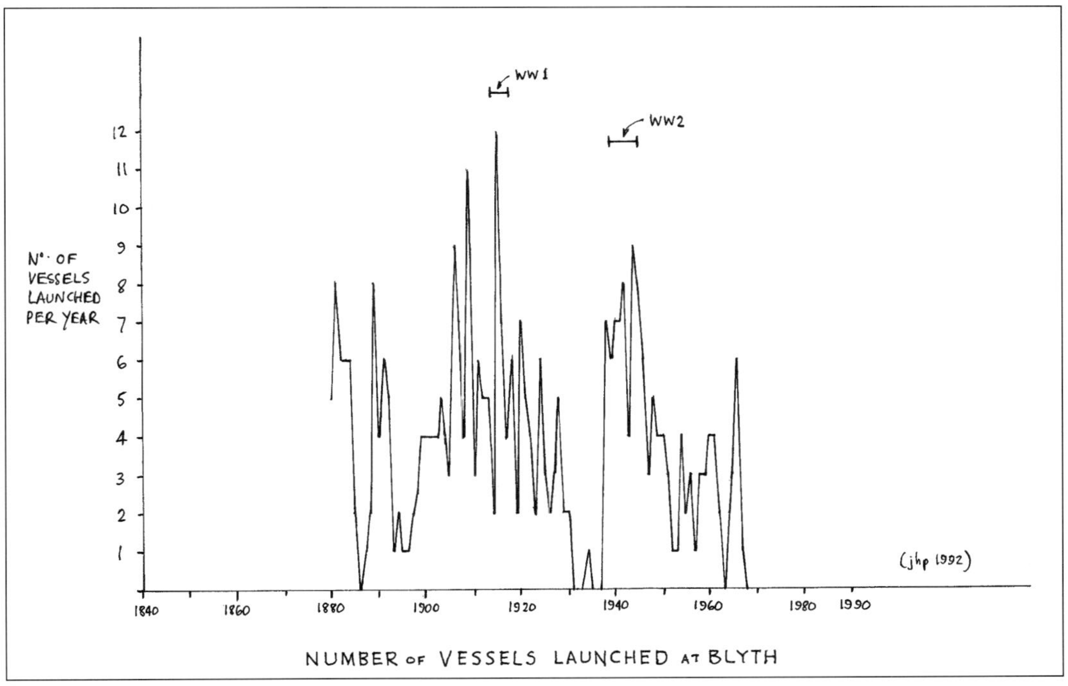

Appendix G to Chapter One

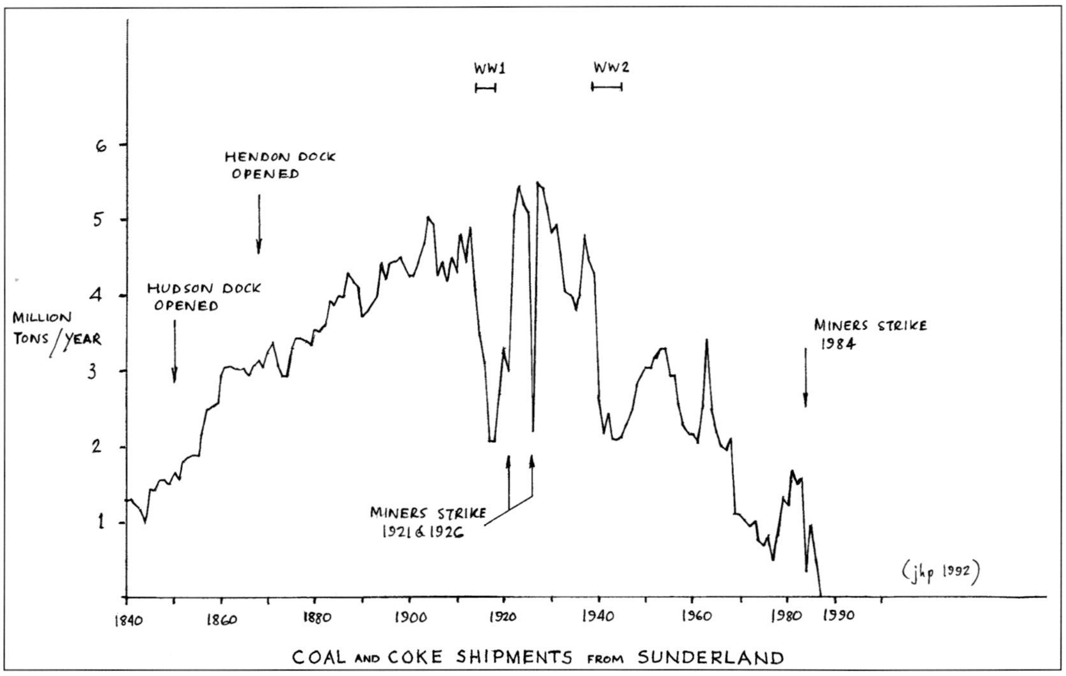

Appendix H to Chapter One

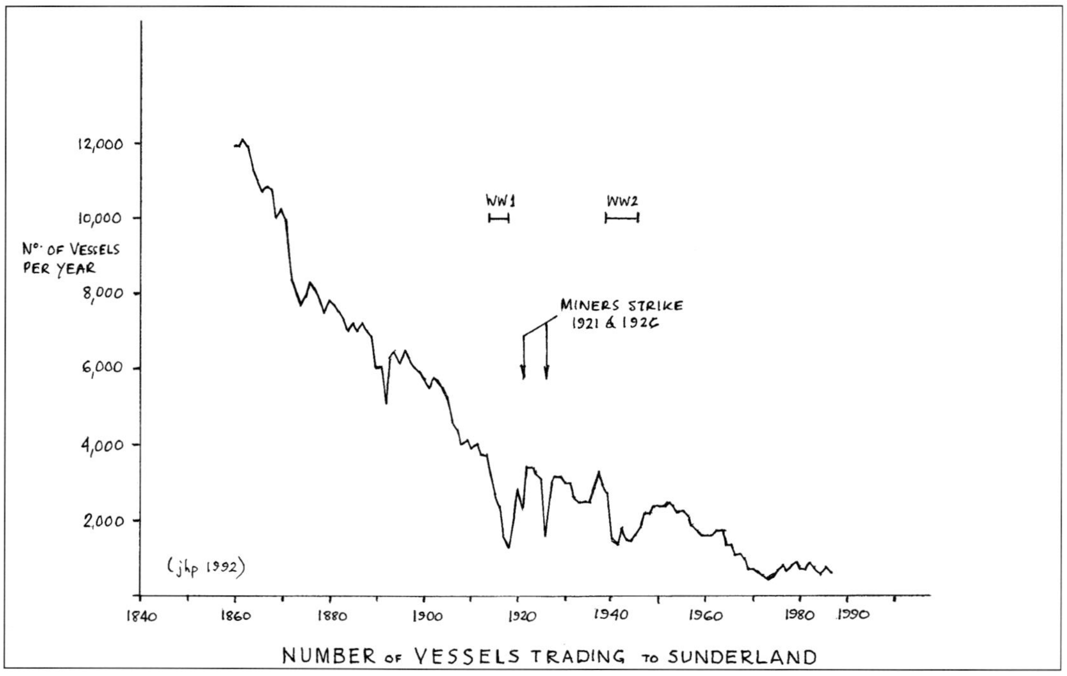

Appendix I to Chapter One

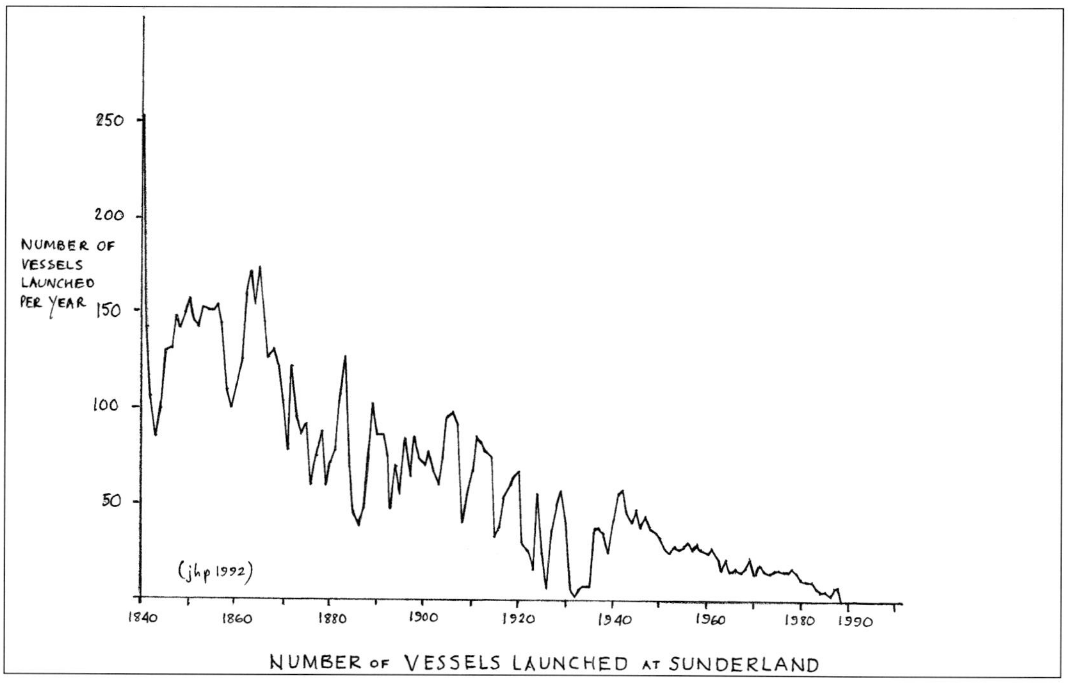

Appendix J to Chapter One

• CHAPTER TWO •

The Lawson Companies

Lawson Tug Company. c.1840 - 1895

In a small booklet, published in the early 1930s and written by Frank Batey managing director of Lawson-Batey Tugs Ltd., it is stated that John Robert Lawson (1822-1901) commenced business on the River Tyne as a tug-owner about 1840 with a small wooden-hulled paddle tug. This vessel is further described as having a single cylinder engine of about 18" diameter and that she was steered by a tiller at the stern. Dimensions of the hull are not given but it can be deduced that tugs of that time must have been quite small and very vulnerable to bad weather. Nevertheless, they did manage to carry out a very useful job, not only in the confines of the river but also outside in the open sea. Unfortunately no details of this first craft have been found in the local registers.

It is known that J.R. Lawson owned, or part owned, the wooden paddle tug *Expert* (52/1856) in 1861 and that his son, Benjamin Dain Lawson (1848-1943), was her skipper in 1867 after having begun work in her in 1862 as fourth hand. At that time there is mention of the ownership of some tugs being shared with Joseph Lawson, who it is believed was another son of John Robert Lawson. It has proved very difficult to follow the Lawson family descent since the same forenames and similar initials were often repeated. About 1865 the partnership of John Robert Lawson and Joseph Lawson purchased the wooden paddle tug *Champion* (1) (61/1858), holding 56 and 8 sixty-fourth shares respectively. Both of the tugs the Lawsons now owned were of a similar size at 75-80 feet long and 16 feet beam. At the age of nineteen, B.D. Lawson joined the partnership in 1867.

A slightly larger paddle tug was purchased in 1869 from Joseph Honeyman of North Shields. She was *William* (2) (80/1865) and was to stay with the Lawsons for about twenty-four years. She was jointly owned by John R. Lawson and B.D. Lawson, the latter at some time also being her skipper. A painting of *William* (2), dated 1873, shows her with funnel colours of a black diamond on a light background with a

William *(2) towing the brig* Brotherly Love *into the Tyne entrance in 1873.* *(Company archives).*

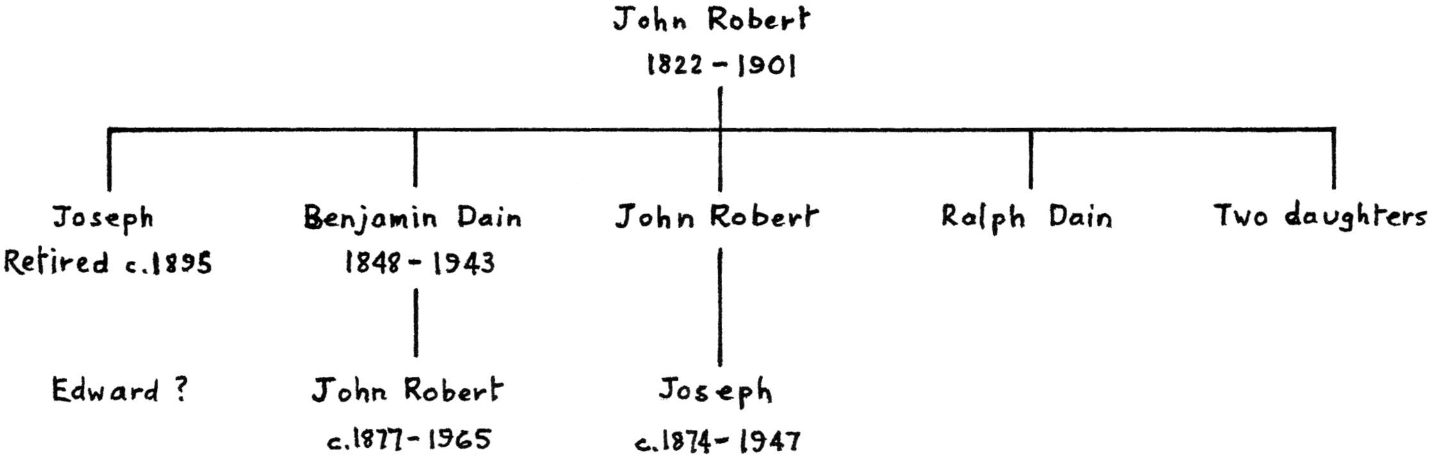

Lawson Family Tree.

black top. Whether Lawson had used this marking any earlier is not known, but this colour scheme with buff as the background to the black diamond was used throughout the rest of the company's existence to 1920 and indeed the company was often referred to as the 'Black Diamond Line of Tugs'. This black diamond also appeared on the paddle boxes. As will be seen in the Lawson-Batey chapter, the funnel mark was carried through for many years into that company's period of existence. In 1871 *Expert* was sold to London owners.

In 1874 *Champion* (1) was sold for further service at Sunderland her place being filled by another wooden paddle tug, *Lion* (75/1858), purchased from North Shields owners and registered in the names of all three Lawson partners. *Ganges* (69/1863) was bought in 1875, again from North Shields owners, in the names of John Robert and Joseph Lawson, and in the following year *Patriot* (91/1867) and in 1878 *Powerful* (89/1865) joined the fleet, which then numbered five tugs. Interestingly, *Patriot* had a hull built of iron. She had previously been in the ownership of William Rundle of Cardiff. *Powerful* had, when built, been owned in South Shields for two years but in the interim had Leith-based owners. About 1879 *Lion* was sold to G. Harker of North Shields, and *Scotia* (1) (78/1868) was purchased. This latter wooden paddle tug had been built and engined at North Shields and was powered with a single-cylinder side-lever engine. Her cylinder dimensions were 32" diameter x 48" stroke and her hull was almost 83 feet long.

In 1881 *Crusader* (67/1875) was purchased from C. Havelock of Sunderland and *Lizard* (82/1869) was bought from C.W. Christie of Cardiff. Both had wooden hulls. About the same time, *Patriot* was sold for further service at Scarborough. *Wild Rose* (75/1874) joined the fleet in 1882 from Gloucester owners, and in the same year *Vivid* (89/1863) came to the Tyne after service on the River Thames and at Kings Lynn. *Vivid* spent a very short period on the north-east coast, being sold to the Irvine Harbour Trust in 1883. *Monarch* (99/1871), *Guide* (104/1869) and *Scotland* (77/1877) were all purchased by the Lawson partnership in 1883. By comparison, *Guide* was a very large wooden paddle tug being 100 feet long x 20 feet beam and much bigger than anything else previously owned by Lawson. The fleet had by now reached ten in number:-

Vessel name	Gross tonnage/year	Hull construction	Means of propulsion
William (2)	(80/1865)	Wood	Paddle
Ganges	(69/1863)	Wood	Paddle
Powerful	(89/1865)	Wood	Paddle
Scotia (1)	(78/1868)	Wood	Paddle
Crusader	(67/1875)	Wood	Paddle
Lizard	(82/1869)	Wood	Paddle
Wild Rose	(75/1874)	Wood	Paddle
Monarch	(99/1871)	Iron	Paddle
Guide	(104/1869)	Wood	Paddle
Scotland	(77/1877)	Wood	Paddle

- with an average age of just over thirteen years. Of the fifteen tugs owned to date it is notable that all but one of them, even though some were purchased from other areas, had been built on the Tyne.

In 1883 *Crusader* was sold to owners at Rye and *Scotia* (1) was lost off Scarborough on 29th June 1884. At 2am. on 24th June 1885 *Guide* blew up at the Penny Ferry landing, South Shields, and one man, Richard Spoors, was killed when he was blown into the beam of the engine by the force of the explosion. The tug ended up, tilted over, on Alum House Ham. Her master at the time of the incident was William Stewart. Badly damaged though she was, she was later sold to J. & D. Morris of Newcastle who converted her into a screw-propelled vessel and put the tug back into service until she was finally lost in 1897.

In 1885 the iron paddle tug *Ulysses* (118/1874) was purchased from Thames owners and was to stay with the Lawson company and its successors until 1937. A serious fire occurred on *William* (2) on 25th June 1886 whilst she was moored overnight at her berth as South Shields - a not uncommon event on wooden-hulled tugs. She was taken to Eltringham's Quay where the fire was fought by the tug *Great Britain* (1). Even though her sides were almost burnt through *William* was repaired and put back into service. *Powerful* was sold, about 1886, to W. Minto of South Shields who continued to use her on the River Tyne until 1893 when she went to Alloa. The iron paddle tug *Cruiser* (1) (98/1874) joined the fleet from Gravesend in 1887. Like *Ulysses,* she was to give long and useful service before being sold for scrap in 1948.

By the 1880s it is known that the Lawson partnership was using 27 Baring Street, South Shields as its company address. Whether this really was an office, a private residence, or both is not known. Previously it was very likely that private residences were used in Bath Street and Winchester Street, South Shields.

On 22nd March 1889 the tug *Scotland* was reported missing having been last sighted four miles ENE of the Tyne entrance on the night of 20th March during a severe NE gale. To the great relief of all concerned, Captain Hopper and his crew arrived home on the morning of 23rd March - by train! *Scotland* had been out 'seeking' before the onset of the severe weather and had tried, along with other tugs, to return to the Tyne but was unable to make her way across the bar. She was forced to turn seawards again and was being completely swamped by the waves. Every moveable object was swept away, the paddle casings smashed and the bulwarks carried away. She was then about twenty miles east of Blyth and settling fast with little hope of the crew surviving in a small boat. Fortunately, the steamer *Libra,* on a voyage Aberdeen to London, appeared on the scene and was able, albeit with difficulty and considerable risk, to rescue the exhausted tug crew. *Scotland* foundered shortly afterwards. The rescued men were landed at Gravesend from whence they were able to make their way home.

Lizard was sold to Middlesbrough owners in 1889 and in 1890 Joseph Lawson purchased the North Eastern Railway Co.'s wooden paddle tug *Tyne Dock No.1* (77/1859) and renamed her *Scotia* (2). Strangely, although built in 1859, she was not registered at the Custom House until 1881. Why this was so is not known, but it is possible that, if she had only worked within the privately-owned Tyne Dock, registration would not have been necessary. *Wild Rose* was sold in 1891 and was replaced by *Tyne Dock No.2* (108/1866) again bought from the NER. Like *Tyne Dock No.1* she had not been registered until 1881. The Lawson company immediately renamed her *Champion* (2).

About 1892 John Robert Lawson Jnr., the son of B.D. Lawson, joined the company and he was to serve throughout the remainder of its independent existence and into the

Lawson-Batey era until retiring in 1944. *Ganges* left the fleet in 1892 and on 12th January 1893 the relatively large iron paddle tug *Hercules* (145/1877) was purchased from the Hull Dock Co. As will be seen later she was to have a tragic end to her service on the Tyne, in 1931. *William* (2) moved in the other direction in 1894 when she was sold to F.T. Munkman of Hull for further service.

A complaint was made in the press against the Tyne tugowners that, when on 22nd December 1894 the brigantine *Northern Star* of South Shields was off the bar for two hours on the eve of the great gale, there was no tug to be had. The sailing vessel had, in consequence, been out in all the subsequent terrible weather and had only been able to return on 20th January 1895. The complainant added that tugs not being available after dark was becoming so general that it was time for the Tyne Improvement Commissioners to provide a tug during the hours of darkness. Joseph Lawson was so incensed about this criticism that he felt bound to reply to what he described as a gross libel. This, in part, is what he wrote:-

> "Anyone with the slightest knowledge of the Tyne and its tug services would not make such a statement. The Tyne tug service stands pre-eminently at the head of all ports in the world. The incessant watching and bravery displayed by the Tyne tugmen in being at sea in bad weather night and day amount to rashness and contempt of danger. The Tyne tugs navigate and search about the North Sea like stormy petrels. The Tyne men and tugs are a subject of eulogy and admiration by all nationalities and it is a great injustice to this courageous body of men to place such a slander in print. Moreover, on the date mentioned my own tugs were in the vicinity of the Tyne bar with others on the look-out for vessels wishing to enter. Also the pilots, particularly in bad weather, keep a sharp watch for vessels."

The Lawson Steam Tug-boat Co.Ltd. 1895 - 1920

In 1895 it was decided that the family partnership should become a private limited company. On 1st March the Lawson Steam Tug-boat Company Ltd. was formally incorporated, with a capital of £20,000 represented by four hundred £50 shares. At this time the fleet consisted of six tugs with an average age of about twenty-five years:-

Monarch	(99/1871)	Iron	Paddle
Ulysses	(118/1874)	Iron	Paddle
Cruiser (1)	(98/1874)	Iron	Paddle
Scotia (2)	(77/1859)	Wood	Paddle
Champion (2)	(108/1866)	Wood	Paddle
Hercules	(145/1877)	Iron	Paddle

Shortly after the change in company status Joseph Lawson retired, on the advice of his doctor. He had been connected with the Lawson business since the age of eleven when he worked on his father's tugs as a 'utility boy'. He had served an apprenticeship with Thomas Hepple & Sons at Low Walker and eventually became a certificated chief engineer. He first took a financial interest in tug ownership when he was twenty-one years of age.

In February 1896 the Lawson company purchased the Great Western Railway Co.'s tender *Sir Walter Raleigh* (151/1876). Just why she was purchased is not understood since she would have been a rather unwieldy vessel for work on the River Tyne. She was, however, soon chartered to Gravesend owners so perhaps she never came to the Tyne. One year later she was sold to Spain and renamed *Nervion*.

A tug which was to give long and useful service on the Tyne was purchased in 1897 after having served at Great Yarmouth, Liverpool and Sunderland. She was the iron-hulled paddle tug *Comet* (132/1876) and had a two-cylinder side-lever engine. *Scotia* (2) was sold in 1898, to be broken up, and, in 1899, *Monarch* was disposed of similarly.

On 15th June 1899 the company's capital was increased to £30,000 by the creation of two hundred new shares of £50 each. The registered office at this time was still 27 Baring Street, South Shields. The current fleet of five tugs was, for the next eighteen years, to be steadily increased in number. The turn of the century saw the addition of *Taliesin* (79/1883), the company's first screw tug which had worked at Connah's Quay for the seventeen years since being built at Cardiff by the curiously named Tyneside Engineering Works Co.! Her twin screws were powered by two tandem compound engines which had the high- and low-pressure cylinders mounted one above the other, instead of the more usual side-by-side configuration. A second screw tug, *Flying Foam* (126/1877), joined the fleet in December 1900 having, when

The complete Lawson fleet c.1905, right to left: Taliesin, Ulysses, Cruiser *(1),* Comet, Royal Briton, Nestor, Mentor, *and* Hercules. *The identity of the tug at far left is unknown.* *(Local History Library, South Shields Central Library).*

built, been owned by the Clyde Shipping Co.Ltd. and later the Union Steam Towing Co. of Belfast. Unfortunately, she had a short life with the company for, on 24th August 1903, she was wrecked on Whitburn Steel.

Alderman John Robert Lawson died on 17th February 1901 at the age of seventy-nine, after a long illness. He had been a student of mathematics, astronomy and practical engineering, and in his 'teens had been given charge of the engine and boiler department of Barnes Forge & Ironworks at Jarrow Slake. Shortly thereafter he began his long involvement with the towage business. At one stage he was manager of the 'Red Star Line' of passenger boats which ran between Shields and Newcastle.

Champion (2) also ended her days by being lost when she foundered seven miles off the Tyne entrance on 13th February 1902 after having been with the company for eleven years. In the same year the screw tug *Royal Briton* (99/1885) was purchased from Cardiff owners.

The iron screw vessel *Africa* (122/1883) had been built on the Tyne as a North Shields trawler and later was owned at Swansea. Whilst there, under the nominal ownership of the African Tug Co.Ltd., she had been managed by, amongst others, Joseph Lawson. By about 1903 *Africa* had been bought by the Lawson Steam Tug-boat Co.Ltd., and when she came to the Tyne in about 1906 she was renamed *Nestor.* In 1907, with *Royal Briton*, she towed the Norwegian steamer *Thordis* from Bodø to the Tyne, in appalling weather, in eight days. Another tug purchased in 1903 was the twin-screw *Brock* (141/1883), which was renamed *Mentor.* In 1903 the tugs *Hercules* and *Ulysses* were chartered for a short time to the Manchester Ship Canal Co. so that they could assess the usefulness or otherwise of tugs with disconnecting engines. The outcome was that *Old Trafford* (later Ridley's *Reliant*) and her sisters were built for the canal company.

In 1908 the company set up office at 15 Thrift Street, South Shields. This was a very narrow street of shops and offices and included the entrances to the Tyne Dock Engineering Company's shiprepair yard. It would appear that, before the company bought the premises, they had in fact been a shop. The site of this street, and several others, are now inside the Tyne Dock Engineering and former Brigham & Cowan yards and one cannot tell, other by reference to old maps and photographs, that shops and houses ever existed in that locality. The company was still referred to as the 'Black Diamond Line of Tugs' and in its advertisements reminded would-be users of the tugs that the signal for Lawson tugs was two long, four short and one long blast on the steam whistle.

On 22nd September 1909 the twin-screw tug *Plover* (245/1881) was purchased from a Liverpool salvage company. She had started life as *Fylde* working for the railway companies using Fleetwood harbour. She was a powerful tug, her engines producing 850 IHP, such that she was able, in 1913, to tow the four-masted German barque *Schiffbek* (2,663/1886) from the Tyne to Hamburg, and later, the five-masted *R.C. Rickmers* (5,548/1906) from the Tyne to the English Channel. Towage of large sailing ships was still a common occurrence, and, at about the same period, *Nestor* towed the French barque *Marie* (2,192/1899) to Beachy Head without a stop in 3 days 21 hours.

Other notable tows during 1910 and 1911 had been:-

Tug	Tow	Voyage
Nestor	British barque *Crown of India* (2,057/1885)	Leith to Frederikstad in severe weather.
	Norwegian steamer *Parran* (1,977/1889)	Bergen to J.J. King, Gateshead for breaking-up
	British barque *Battle Abbey* (1,606/1875) bound for San Francisco	Tyne to Beachy Head in three days.
Plover	A.D.Bordes & Son's *Cambronne* (1,863/1897)	Hull to Tyne.
	German barque *Ostara* (1,952/1896)	Ardrossan to Tyne.
	French barque *Edouard Detaille* (2,185/1901)	Grimsby to Shields and Shields to Brest.
	Four-masted barque *African*	Hamburg to Tyne with part cargo. Three days in poor weather.
	German barque *Athene* (2,470/1892)	Tyne to Port Talbot. Twice had to turn back and shelter under the Isle of Wight in a hurricane.
	French barque *La Parouse* (2,186/1901)	Liverpool to Tyne in five days.

TYNE TOWAGE.

Telegraphic Address—
"Lawson, South Shields."

Telephones—
Nat. 114 & 334.

THE . .
LAWSON STEAM TUG-BOAT Co., Ltd.,
15 Thrift Street,
SOUTH SHIELDS.

OWNERS OF THE . .
Black Diamond Line of Tugs.

TUGS :—'Mentor,' 'Nestor,' 'Plover,' 'Royal Briton,' 'Taliesin,' 'Cruiser,' 'Comet,' 'Hercules,' 'Ulysses.'

Steam Signals :—Two Long, Four Short, One Long Blast.

Advertisement on back of a postcard, c.1912. *(Company archives).*

For the first time the Lawson company was to place an order for two new screw tugs. The first, delivered in May 1915, was *Homer* (157/1915). She was notable for another reason because she was the first Lawson-owned tug to have a steel hull. She had a triple-expansion steam engine

Hercules pulling hard during the First World War. *(Local History Library, South Shields Central Library)*

developing 850 IHP and had been built at Ardrossan. The second to arrive, exactly one year later, was *Joffre* (260/1916). She had come from the same shipyard but was considerably larger than *Homer.* She was named in honour of the French General, because of the many contracts which Lawson held for the towage of merchant vessels owned by French companies. One of her earliest jobs was to tow the old wooden HMS *Britannia* from Dartmouth to Blyth for breaking up.

In 1917 three second-hand iron paddle tugs were purchased, namely, *Criterion* (99/1876) from the Duke of Buccleuch, *Admiral* (102/1876) from Charlestown, and *Privateer* (117/1883) from Middlesbrough, bringing the fleet to its largest size yet. However, during the First World War, several of the company's tugs were on hire to the Royal Navy and thus were not always available for work which the Lawson management would have wanted for them. Those tugs involved were:-

Ulysses	22.05.1917 to 07.07.1917
Cruiser (1)	08.07.1917 to 01.09.1919
Hercules	28.06.1915 to 12.03.1919
Taliesin	24.09.1914 to 07.08.1919
Royal Briton	31.05.1916 to 31.08.1916
Nestor	20.02.1918 to 18.12.1918
Plover	26.09.1914 to 08.03.1920
Homer	01.11.1915 to 28.08.1919
Joffre	23.11.1917 to 15.03.1919

During the period of hire, to avoid any confusion of names, *Hercules* was renamed *Nemea; Nestor* became *Velmar* and *Plover* changed to *Peewit.* All reverted to their original names when the hire ended. *Homer* was fitted with a bow-mounted gun, and, on 8th April 1915, fought off a German submarine attack whilst towing the French barque *General de Sonis* (2,190/1901) from Queenstown to Sunderland. This attack, by gunfire and torpedo, took place 25 miles south-west of the Owers Light Vessel. On 2nd July 1918 *Admiral,* when towing two French lighters from Shields to Boulogne, was captured by a German submarine two miles north of Flamborough Head and sunk by gunfire. Her skipper was John McRitchie. Fortunately, there were no casualties. Less seriously, Benjamin Newton, the skipper of *Joffre* was fined £2 in court for using a tow rope longer than 20 fathoms within the River Tyne! Just after the war, in November 1918, *Ulysses* was in collision with the British submarine H.41, an event which later ended up as a court claim for damages. A month later, *Privateer* ran ashore at Equitien near Boulogne on a voyage from Le Tréport, and became a total loss.

The chairman of the company, T.N. Alexander, died in March 1918 and was succeeded by Matthew Coulson James, who had been a director of the company since May 1909. T.N. Alexander, with Thomas Whitfield, another Lawson director, were partners in Tyne Dock Engineering Co. (TDE). This may be why the Lawson office in Thrift Street came to be taken up, since it adjoined the TDE site. Another connection with TDE was that, in the early 1900s, they are recorded as being the registered office of the African Tug Co.Ltd. R.J. Thompson, a director of Furness, Withy & Co.Ltd., was appointed to the board in August 1918.

After the First World War in 1918, when the Lawson company was clearly the owner of the largest fleet of tugs on the River Tyne, suggestions had been made that co-operation, and possibly amalgamation, with J. Batey & Son Ltd. would be to the two companies' mutual benefit. The outcome was that the Lawson and Batey companies decided to merge in 1920. On 28th January The Lawson Steam Tug-boat Co.Ltd. was converted into a public limited company, and in the following month the company's shares were sub-divided into 30,000 of £1 each. On 17th April 1920 The Lawson Steam Tug-boat Co.Ltd. formally changed its name to Lawson-Batey Tugs Ltd. and thus absorbed the business of J. Batey &

A superb letter head from the last days of the independent existence of the Lawson company. (The Softley family).

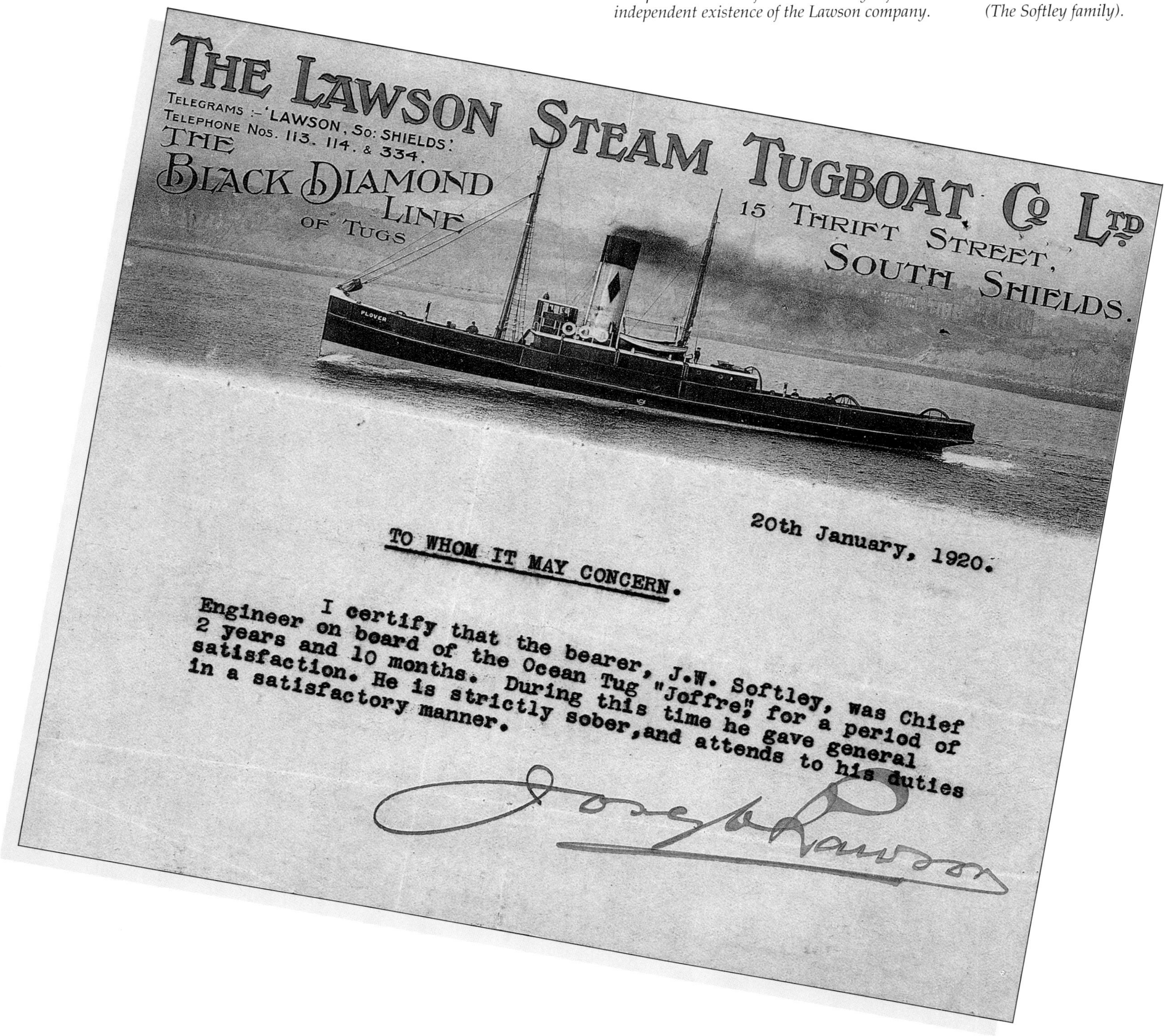

THE LAWSON STEAM TUGBOAT Co LTD

Telegrams:-'Lawson, So: Shields.'
Telephone Nos. 113. 114. & 334.

The Black Diamond Line of Tugs

15 Thrift Street, South Shields.

20th January, 1920.

TO WHOM IT MAY CONCERN.

I certify that the bearer, J.W. Softley, was Chief Engineer on board of the Ocean Tug "Joffre" for a period of 2 years and 10 months. During this time he gave general satisfaction. He is strictly sober, and attends to his duties in a satisfactory manner.

Joseph Lawson

Son Ltd. The share capital of the combined company was £150,000 in units of £1.

The Lawson company, throughout its eighty-year history, had owned thirty-three different tugs and at the time of the merger it contributed the following tugs to the combined Lawson-Batey fleet:-

Ulysses	(118/1874)	Iron	Paddle
Cruiser (1)	(98/1874)	Iron	Paddle
Hercules	(145/1877)	Iron	Paddle
Comet	(132/1876)	Iron	Paddle
Taliesin	(79/1883)	Iron	Twin Screw
Royal Briton	(99/1885)	Iron	Screw
Nestor	(122/1883)	Iron	Screw
Mentor	(141/1883)	Iron	Twin Screw
Plover	(245/1881)	Iron	Twin Screw
Homer	(157/1915)	Steel	Screw
Joffre	(260/1916)	Steel	Screw
Criterion	(99/1876)	Iron	Paddle

The twelve tugs had an average age of almost thirty-five years.

The individual Lawson company name may have disappeared in 1920 but their distinctive 'black diamond' funnel marking was to continue in use until after the Second World War.

APPENDIX A TO CHAPTER TWO

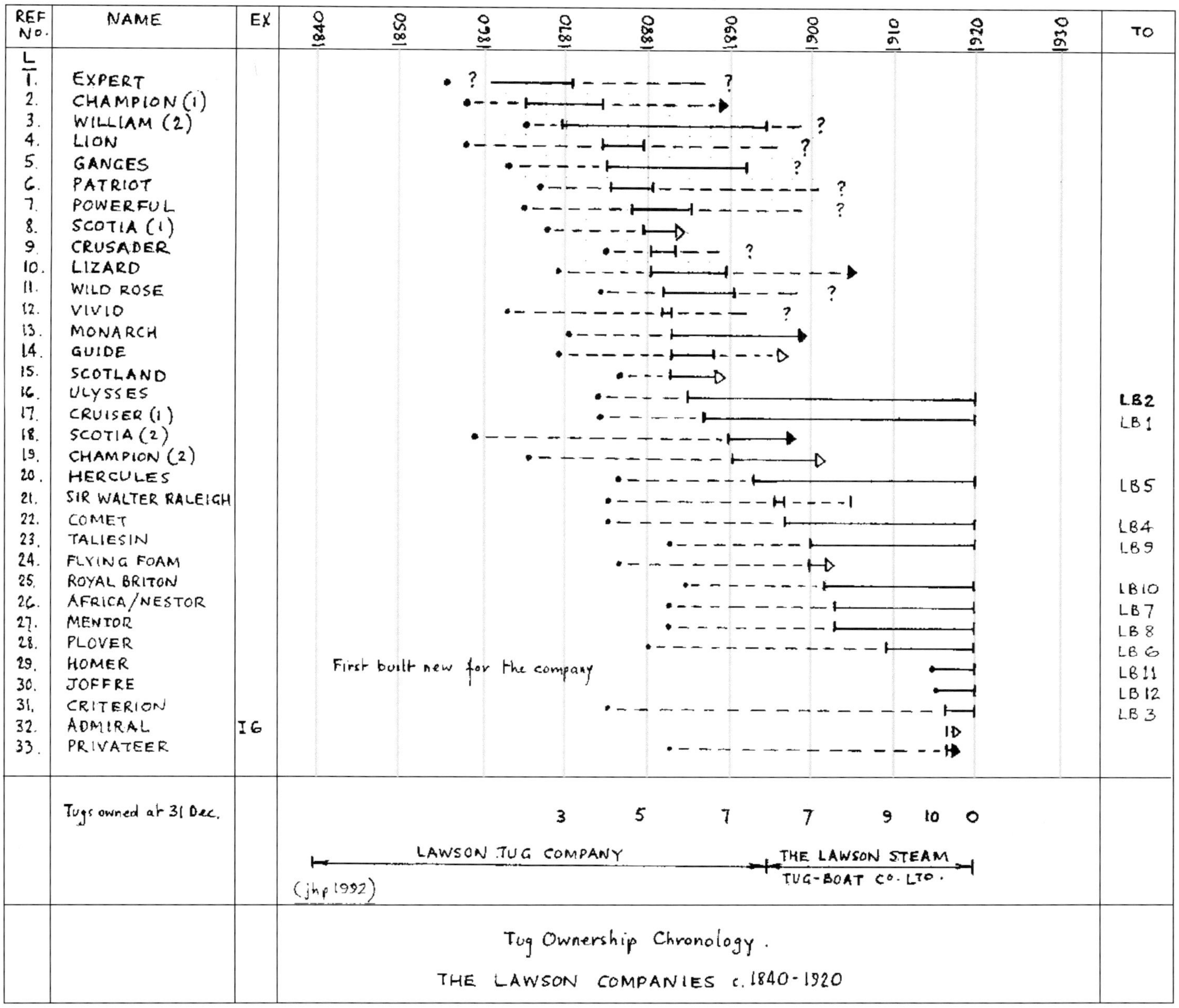

APPENDIX B TO CHAPTER TWO

The Lawson Companies Fleet List

L 1 **EXPERT** **(By 1861-1871)**
ON 17044 52 g. 75.2 x 15.3 x 8.2 ft. Wood Paddle
B: At North Shields.
E: 25 NHP.

PR: 1856 Shields.

1856:	Completed. Owner not known.
By 1861:	Purchased by John Robert Lawson, South Shields.
1871:	Sold to S.T. Freeman, London.
1887:	Not in register

L 2 **CHAMPION (1)** **(c.1865-1874)**
ON 21457 61 g. 79.6 x 15.8 x 8.7 ft. Wood Paddle
B: Hepple & Landells, North Shields.
E: No details available. 30 NHP. By builder.

PR: 1858 Shields; ?? North Shields; 1874 Sunderland

1858:	Completed for R., J., W., J.& R. Ray, North Shields.
1864:	Sold to R. Rowland, G. Lamb, W. Milburn, E. Storey & T. Watson.
c.1865:	Purchased by John Robert and Joseph Lawson, South Shields.
10.1874:	Sold to Margaret Thwaite and J. Wilkinson, Sunderland.
11.1880:	Owned by Margaret Thwaite only.
02.1887:	Sold to F. Wilton et al, Sunderland.
.1889:	Broken up.

William *(2), from a painting made c.1873 (John Clarkson, Longton)*

L 3 **WILLIAM (2)** **(1869-1894)**
ON 53437 80 g. 85.7 x 17.4 x 9.2 ft. Wood Paddle
B: Richard Stobbs, North Shields.
E: L. 1 Cy. 32½" x 46". 34 NHP. William Scott, North Shields.

PR: 1865 North Shields; 1872 North Shields; 1894 Hull.

.1865:	Completed for Joseph Honeyman, North Shields.
.1869:	Purchased by John R. Lawson and B.D. Lawson.
.1894:	Sold to F.T. Munkman, Hull.
Post 1898:	History not known.

L 4 **LION** **(1874-c.1879)**
ON 20438 75 g 85.4 x 16.9 x 9.1 ft. Wood Paddle
B: At South Shields.
E: 37 NHP

PR: 1858 Shields

.1858:	Completed for T.B. & C.D. Barker, North Shields.
.1874:	Purchased by John R. Lawson, B.D. Lawson & J. Lawson.
c.1879:	Sold to G. Harker, North Shields.
c.1880:	Sold to C. Lawrence, Sunderland.
c.1881:	Sold to T. Appleby, Hepworth.
c.1884:	Sold to G.R. Robson & J. Robson, Bill Quay.
c.1896:	Not in register

L 5 **GANGES** **(1875-1892)**
ON 45622 69 g 81.8 x 16.9 x 9.1 ft. Wood Paddle
B: At Willington Quay.
E: 35 NHP.

PR: 1863 Shields.

.1863:	Completed for P. Clendon, North Shields.
.1874:	Sold to J.T. Davison, North Shields.
.1875:	Purchased by John Robert Lawson & Joseph Lawson, South Shields.
.1892:	Not in register

L 6 **PATRIOT** **(1876-1881)**
ON 55205 91 g 86.4 x 17.6 x 9.3 ft. Iron Paddle
B: J. Hepple & Son, Low Walker
E: L. 1 Cy. 33¼" x 50". 40 NHP.

PR: 1867 Cardiff; 1870 Cardiff; 1876 North Shields;1881 Scarborough; 1883 North Shields.

04.1867:	Completed for W. & C.O. Rundle, Cardiff.
1876:	Purchased by J.R. Lawson, J. Lawson, B.D. Lawson and two others.
03.1881:	Sold to W.H. Shawcross, Scarborough.
10.1883:	Sold to J.Gray and J. Murray, South Shields. Later J. Gray alone.
02.1901:	Register closed

L 7 **POWERFUL** **(1878-c.1886)**
ON 53201 89 g 89.9.x.18.2 x 9.5 ft. Wood Paddle
B: At Coble Dene.
E: L. 1 Cy. 34" x 48". 34 NHP. T. Hepple, Low Walker.

PR: 1866 South Shields; 1867 Leith; 1876 North Shields; 1893 Alloa.

1865:	Completed for W. Best, South Shields.
1867:	Sold to H.P. Hanson, Leith.
02.1876:	Sold to R. Oats.
12.1878:	Purchased by J.R. Lawson and B.D. Lawson. Later B.D. Lawson alone.
c.1886:	Sold to W. Minto, South Shields.
1893:	Sold to A.A.T. Moyes, Alloa.
Post 1898:	History not known.

L 8 **SCOTIA (1)** **(1879-1884)**
ON 60484 78 g 82.6 x 17.5 x 9.2 ft. Wood Paddle
B: Brodie, North Shields.
E: L. 1 Cy. 32" x 48". 34 NHP. William Scott, North Shields.

PR: 1868 North Shields.

03.1868:	Completed for Andrew Bain, North Shields.
01.1873:	Sold to W. Williams, Newport, Mon.
1879:	Purchased by John Robert and Joseph Lawson, South Shields.
06.1884:	Foundered off Scarborough.

L 9 **CRUSADER** **(1881-1883)**
ON 69867 67 g 80.7 x 17.0 x 8.7 ft. Wood Paddle
B: C.W. Dodgin & Co., North Shields.
E: L. 1 Cy. 29½" x 42". 30 NHP. Manchester, Sheffield & Lincolnshire Railway Co., New Holland.

PR: 1876 South Shields; 1876 Sunderland; 1884 Rye.

11.1875:	Completed for Christian Havelock et al, Sunderland.
05.1881:	Purchased by John R. Lawson & B.D. Lawson, South Shields.
11.1883:	Sold to K. Reeves, Rye.
Post 1888:	History not known.

L 10 **LIZARD** **(1881-1889)**
ON 60524 82 g 83.3 x 17.8 x 9.4 ft. Wood Paddle
B: T. Hepple & Co., Low Walker.
E: L. 1 Cy. 31" x 48". 32 NHP. By Builder.

PR: 1869 North Shields; 1891 Middlesbrough.

05.1869:	Completed for C.O. Young et al, Cardiff.
1877:	Sold to C.W. Christie, Cardiff.
1881:	Purchased by John R. Lawson et al, South Shields.
1889:	Sold to Diana L. Sutherland, Middlesbrough.
09.1891:	Sold to The Lizard Steam Tug Co.Ltd., Middlesbrough.
05.1892:	Sold to W. Barnes, Hull.
1906:	Broken up.

Wild Rose *lying off Jarrow Slake. (Local History Library, South Shields Central Library).*

L 11 WILD ROSE (1882-1891)

ON 70378 75 g 85.0 x 17.6 x 9.2 ft. Wood Paddle
B: Richard Stobbs, North Shields.
E: L. 1 Cy. 32″ x 48″. 34 NHP. T.R. Scott & Co., North Shields.

PR: 1874 North Shields; 1875 Gloucester; 1882 North Shields; 1897 Leith.

1874:	Completed for G. Wharrier, North Shields.
1875:	Sold to Hartley Hodder, Purton, Gloucester.
1882:	Purchased by Joseph Lawson, Robert Dain Lawson et al, South Shields.
1891:	Sold to J. Minto, Leith.
Post 1898:	History not known.

L 12 VIVID (1882-1883)

ON 48649 89 g 89.0 x 17.8 x 9.2 ft. Iron Paddle
B: At Low Walker.
E: 50 NHP.

PR: 1864 London; 1880 Lynn; 1883 Irvine.

1863:	Completed for James Walker, Gravesend.
c.1872:	Sold to John Gillett, Gravesend.
1880:	Sold to The Vivid Tug Co. of Kings Lynn, Lynn.
1882:	Purchased by Joseph Lawson.
1883:	Sold to Irvine Harbour Trust.
c.1890:	Sold to James Douglass, Belfast.
Post 1892:	History not known.

L 13 MONARCH (1883-1899)

ON 65832 99 g 92.8 x 18.1 x 9.1 ft. Iron Paddle
B: J.T. Eltringham, South Shields.
E: L. 1 Cy. 24½″ x 54″. 50 NHP. J.P. Rennoldson, South Shields.

PR: 1871 Grimsby; 1880 Sunderland.

1871:	Completed for T. & S. Oates, Grimsby.
07.1880:	Sold to Martha Nicholson, Sunderland.
1883:	Purchased by John R. Lawson.
1895:	Transferred to Lawson Steam Tug-boat Co.Ltd.
1899:	Broken up.

L 14 GUIDE (1883-1888)

ON 60110 104 g 100.0 x 19.5 x 9.2 ft. Wood Paddle
111 g 97.7 x 19.8 x 9.0 ft. (from 1888) Screw from 1888
B: Harvey & Co., Hayle.
E: L. 1 Cy. 38″ x 54″. 50 NHP. By Builder.

PR: 1869 Dartmouth; 1877 Milford; 1888 Newcastle.

09.1869:	Completed for Dartmouth Steam Packet Co.Ltd.
1877:	Sold to T.T. Jackson, Milford and later transferred to Mrs. C. Jackson.
1883:	Purchased by Joseph Lawson.
06.1885:	Explosion on board at 2am. killed one man. Later repaired.
04.1888:	Sold to J. & D. Morris, Newcastle, renamed JUBILANT.
1888:	New engine fitted and converted to screw propulsion. Engine now C. 2 Cy. 13″ x 23½″ x 20″. 30 NHP, by J. Abbot & Co., Gateshead.
11.1897:	Vessel reported missing on voyage Maldon to River Tyne.

L 15 SCOTLAND (1883-1889)

ON 76647 77 g 84.0 x 17.2 x 9.3 ft. Wood Paddle
B: Driver & Hughes, Willington Quay.
E: L. 1 Cy. 30⅝″ x 48″. 93 NHP. T.R. Scott & Co., North Shields.

PR: 1877 North Shields.

09.1877:	Completed for J. Spence & T. Spence, North Shields.
02.1879:	Sold to W. Murray, North Shields.
1882:	Sold to T. Thompson, St. Peters.
1883:	Purchased, for £1,570, by B.D. Lawson, South Shields.
03.1889:	Vessel foundered in a gale, 20 miles east of Blyth.

L 16 ULYSSES (1885-1920)

ON 70605 118 g 100.6 x 18.0 x 9.4 ft. Iron Paddle
B: John Readhead & Co., South Shields. (YN 103)
E: L. 2 Cy. 26½″ x 48″. 450 IHP. By Builder.

PR: 1874 London; 1895 South Shields.

1874:	Completed for T.A.F. Banks et al, Gravesend.
1879:	Sold to C. Brownfield et al, Gravesend.
1885:	Purchased by Joseph Lawson.
03.1895:	Owners became Lawson Steam Tug-boat Co.Ltd.
05.1917:	Hired by Royal Navy. 07.1917: Returned to owners.
06.1920:	Transferred to Lawson-Batey Tugs Ltd.
06.1937:	Sold, for £445, to Clayton & Davie Ltd., Dunston, for breaking up.

Crusier *(1)*

L 17 CRUISER (1) (1887-1920)

ON 68532 98 g. 96.3 x 18.2 x 9.2 ft. Iron Paddle
B: John Softley & Co., South Shields.
E: L. 2 Cy. 25⅝″ x 46″. 300 IHP. Hepple & Co., South Shields.

PR: 1874 London; 1891 South Shields.

1874:	Completed for Daniel S. Mitchell, Gravesend.
1887:	Purchased by Joseph Lawson et al.
03.1895:	Owners became Lawson Steam Tug-boat Co.Ltd.
07.1917:	Hired by Royal Navy. 09.1919: Returned to owners.
06.1920:	Transferred to Lawson-Batey Tugs Ltd.
1945:	Chartered to Blyth Tug Co.Ltd.
11.1948:	Sold to British Iron & Steel Corporation for £275. Allocated to J.J. King & Co.Ltd., Gateshead for breaking up.

L 18 SCOTIA (2) (1890-1898)

ON 81972 77 g. 82.0 x 19.1 x 9.8 ft. Wood Paddle
B: Edward Oliver, South Shields.
E: L. 1 Cy. 24″ x 42½″. 50 NHP. J.P. Almond, North Shields.

PR: 1881 South Shields (1st Registry).

11.1859:	Completed, as TYNE DOCK No.1 (1), for Tyne Dock Co.
05.1881:	Transferred to North Eastern Railway Co.
03.1890:	Purchased by Joseph Lawson, renamed SCOTIA.
03.1895:	Transferred to The Lawson Steam Tug-boat Co.Ltd.
02.1898:	Vessel broken up.

Champion *(2) lying at South Shields buoys. (Local History Library, South Shields Central Library).*

L 19 CHAMPION (2) (1891-1902)

ON 81973 108 g. 87.0 x 19.0 x 9.2 ft. Wood Paddle
B: Richard Stobbs, North Shields.
E: L. 2 Cy. 25″ x 41¾″ 42NHP. J.P. Almond, North Shields.

PR: 1881 South Shields (1st Registry).

1886:	Completed, as TYNE DOCK No.2, for Tyne Dock Co.
05.1881:	Transferred to North Eastern Railway Co.
1891:	Purchased by Joseph Lawson, renamed CHAMPION.
03.1895:	Transferred to Lawson Steam Tug-boat Co.Ltd.
02.1902:	Foundered 7 miles off Tyne entrance.

Hercules at launch of Chilean warship Latorre in 1913. (George Scott).

L 20 HERCULES (1893-1920)

ON 77489 145 g. 105.3 x 19.2 x 10.6 ft. Iron Paddle
B: J.T. Eltringham, South Shields. (YN 62)
E: L. 2 Cy. 34" x 60". 80 NHP. 13 K. J.P. Rennoldson, South Shields.

PR: 1877 Hull; 1893 South Shields.

08.1877:	Completed for Dock Company of Kingston-upon-Hull.
01.1893:	Purchased by Joseph Lawson.
03.1895:	Transferred to Lawson Steam Tug-boat Co.Ltd.
06.1915:	Hired by Royal Navy, renamed NEMEA.
03.1919:	Returned to owners and reverted to name HERCULES.
06.1920:	Transferred to Lawson-Batey Tugs Ltd.
01.1931:	Sank after collision with NAPIER STAR. Beached inverted on Jarrow Slake.
1931:	Vessel broken up.

Sir Walter Raleigh when owned by the Great Western Railway Co. (Local History Library, South Shields Central Library).

L 21 SIR WALTER RALEIGH (1896-1897)

ON 74618 151 g. 110.1 x 19.1 x 10.0 ft. Iron Paddle
B: W. Allsup & Sons, Preston. (YN 37)
E: D. 2 Cy. 30" x 54". 85 NHP. By Builders.

PR: 1876 Plymouth; 1896 South Shields; 1897 Bilbao.

1876:	Completed for Great Western Railway Co. as tender at Plymouth.
02.1896:	Purchased by Lawson Steam Tug-boat Co.Ltd.
1896:	Chartered to Sandford, Gravesend.
1897:	Sold to Cia de Remolcadores del Cantabrico, Spain, renamed NERVION.
1905:	Reduced to a lighter for use at Bilbao.

Comet lying at buoy off South Shields groyne. (Local History Library, South Shields Central Library).

L 22 COMET (1897-1920)

ON 74712 132 g. 106.7 x 19.4 x 9.8 ft. Iron Paddle
B: J. & M. Gunn & Co., Cardiff.
E: L. 2 Cy. 28" x 48". 500 IHP. By Builders.

PR: 1876 Great Yarmouth; 1880 Liverpool; 1882 Sunderland; 1900 South Shields.

1876:	Completed for Great Yarmouth Star Steam Tug Co.Ltd., Great Yarmouth.
1880:	Sold to J. Corkhill, Liverpool.
1882:	Sold to R.M. Hudson & Son, Sunderland.
05.1897:	Purchased by Lawson Steam Tug-boat Co.Ltd.
06.1920:	Transferred to Lawson-Batey Tugs Ltd.
04.1952:	Transferred, for £1,650, to Blyth Tug Co.Ltd.
03.1953:	Sold, for £950, to British Iron & Steel Corporation. Allocated to Clayton & Davie Ltd., Dunston, for breaking up.

Taliesin outward bound at the mouth of the Tyne. (Company archives).

L 23 TALIESIN (1900-1920)

ON 88701 79 g. 85.6 x 18.2 x 8.0 ft. Iron Twin screw
B: Tyneside Engineering Works Co., Cardiff.
E: Two-C. 2 Cy. 12" & 24" x 14" (Tandem). 50 NHP. By Builder.

PR: 1883 Chester; 1900 South Shields.

05.1883:	Completed for Coppack, Carter & Co., Connah's Quay.
1900:	Purchased by Lawson Steam Tug-boat Co. Ltd.
09.1914:	Hired by Royal Navy. 08.1919: Returned to owners.
06.1920:	Transferred to Lawson-Batey Tugs Ltd.
09.1920:	Sold, for £5,000, to Edwin Cant, Southgate, (J.W. Coppack, Manager).
11.1930:	Sold to J.W. Coppack, Connah's Quay.
09.1931:	Sold to Allan M. Rouse, Neyland, Pembrokeshire.
1945:	Sold to British Iron & Steel Corporation for breaking up.

L 24 FLYING FOAM (1900-1903)

ON 76739 126 g. 109.2 x 18.1 x 9.2 ft. Iron Screw
B: Murdoch & Murray, Port Glasgow. (YN 21)
E: C. 2 Cy. 20" & 40" x 24" 70 NHP. Rait & Lindsay, Glasgow.

PR: 1877 Glasgow; 1901 South Shields.

1877:	Completed for Clyde Shipping Co.Ltd.
1889:	Sold to Union Steam Towing Co., Belfast.
12.1900:	Purchased by Lawson Steam Tug-boat Co.Ltd.
08.1903:	Wrecked on Whitburn White Steel. Total loss.

Royal Briton *shortly after arriving on the Tyne, c.1902. (Local History Library, South Shields Central Library).*

L 25 **ROYAL BRITON** **(1902-1920)**
ON 89198 99 g. 89.6 x 18.0 x 9.8 ft. Iron Screw
B: Finch & Co., Chepstow.
E: C. 2 Cy. 18" & 36" x 24". 500 IHP. Tyneside Engineering Works Co.Ltd., Cardiff.

PR: 1885 Cardiff.

06.1885:	Completed for P. Gibbs, G. Lee & E. Finch, Cardiff.
03.1902:	Purchased by Lawson Steam Tug-boat Co.Ltd.
05.1916:	Hired by Royal Navy. 08.1916: Returned to owners.
06.1920:	Transferred to Lawson-Batey Tugs Ltd.
06.1920:	Sank at Boulogne after collision with barge. Raised and repaired at Grimsby.
06.1934:	Sold, for £500, to Jarrow Dredging & Salvage Co. Ltd.
03.1942:	Sold to Edmund Handcock (1929) Ltd., Cardiff.
04.1949:	Broken up by local scrap merchants after hitting quay wall and sinking in Cardiff Docks. Was raised and beached outside the docks.

Nestor. *(Word Ship Photo Library).*

L 26 **AFRICA/NESTOR** **(c.1903-1920)**
ON 84878 122 g. 100.0 x 19.8 x 10.1 ft. Iron Screw
B: Sir W.G. Armstrong, Mitchell & Co.Ltd., Low Walker. (YN 452)
E: C. 2 Cy. 21" & 38" x 24". 600 IHP. Ross & Duncan, Glasgow.

PR: 1883 North Shields; 1891 Swansea; 1906 South Shields.

01.1883:	Completed, as AFRICA, for C. Dyble, North Shields and operated as a trawler.
1891:	Sold to C. Hooper, Swansea.
c.1896:	Sold to African Tug Co.Ltd., Swansea.
c.1903:	Purchased by Lawson Steam Tug-boat Co.Ltd.
c.1906:	Renamed NESTOR.
1915:	Chartered to William Watkins Ltd., London.
02.1918:	Hired by Royal Navy and renamed VELMAR.
12.1918:	Returned to owners and became NESTOR again.
06.1920:	Transferred to Lawson-Batey Tugs Ltd.
06.1937:	Sold, for £460, to Clayton & Davie Ltd., Dunston, for breaking up.

L 27 **MENTOR** **(1903-1920)**
ON 89701 141 g. 95.2 x 19.1 x 9.0 ft. Iron Twin screw
B: T.B. Seath & Co., Rutherglen. (YN 236)
E: Two-C. 2 Cy. 18" & 34" x 24". 500 IHP. Rankin & Blackmore Ltd., Greenock.

PR: 1883 Fleetwood; 1904 South Shields; 1937 Dundee

1883:	Completed, as BROCK, for Lancashire & Yorkshire and London & North Western Railway Companies.
c.1902:	Sold to G.E.J. Moody, Cleethorpes.
10.1903:	Purchased by Lawson Steam Tug-boat Co.Ltd., renamed MENTOR.
06.1920:	Transferred to Lawson-Batey Tugs Ltd.
03.1937:	Sold, for £570, to V.A. Cappon Tugs Ltd., Dundee.
03.1940:	Transferred to Dundee, Perth & London Shipping Co.Ltd.
07.1946:	Broken up in Belgium.

L 28 **PLOVER** **(1909-1920)**
ON 73764 245 g. 122.4 x 22.5 x.10.8 ft. Iron Twin screw
B: J. Reid & Co., Port Glasgow. (YN 6q.)
E: Two-C. 2 Cy. 25" & 48" x 27". 850 IHP. 11K. Rankin & Blackmore Ltd., Greenock.

PR: 1881 Fleetwood; 1904 Liverpool; 1909 South Shields.

1881:	Completed, as FYLDE, for Lancashire & Yorkshire and London & North Western Railway Companies.
1904:	Sold to Liverpool Association for Protection of Commercial Interests as respect wrecked and damaged property, renamed PLOVER.
09.1909:	Purchased by Lawson Steam Tug-boat Co.Ltd.
09.1914:	Hired by Royal Navy, renamed PEEWIT.
03.1920:	Returned to owners and became PLOVER again.
06.1920:	Transferred to Lawson-Batey Tugs Ltd.
01.1953:	Damaged when run down by BLENHEIM at Stanhope Buoys, South Shields.
04.1957:	Sold, for £4,500, to British Iron & Steel Corporation for breaking up at C.W.Dorkin & Co.Ltd., Gateshead.

Homer *when on Admiralty service with two masts and fitted with a bow gun. (Local History Library, South Shields Library*

L 29 **HOMER** **(1915-1920)**
ON 135728 157 g. 95.3 x 22.1 x 10.9 ft. Steel Screw
B: Ardrossan Drydock & Shipbuilding Co.Ltd. (YN 264)
E: T. 3 Cy. 13", 21" & 34" x 24". 850 IHP. Aitchison, Blair Ltd., Glasgow.

PR: 1915 Ardrossan.

03.1915:	Completed for Lawson Steam Tug-boat Co.Ltd.
11.1915:	Hired by Royal Navy. 08.1919: Returned to owners.
06.1920:	Transferred to Lawson-Batey Tugs Ltd.
1931:	Chartered to Gamecock Steam Towing Co.Ltd. for a short period.
08.1939:	Hired by Director of Sea Transport. 04.1941: Returned to owners.
12.1958:	Transferred, for £8,000, to Blyth Tug Co.Ltd.
03.1968:	Sold, for £2,950, to Hughes Bolckow Ltd., Blyth, for breaking up.

Joffre. *(A. Duncan)*

L 30 **JOFFRE** **(1916-1920)**
ON 135730 260 g. 115.0 x 26.1 x 11.7 ft. Steel Screw
B: Ardrossan Drydock & Shipbuilding Co.Ltd. (YN 265)
E: T. 3 Cy. 18", 28½" & 46" x 30". 1140 IHP. 11 K. J.G. Kincaid & Co. Ltd., Greenock.

PR: 1916 Ardrossan; 1925 Newcastle.

05.1916:	Completed for Lawson Steam Tug-boat Co.Ltd. after default of original purchasers.
11.1917:	Hired by Royal Navy. 03.1919: Returned to owners.
06.1920:	Transferred to Lawson-Batey Tugs Ltd.
05.1925:	Stranded near Bolt Head, Devon.
05.1925:	Refloated and beached at Salcombe, later towed to Falmouth.
06.1925:	Towed to Tyne for repair. Recommissioned 09.1925.
01.1953:	Damaged when run down by BLENHEIM at Stanhope Buoys, South Shields.
01.1959:	Chartered to Tyne Tugs Ltd.
03.1966:	Sold, for £4,075, to P.W. MacLellan, Bo'ness, for breaking up.

L 31 **CRITERION** **(1917-1920)**

ON 68946 99 g. 98.7 x 18.2 x 9.2 ft. Iron Paddle

B: Softley Brothers, North Shields.

E: 1876: L. 1 Cy. 35" x 52". R.J. Marshall, South Shields. 40 NHP.
1925: L. 2 Cy. 27" x 45". J.P. Rennoldson & Sons Ltd., South Shields. 50 NHP.

PR: 1876 Sunderland; 1878 Granton.

11.1876:	Completed for C. Havelock & T.C. Wake, Sunderland.
05.1878:	Sold to The Duke of Buccleuch, Edinburgh.
1917:	Purchased by The Lawson Steam Tug-boat Co.Ltd.
06.1920:	Transferred to Lawson-Batey Tugs Ltd.
05.1929:	Sold, for £290, to Malcolm Thompson, Bo'ness, for breaking up.

L 32 **ADMIRAL** **(1917-1918)**

See Anchor Steam Tug fleet list, AST 5.

L 33 **PRIVATEER** **(1917-1918)**

ON 78447 117 g. 96.5 x 18.3 x 9.2 ft. Iron Paddle

B: T. & W. Toward, Newcastle.

E: L. 1 Cy., 35" x 54". 65 NHP. G.P. Hepple, North Shields.

PR: 1883 Swansea; 1895 Boston; 1915 Middlesbrough.

1883:	Completed for W.P. Ching, Swansea.
1890:	Sold to J.M. Thomas, Swansea.
1895:	Sold to Boston Steam Tug Co.Ltd.
1914:	Sold to F. Palmer, Boston.
1915:	Sold to The Robinson Tug Co. Ltd., Middlesbrough.
02.1917:	Sold, for £1,450, to Charles Duncan & Sons Ltd., Middlesbrough.
12.1917:	Purchased by The Lawson Steam Tug-boat Co.Ltd., South Shields.
12.1918:	Vessel total loss. Ashore Equitien, nr. Boulogne, on voyage Le Tréport to Boulogne.

Privateer *giving pleasure trips from Bideford when owned at Swansea.* *(North Devon Museum Trust).*

• CHAPTER THREE •

The Batey Companies

John Batey. c.1840 - c.1874

About the same time as John R. Lawson became involved in the towage business on the River Tyne, c.1840, John Batey, at the age of thirty-one, was also the owner of a small wooden paddle tug. As was the case with the earliest tugs owned by Lawson, no details have been discovered of the vessel first owned by John Batey. It is likely that, to begin with, he would have been only a part owner with another or even several other persons. The early Batey tugs must have been quite small but nevertheless they are on record as having gone 'seeking' jobs as far south along the coast as Flamborough Head. John Batey, who was born at Willington Quay in 1809, had begun his involvement in the towage business at North Shields but it was not long before he was living in, and working from, 3 Gibson Street not far behind the Quayside, in Newcastle. In a Lawson-Batey booklet published in the 1930s, John Batey's grandson Frank records that, in 1854, John Batey was instrumental in saving several sailing vessels at Newcastle Quayside. There had been a disastrous explosion at a gunpowder factory across the river at Gateshead which had been severe enough to lay waste, and set fire to, the buildings and vessels at Newcastle Quay. John Batey was able to tow those vessels at risk to places of safety.

The earliest evidence which has been discovered of a vessel in which John Batey had a financial interest concerns *William* (1) which had been built on the Tyne in 1842. He was part owner, along with W. Charlton, and also her skipper in 1854. However, by 1857, although W.Charlton, along with several others, was still a part owner, John Batey was not. This was probably due to the fact that he had had built for himself *Amazon* (42/1856) for operation at Montrose on the east coast of Scotland, to assist the many large sailing vessels bringing grain and flax from Australia and sailing outwards with dressed timber and general woollen and linen goods. According to the Montrose Directory Almanac for 1897 *Amazon* was the first tug ever employed at Montrose. If so, then John Batey had chosen a location where it could have been expected that competition would be far less than on the River Tyne. In 1862 *Amazon* was sold to a North Shields owner and replaced by another newly-built wooden paddle tug, appropriately named *Montrose* (51/1862).

About 1862 or 1863, when competition began to appear at Montrose, a local company or partnership was set up in which John Batey was involved. This new enterprise soon became known as the Montrose Steam Tug Co.Ltd., but it did not last very long since, in the early 1870s, John Batey, James Mitchell and others formed a new partnership at Montrose. What name or title they traded under has not been discovered, but it could be that under different ownership they retained the original title.

John Batey. 1809-1897 (Company archives).

In 1864 John Batey's son, Francis, at the age of twenty-two, became a partner with his father. The wooden paddle tug *Rosamond* (47/1859) was acquired about that time, in the names of John Batey, Francis Batey and Robert Armstrong. It is presumed that this vessel operated on the Tyne. These early tugs owned by the Batey concern were relatively small, as is indicated by their low gross tonnage, and were only about seventy feet long. *Venus* (69/1867) was built new in the names of John and Francis Batey and also Thomas Hunter Batey, another member of the family. Francis is recorded as being skipper of this tug in 1869. In that year *Montrose* was sold to T.Hepple of Newcastle, presumably acting as broker, and within a few months had been sold on to Russian owners. The following year *Rosamond* was sold to Tyne owners and later went to Grimsby.

Francis Batey. 1842-1915 (Company archives).

Providence (58/1871) was built new, again in the names of the Bateys, father and son, and Robert Armstrong. Also in 1871, the same owners purchased *Sir Colin Campbell* (47/1858) from G.R. Turnbull of St.Anthony's.

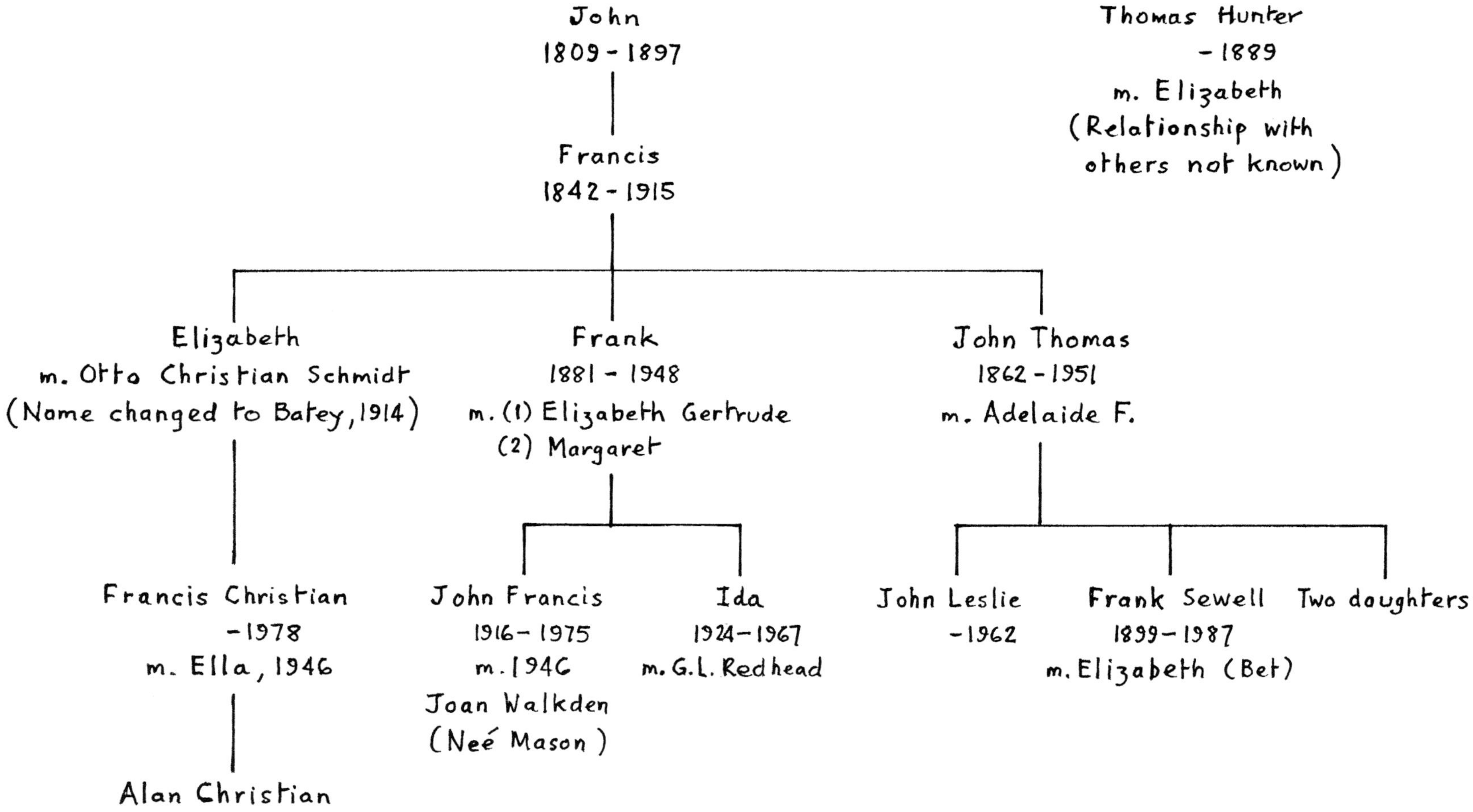

Batey Family Tree.

The large increase in the cost of coal in the early 1870s was causing serious concern to the Tyne tugowners. Since neither the Board of Trade nor the Tyne Improvement Commission would allow an increase in towage rates, and following protracted discussions, the tugowners laid off all of their tugs on 8th February 1873 claiming that they were unable to make a profit on the current rates. Following an enquiry certain increases were permitted for a trial period and the tugs returned to work after a stoppage of eleven days. It is of interest to note that prior to 1865 the towage rates were charged as individual tug companies thought fit, but thereafter what was termed the compulsory or Parliamentary rate, was fixed by agreement with the Tyne Improvement Commission. Even so the tug companies were in the habit of allowing substantial discounts to local and favoured shipowners. After the 1873 dispute the tugowners re-affirmed that they wished to stay with the Parliamentary system.

J. Batey & Son. c.1874 - 1906

About 1874 the partnership was formally registered as J. Batey & Son, although it would appear that Robert Armstrong, in particular, and others were still closely involved. It has already been noted that Batey's *Montrose* went to Russia in 1869. During the 1870s several other Tyne-owned tugs were sold to Black Sea ports, all manned on their delivery voyage by Tyne tugboatmen, amongst them Francis Batey. These lengthy voyages must have been quite a test for the low-powered single-engined paddle tugs of the day. A somewhat larger tug than hitherto was purchased from South Shields owners in December 1874. She was *Storm King* (80/1866), built of wood and almost ninety feet long, and was sent to work at Montrose. J. Batey & Son, at the time of the company's formation, owned the following fleet of four tugs:-

Vessel name	Gross tonnage/year	Hull construction	Means of propulsion
Venus	(69/1867)	Wood	Paddle
Providence	(58/1871)	Wood	Paddle
Sir Colin Campbell	(47/1858)	Wood	Paddle
Storm King	(80/1866)	Wood	Paddle

– with an average age of eight and a half years; surprisingly low when compared with the fleet ages which were to apply in the next century.

In 1876, the wooden paddle tug *Warrior* was purchased from B.H. Nichols of Blakeney and was renamed *Star* (62/1866) on arrival in the Tyne. About twelve months later *Sir Colin Campbell* was sold to James Hunter of Willington Quay for further local service. The paddle tug *Bob Chambers* (67/1862) since being built, had been operated by the Montrose Steam Tug Co. in which John Batey had an interest. From 1879 this tug was registered in the names of Francis Batey, James Mitchell and others, and continued to work at Montrose. In the following year *Flying Scotchman* (80/1876) was purchased from owners at Dundee and was then employed at Montrose. In 1884 a much larger iron-hulled paddle tug was completed for J. Batey & Son by Hepple & Co. at North Shields. She was *John Batey* (109/1884) and was also sent to work at Montrose. It had been intended that, when the towage business was slack, she would be employed as a trawler but after a short while she operated as a tug only.

Shortly after *John Batey* commenced service, *Storm King* stranded at the entrance to Montrose Harbour on 10th October 1884 and became a total loss. To replace this loss,

Rapid (1) (90/1868) was purchased in 1885 and she, together with *John Batey*, carried out the towage at Montrose for a number of years under the management of James Mitchell. *Bob Chambers* was then only employed in emergencies. In 1888 *Titan* (102/1865) was purchased from William Watkins of London, she having been built and owned earlier at Dundee, *Providence* went to Sunderland owners in 1892 and *Harwich* (125/1877) was purchased in October 1896 from R. & W. Paul Ltd. of Ipswich.

Large sailing ships, when gathering cargo from British and European ports, before making trans-world voyages would commonly use tugs to expedite their movements rather than wait for favourable winds. One such occasion involved *John Batey* which left the Tyne on 1st October 1894 for Hamburg where she arrived two days later. Early on 5th October she left on the return journey towing the iron sailing vessel *Jupiter* which was to complete loading in the Tyne. The pair arrived late on 7th October after an uneventful journey which had taken seventy hours.

John Batey, the founder of the company, died in 1897 at the age of eighty-seven years. He had built up a fleet of eight tugs from nothing, over a period of about fifty-seven years. At the time of his death three of those tugs were at Montrose with the others working in and around the River Tyne. Others had, of course, been involved in the partnership over the years, but John and, later, his son Francis had been those responsible for the operation of the towage business. About 1897 John Thomas Batey, a son of Francis Batey, became a partner, but he was not involved in the day-to-day running of the company since he was employed in the shipbuilding industry on the Tyne. The level of business at Montrose had been declining due to the very much reduced number of sailing vessels using the port. Thus it was that *Bob Chambers* was sold for further work at Leith and *John Batey* returned to the Tyne. This left only *Rapid* (1) at Montrose, which must have made the Batey operations there uneconomical. Since she was still needed at the port she was hired to the Montrose Harbour Trustees for a time. By 1904 all Batey involvement at Montrose had ceased and work was concentrated on the Tyne.

Star had been disposed of by 1899 and the company address was then 87 Quayside, Newcastle, after having been for very many years located at 3 Gibson Street. This move gave the company a much better presence amongst the maritime community based on the Quayside. Frank Batey, John's grandson and Francis' son, joined the company in 1904 at the age of twenty-three. The thirty-three year old iron-hulled paddle tug *Charles Dickens* (121/1872) was purchased from Middlesbrough owners in 1905. She had earlier served on the Thames after being built at South Shields by Readhead, Softley & Co. *Venus*, after having been in Batey service since 1867, was sold for further service.

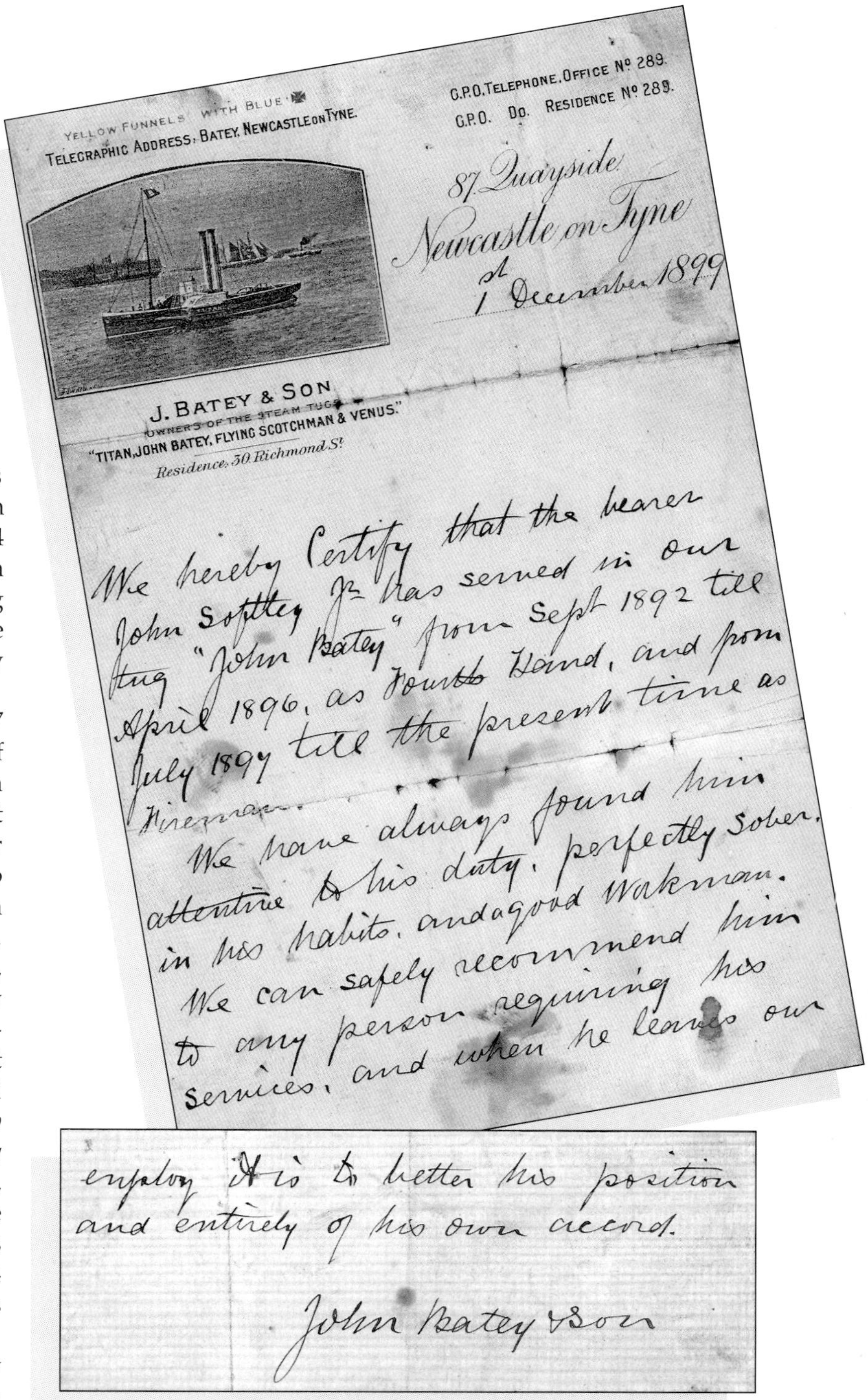

YELLOW FUNNELS WITH BLUE
TELEGRAPHIC ADDRESS: BATEY, NEWCASTLE ON TYNE.
G.P.O. TELEPHONE, OFFICE Nº 289.
G.P.O. Do. RESIDENCE Nº 289.

87 Quayside
Newcastle on Tyne
1st December 1899

J. BATEY & SON
OWNERS OF THE STEAM TUGS
"TITAN, JOHN BATEY, FLYING SCOTCHMAN & VENUS."
Residence, 30 Richmond St.

We hereby Certify that the bearer John Softley Jr has served in our tug "John Batey" from Sept 1892 till April 1896, as Fourth Hand, and from July 1897 till the present time as Fireman.

We have always found him attentive to his duty, perfectly sober in his habits, and a good workman. We can safely recommend him to any person requiring his services, and when he leaves our employ it is to better his position and entirely of his own accord.

John Batey & Son

Letter head 1899 with illustration of Titan *and reference for John Softley Jnr.* *(The Softley family).*

Titan *alongside the newly-completed Great Lakes steamer* Turbinia *(1,064/1904). Note the letter 'B' on the paddle boxes.* *(Swan Hunter Ltd.).*

From time to time the Royal Navy has visited ports in Great Britain and elsewhere for the purpose of 'showing the flag'. In the 'big ship' era many of the warships were too large to enter harbour and it was common practice for the fleet to anchor off the entrance to ports or mouths of rivers. In September 1904 the Channel Fleet visited the Tyne and some of the ships were anchored offshore. The presence of the fleet attracted tugs from far and wide to the Tyne, with the intention of carrying passengers for a trip around the fleet. Not all tugs had passenger certificates which meant that

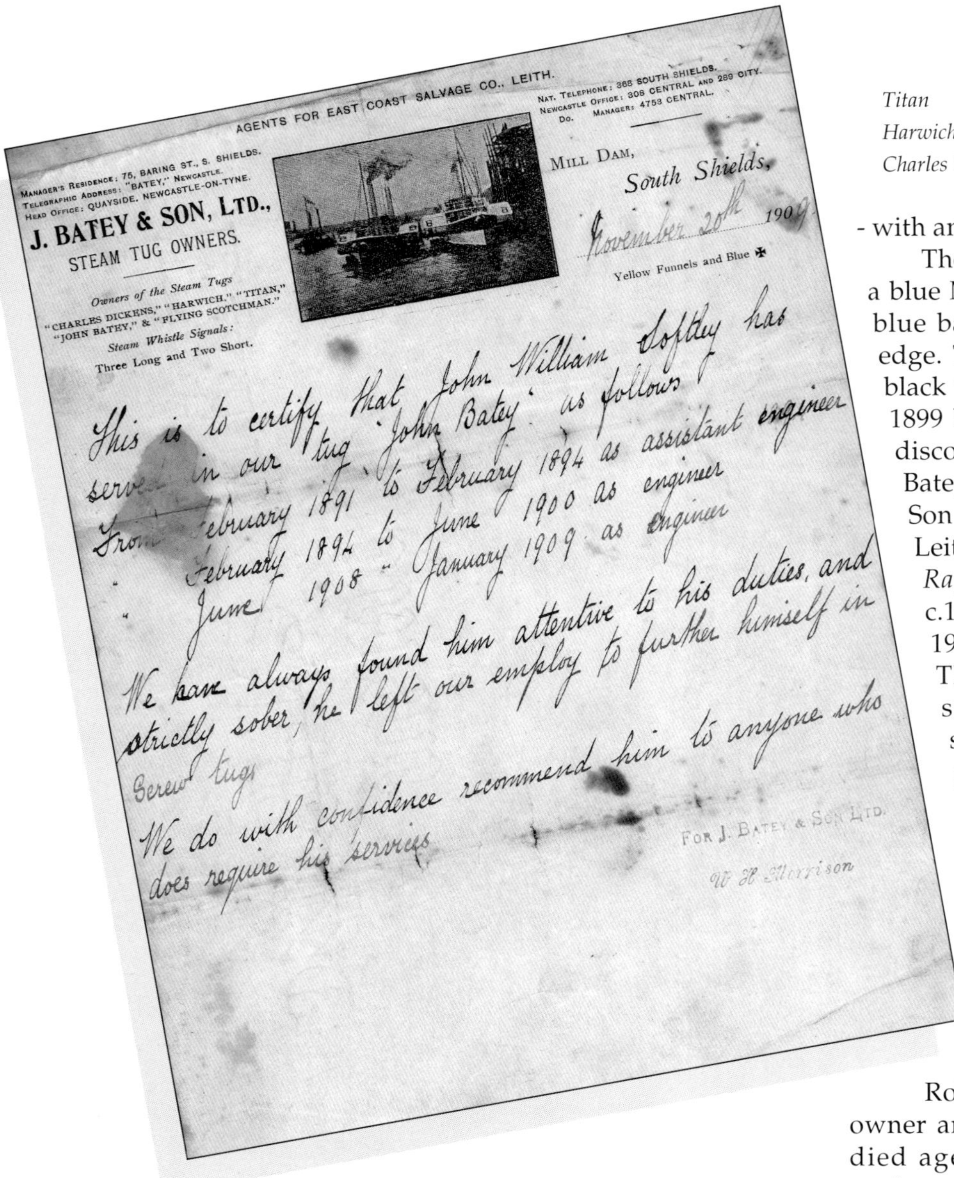

AGENTS FOR EAST COAST SALVAGE CO., LEITH.

Manager's Residence: 75, Baring St., S. Shields.
Telegraphic Address: "Batey," Newcastle.
Head Office: Quayside, Newcastle-on-Tyne.

J. BATEY & SON, LTD.,
STEAM TUG OWNERS.

Owners of the Steam Tugs
"Charles Dickens," "Harwich," "Titan,"
"John Batey," & "Flying Scotchman."
Steam Whistle Signals:
Three Long and Two Short.

Nat. Telephone: 366 South Shields.
Newcastle Office: 309 Central and 289 City.
Do. Manager: 4759 Central.

Mill Dam,
South Shields,
November 20th 1909

Yellow Funnels and Blue ✠

This is to certify that John William Softley has served in our tug 'John Batey' as follows
From February 1891 to February 1894 as assistant engineer
February 1894 to June 1900 as engineer
June 1908 - January 1909 as engineer

We have always found him attentive to his duties, and strictly sober, he left our employ to further himself in Screw tugs

We do with confidence recommend him to anyone who does require his services

For J. Batey & Son Ltd.
W. H. Morrison

Letter of 1909 giving a reference for John William Softley. (The Softley family).

a Board of Trade survey was necessary to obtain the required authority. One Shields shiprepair company had a queue of eighteen tugs awaiting survey for this purpose. The number of tugs involved can be imagined when it is realised that at this time there were seventy-eight tugs owned on Tyneside alone. Not all of these, of course, come within the scope of this history.

J. Batey & Son Ltd. 1906 - 1920.

On 26th June 1906, J. Batey & Son was converted into a limited liability company entitled J. Batey & Son Ltd., with their registered office described as at 'Three Indian Kings' Court, Newcastle. The first subscribers were Francis Batey, Robert Armstrong, Frank Batey, John Thomas Batey, Adelaide F. Batey, Margaret Batey and Elizabeth Schmidt; the ladies being wives and a sister respectively of John Thomas and Frank Batey. Francis Batey was the company chairman and also the designated manager at a salary of £1 per week per tug. Frank and John Thomas Batey were appointed directors, the former also acting as secretary. The initial capital of the new company was £15,000 in £1 shares. The fleet at the time that J. Batey & Son Ltd. was incorporated was:-

Flying Scotchman	(80/1876)	Wood	Paddle
John Batey	(109/1884)	Iron	Paddle
Rapid (1)	(90/1868)	Wood	Paddle
Titan	(102/1865)	Iron	Paddle
Harwich	(125/1877)	Wood	Paddle
Charles Dickens	(121/1872)	Iron	Paddle

- with an average age of over thirty-two years.

The Batey tugs had a buff (or light yellow) funnel with a blue Maltese cross. Their house flag had a white 'B' on a blue background with a narrow red border around the edge. The forward side of the paddle boxes bore a large black 'B'. It is known that this colour scheme was in use in 1899 but just when it was first introduced has not been discovered. The steam whistle signal code for calling a Batey tug was three long and two short blasts. J. Batey & Son Ltd. were agents for the East Coast Salvage Co. of Leith.

Rapid (1) was sold to Sunderland Towage Co.Ltd. c.1906. The company obtained their first screw tug in 1909 when they purchased *Langton* (157/1892) from The Alexandra Towing Co.Ltd. of Liverpool. This steel-hulled vessel was powered by a compound steam engine developing almost 600 IHP. She was to give fifty-four years' service to the company and its successors. On 11th December 1909 *John Batey*, skippered by Charles Edward Jopling, ran aground on Bondi Carrs at Hauxley on the Northumberland coast whilst on a voyage from the Tyne to St. Abbs Head. The crew had to abandon ship and was rescued by local fishermen who ultimately received an award of £80 for their services. The tug was later refloated and taken to Amble for temporary repairs.

Robert Armstrong, who had been a tug and wherry owner and associated with the Batey firm since the 1860s, died aged 71 at Amble in January 1911. Frank Batey's nephew, Francis Christian Batey, was introduced into the company in 1912. In that year the iron-hulled screw tug *Bulldog* (72/1884) was purchased from the Boston Steam Tug Co. and *Flying Scotchman* was sold to Tyneside owners for further service.

During 1911, through 1912 and into 1913, there were serious disputes between the tugowners and the tug crews over wages and conditions. It was claimed that there had been no increase of basic pay since 1888 when the weekly rates were fixed at Master £2, Engineman £1-16-0 (£1.80), Fireman £1-10-0 (£1.50). Current earnings including overtime, averaged over eight months, were Master £3, Engineman £2-15-0 (£2.75), Fireman £2-9-0 (£2.45) and Lads £1-3-0 (£1.15). The most serious part of the dispute concerned overtime payments and, arising from this, the definition of normal working hours. This is not surprising since the rules for deciding on these matters were complicated beyond belief and with hindsight arguments must have been inevitable. It is worth quoting from some of these conditions drawn up by the Tyne Tugowners' Association in 1912 following arbitration:-

'When a tug is ordered to look out for a particular vessel during overtime hours, the crew are to be paid two shillings (10p.) per man per night, whereas previously they had received two shillings per ship even if they waited two nights.

When steamers of 500 tons and upwards are towed during overtime hours, overtime is to be paid at the rate of one shilling (5p.) per man per ship instead of sixpence (2.5p) previously.

With regard to Sunday pay, any twelve hours consecutive work between Saturday midnight and Sunday midnight shall be regarded as one day and that any time over twelve hours shall be regarded as a quarter day extra.

If a crew is detained on board a tug, waiting for a particular vessel, and the services of the tug are not required, overtime is to be paid for the vessel waited for.

No change is made for the proportion of salvage money allowed to tug crews (20%).

When tugs are engaged towing vessels to the Tees or Hartlepool or any neighbouring ports, the usual overtime is to be paid, in addition to trip money.

In respect of claims for overtime for getting a tug ready for 6am. and in the case when a tug is not moored at 5pm. and where a tug is in attendance on a vessel after midnight, and is ordered to attend another steamer before 7am. no change is made in the present practice.

Overtime on Saturday is to commence at 2pm. instead of 4pm., any of the crew sent on board on Sunday, to attend to the fires etc., is to be paid a day's wage. Under the old conditions this was considered to be part of the ordinary week's work.'

These are only a few extracts but they show how complex the working rules had become in an attempt to cope with the unpredictability of the life of a tugboatman. Arguments, involving 40 tugs and some 250 personnel, continued between the owners and the men. In 1913 a further demand had been for the crew's share of salvage awards to be increased to 24%. When agreement could not be reached the tugboatmen went on strike on 13th September 1913 and they stayed out until 20th October when they agreed to further arbitration. The awards subsequently achieved gave the men an increase in basic rates but no change in overtime rates. Salvage rates were revised in a curious manner, giving 22.5% to double-engined tugs, but only 18.5% to those with a single engine.

On 28th October 1913, *John Batey* was seriously damaged when in collision with the Tyne pilot cutter *Protector* (200/1907). Her paddle casing and sponson were completely carried away.

A new steel screw tug was delivered to the company in September 1914 by Hepple & Co.Ltd. of South Shields. She was *Francis Batey* (151/1914) and was powered by a compound steam engine developing 750 IHP. Almost immediately it was proposed that she should be sold to Turkish owners, but this move fell through, presumably due to the outbreak of the First World War. The cancellation of this transaction was to cause a worry which arose in 1927 and which is referred to in the Lawson-Batey chapter. Within a month of her completion *Francis Batey* was hired to the Royal Navy until May 1919. The latter part of her war-time employment was as a rescue tug at Milford Haven. In August 1915 *Langton* was also hired by the Royal Navy but in this instance she was to travel much further from home than did *Francis Batey*. She was involved, with *Joseph Crosthwaite* of the North Shields-based Crosthwaite company, in the Dardanelles, and assisted in the evacuation from 'Clyde' Beach at Gallipoli. She was not returned to the control of her owners until March 1920, sixteen months after the end of the war.

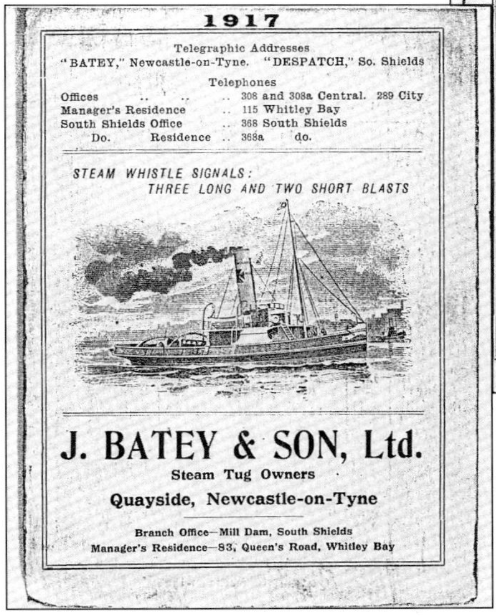

1917

Telegraphic Addresses
"BATEY," Newcastle-on-Tyne. "DESPATCH," So. Shields

Telephones

Offices	308 and 308a Central. 289 City
Manager's Residence ..	115 Whitley Bay
South Shields Office ..	368 South Shields
Do. Residence ..	368a do.

STEAM WHISTLE SIGNALS:
THREE LONG AND TWO SHORT BLASTS

J. BATEY & SON, Ltd.
Steam Tug Owners
Quayside, Newcastle-on-Tyne

Branch Office—Mill Dam, South Shields
Manager's Residence—83, Queen's Road, Whitley Bay

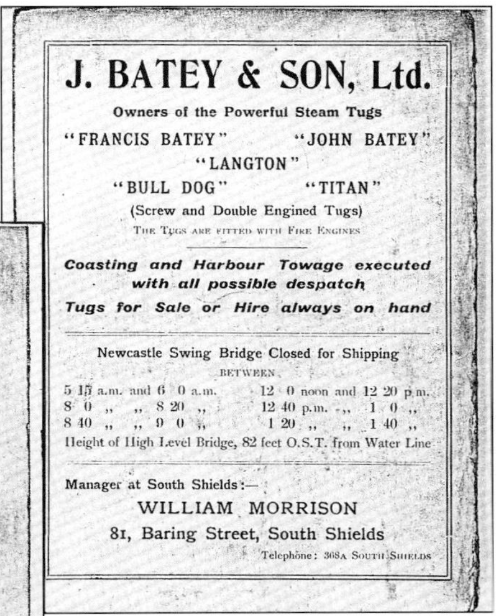

J. BATEY & SON, Ltd.
Owners of the Powerful Steam Tugs
"FRANCIS BATEY" "JOHN BATEY"
"LANGTON"
"BULL DOG" "TITAN"
(Screw and Double Engined Tugs)
The Tugs are fitted with Fire Engines

Coasting and Harbour Towage executed with all possible despatch
Tugs for Sale or Hire always on hand

Newcastle Swing Bridge Closed for Shipping
BETWEEN

5 15 a.m. and 6 0 a.m.	12 0 noon and 12 20 p.m.
8 0 ,, ,, 8 20 ,,	12 40 p.m. ,, 1 0 ,,
8 40 ,, ,, 9 0 ,,	1 20 ,, ,, 1 40 ,,

Height of High Level Bridge, 82 feet O.S.T. from Water Line

Manager at South Shields:—
WILLIAM MORRISON
81, Baring Street, South Shields
Telephone: 368a South Shields

Cover of booklet issued in 1917. (Company archives).

Francis Batey, the son of the founder of the company, died in 1915 at the age of seventy-three. He was succeeded as chairman and managing director by his son, Frank, then aged thirty-four. When *Charles Dickens* was sold, in 1915, for breaking up on the Tyne the company was left with only four tugs available for operation on their home river. About 1916, *Harwich* was sold to Sunderland Towage Co.Ltd. and, to make up for the depletion of the fleet, the Tyne Tees Steam Shipping Co.Ltd.'s screw tug *W.D.S.* was taken on charter. This arrangement lasted until July 1919 when *W.D.S.* (103/1892) was purchased for £5,500 by the Batey company. An advertisement published in 1917 indicated that the company had offices at the Quayside, Newcastle and at the Mill Dam, South Shields, and that the South Shields manager was William Morrison.

Following overtures from The Lawson Steam Tug-Boat Co.Ltd. agreement was reached that the two companies should merge their operations. This was achieved by the Lawson company changing its name to Lawson-Batey Tugs Ltd. and absorbing the assets of J. Batey & Son Ltd., the latter company then going into voluntary liquidation.

John Batey *leaving Ramsgate Harbour, c.1913, when she was standing-in for the local tug* Aid. *(P.N. Thomas).*

John Batey *towing* Birma *(4,588/1894) owned by East Asiatic Co. c.1918.* *(Company archives/Parry).*

All of these changes were made effective from 17th April 1920. The Batey company throughout its eighty-year history had owned twenty different tugs. At the time of the merger they contributed the following six tugs to the combined Lawson-Batey fleet:-

John Batey	(109/1884)	Iron	Paddle
Titan	(102/1865)	Iron	Paddle
Langton	(157/1892)	Steel	Screw
Bulldog	(72/1884)	Iron	Screw
Francis Batey	(151/1914)	Steel	Screw
W.D.S.	(103/1892)	Steel	Screw

They had an average age of over thirty-one years.

Although the individual Batey company title had disappeared in 1920 the blue Maltese cross funnel mark continued to be used for the ex-Batey tugs. Even when the first additional tug was brought into the Lawson-Batey fleet in 1946 the Batey colours were retained, albeit with a different house flag, and, even at the time of writing, these colours are still in use. There was, however, to be a gap in the use of this colour scheme, from 1959 to 1986, when the Lawson-Batey tugs were chartered to Tyne Tugs Ltd.

APPENDIX A TO CHAPTER THREE

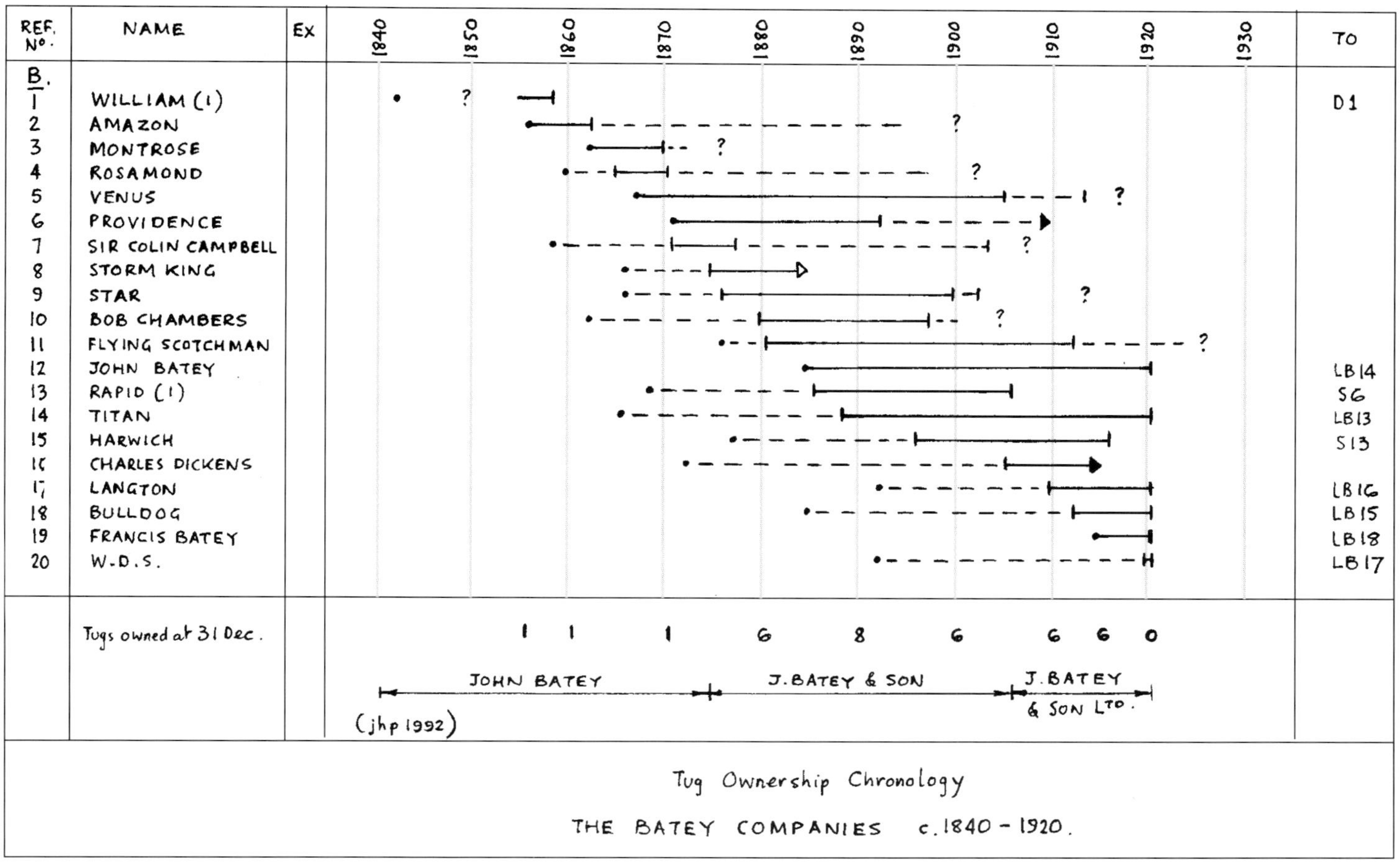

APPENDIX B TO CHAPTER THREE

The Batey Companies Fleet List

B 1 **WILLIAM (1)** **(By 1854-1857)**
ON 24495 Wood Paddle
B: On River Tyne.
E: 15 NHP.

PR: 1842 ??; By 1855 Newcastle.

1842:	Completed, owner not known.
By 1854:	Purchased by W. Charlton and J. Batey, Newcastle.
1857:	Owners now W. Charlton, J. Carr and T.A. Dry.
1858:	Owners now W. Charlton, J. Carr and T. Wilson.
1864:	Owners now W. Charlton and T. Wilson.
1866:	Not in register.

B 2 **AMAZON** **(1856-1862)**
ON 7299 42 g. 72.6 x 13.9 x 7.1 ft. Wood Paddle
B: At Blackwall, Co. Durham.
E: L. 1 Cy. 20 NHP.

PR: 1856: Newcastle; 1862 Newcastle: 1868 Newcastle; 1872 Padstow.

04.1856:	Completed for John Batey, Newcastle.
08.1862:	Sold to H. Jopling, North Shields.
1868:	Sold to Roger Walker, Felling.
1872:	Sold to J.H. Hutchings, Padstow.
1894:	Further history not known.

B 3 **MONTROSE** **(1862-1869)**
ON 43612 51 g. 73.0 x 15.0 x 8.0 ft. Wood Paddle
B: T. Hepple & Co., Low Walker.
E: 20 NHP.

PR: 1862 Newcastle.

05.1862:	Completed for John Batey, Newcastle.
04.1869:	Sold to T. Hepple, Newcastle.
10.1869:	Sold to Russian owners. Further history not known.

B 4 **ROSAMOND** **(c.1864-1870)**
ON 26875 47 g. 71.7 x 14.5 x 7.5 ft. Wood Paddle
B: T. Hepple, Low Walker.
E: L. 1 Cy. 21 NHP. Hepple & Landells, Low Walker.

PR: 1859 Newcastle; 1870 Newcastle; 1885 Grimsby

1859:	Completed, owners not known.
c.1864:	Purchased by John and Francis Batey and Robert Armstrong.
1870:	Sold to John Lakey and son.
12.1885:	Owned by John Lakey Jr., Cleethorpes.
c.1897:	Sold to Thomas Chambers, Newcastle. No further details known.

B 5 **VENUS** **(1867-1905)**
ON 56062 69 g. 81.3 x 16.8 x 8.6 ft. Wood Paddle
B: T. Hepple & Co., Low Walker.
E: L. 1 Cy. 25 NHP.

PR: 1867 Newcastle; 1869 Newcastle.

1867:	Completed for John Batey, Newcastle.
07.1869:	Owners John Batey, Francis Batey and T.H. Batey.
1884:	Owners John Batey, Francis Batey and Robert Armstrong.
1905:	Sold.
1913:	Registry closed.

B 6 **PROVIDENCE** **(1871-1892)**
ON 62389 58 g. 75.8 x 16.3 x 8.1 ft. Wood Paddle
B: T. Hepple & Co., Low Walker.
E: L. 1 Cy. 25½" x 45". 21 NHP. By Builder.

PR: 1871 Newcastle; 1892 Sunderland; 1894 Newcastle.

07.1871:	Completed for Robert Armstrong, John Batey and Francis Batey, Newcastle.
08.1892:	Sold to Henry Wilson, Sunderland.
05.1894:	Sold to J. Mitcheson, Newcastle.
01.1896:	Sold to T.J.R. Hindmarsh, Durham.
04.1897:	Sold to R. Lodge, Howdon.
11.1900:	Sold to E. & R. Jull, St. Anthony's.
1909:	Broken up.

B 7 **SIR COLIN CAMPBELL** **(1871-1877)**
ON 20741 47 g. 73.8 x 14.9 x 7.7 ft. Wood Paddle
B: Hepple & Landells, Low Walker.
E: L. 1 Cy. 25½" x 42". 22 NHP.

PR: 1858 Newcastle; 1871 Newcastle.

1858:	Completed for G.R. Turnbull, St. Anthony's.
1871:	Purchased by John Batey, Francis Batey and Robert Armstrong.
04.1877:	Sold to James Hunter, Willington Quay.
1903:	Register closed.

B 8 **STORM KING** **(1874-1884)**
ON 56106. 80 g. 89.1 x 18.3 x 9.5 ft. Wood. Paddle.
B: Readhead & Softley, South Shields.
E: L. 1 Cy. 32½" x 48". 35 NHP. By Builder.

PR: 1866 South Shields; 1873 South Shields; 1874 Newcastle.

12.1866:	Completed, owner not known.
c.1868:	Sold to G. Jamieson and I. Stoker, and later W. Aitken, South Shields.
12.1874:	Purchased by John & Francis Batey.
02.1880:	Part ownership (14/64) transferred to individuals at Montrose.
10.1884:	Vessel stranded at entrance to Montrose Harbour and became total loss.

B 9 **STAR** **(1876-c.1899)**
ON 56006 62 g. 75.1 x 16.8 x 8.6 ft. Wood Paddle
B: At South Shields.
E: L. 1Cy. 30" x 48". 40 NHP. J.P. Rennoldson, South Shields.

PR: 1866 Grimsby; 1872 Wells; 1876 Newcastle.

1866:	Completed, as WARRIOR, for T.S. Oates, Grimsby.
1872:	Sold to J.E. Turner, Clay, Norfolk.
1875:	Sold to B.H. Nichols, Blakeney.
07.1876:	Purchased by John Batey, Francis Batey & Robert Armstrong and renamed STAR.
c.1899:	Sold.
12.1902:	Registry closed.

B 10 **BOB CHAMBERS** **(1879-1897)**
ON 44298 67 g. 80.6 x 16.8 x 8.8 ft. Wood Paddle
B: At North Shields.
E: 29 NHP.

PR: 1862 Montrose; 1897 Leith

1862	Completed for Tyne owners but immediately sold to Montrose Steam Tug Co.
1879:	Purchased by Francis Batey and James Mitchell et al.
1897:	Sold to J.I. Minto, Leith. Further history not known.

B 11 **FLYING SCOTCHMAN** **(1880-1912)**
ON 63720 80 g. 83.0 x 17.9 x 9.3 ft. Wood Paddle
B: At North Shields.
E: 30 NHP.

PR: 1876 Montrose.

1876:	Completed for James Dykes, Dundee.
1880:	Purchased by J. Batey & Son, Newcastle.
1906:	Transferred to J. Batey & Son Ltd.
1912:	Sold to J. Tulip, Jarrow.
By 1923:	Owned by Robson & Cross, Newcastle. Further history not known.

John Batey. *(Authors' collection).*

B 12 **JOHN BATEY** **(1884-1920)**
ON 63729 109 g. 96.3 x 18.4 x 9.0 ft. Iron Paddle
B: Hepple & Co., North Shields. (YN 380)
E: As built: L. 1 Cy. 34½" x 54". 40 NHP.
c.1920: L. 2 Cy. 28" x 54". R.B. Findley & Son, North Shields.

PR: 1884 Montrose; 1920 Newcastle; 1921 Grangemouth.

1884:	Completed for J. Batey & Son, Newcastle and James Mitchell, Montrose.
1904:	Owners now J. Batey & Son, Newcastle only.
06.1906:	Owners became J. Batey & Son Ltd., Newcastle.
06.1920:	Transferred to Lawson-Batey Tugs Ltd.
10.1921:	Sold, for £3,000, to Grangemouth & Forth Towing Co. Ltd.
1936:	Sold to G.W. Brunton for breaking up.

B 13 **RAPID (1)** **(1885-c.1906)**
ON 56933 90 g. 89.0 x 18.3 x 9.3 ft. Wood Paddle
B: T. Hepple & Co., Low Walker.
E: L. 1 Cy. 33¾" x 50". 40 NHP. By Builder.

PR: 1868 Grangemouth; 1885 Montrose.

01.1868:	Completed for G.G. Mackay, Grangemouth.
1885:	Purchased by J. Batey & Son, Newcastle and James Mitchell, Montrose.
1897:	Hired to Montrose Harbour Trustees for an unknown period.
06.1906:	Ownership restyled J. Batey & Son Ltd.
c.1906:	Purchased, for £400, by Sunderland Towage Co.Ltd.
03.1911:	Sold, for £350, to W. Chambers, Newcastle. Further history not known.

B 14 **TITAN** **(1888-1920)**
ON 45829 102 g. 85.5 x 18.6 x 9.7 ft. Iron Paddle
B: Gourlay Bros. & Co., Dundee. (YN 23)
E: L. 2 Cy. 28" x 45". 300 IHP. 8 K. By Builder.

PR: 1865 Dundee; 1873 London; 1920 Newcastle.

1865:	Completed, as ATLAS, for Dundee, Perth & London Shipping Co.
1873:	Sold to William Watkins, London, and renamed TITAN.
10.1888:	Purchased, for £1,700, by J. Batey & Son, Newcastle.
06.1906:	Owners became J. Batey & Son Ltd.
06.1920:	Transferred to Lawson-Batey Tugs Ltd.
09.1936:	Sold, for £205, to Clayton & Davie Ltd., Dunston, for breaking up.

B 15 **HARWICH** **(1896-1916)**
ON 74437 125 g. 100.6 x 18.9 x 10.4 ft. Wood Paddle
B: J.H. Vaux, Harwich. (YN 24).
E: L. 1 Cy. 27" x 46". 50 NHP. J. Stewart, London.

PR: 1877 Harwich.

1877:	Completed for J.H. Vaux, Harwich.
c.1894:	Sold to R. & W. Paul Ltd., Ipswich.
10.1896:	Purchased by J. Batey & Son, Newcastle.
06.1906:	Owners restyled J. Batey & Son Ltd.
01.1916:	Purchased, for £2,000, by Sunderland Towage Co.Ltd.
05.1918:	Transferred to France, Fenwick Tyne & Wear Co.Ltd.
c.1921:	Purchased by Coulson Tug Co.Ltd.
03.1926:	Purchased, for £1,500, by Blyth Tug Co.Ltd.
11.1932:	Sold, for £42-10-0, to Clayton & Davie Ltd., Dunston, for breaking up.

B 16 **CHARLES DICKENS** **(1905-1915)**
ON 65662 121 g. 103.2 x 18.3 x 10.1 ft. Iron Paddle
B: Readhead, Softley & Co., South Shields. (YN 74)
E: L. 2 Cy. 28" x 48". 55 NHP. By Builder.

PR: 1872 London; 1880 Stockton-on-Tees.

1872:	Completed for J. Martin, London.
c.1879:	Sold to A. Tolhurst, Gravesend.
05.1880:	Sold to Tees Conservancy Commissioners.
12.1903:	Sold to The Florence Tugboat & Salvage Co.Ltd., Middlesbrough.
1904:	Sold to Linthorpe Dinsdale Smelting Co., Middlesbrough.
10.1905:	Purchased by Francis Batey, J.T. Batey and R. Armstrong, Newcastle.
06.1906:	Transferred to J. Batey & Son Ltd.
05.1915:	Broken up on River Tyne.

Titan *at the mouth of the Tyne.* *(Local History Library, South Shields Central Library).*

Langton with tall funnel carrying metal cut-out Maltese crosses. (Company archives).

B 17 **LANGTON** **(1909-1920)**
ON 102076 157 g. 92.0 x 20.1 x 11.3 ft. Steel. Screw
B: S. McKnight & Co., Ayr. (YN 38)
E: C. 2 Cy. 20" & 42" x 27". 580 IHP. 10 K. Muir & Houston, Glasgow.

PR: 1893 Liverpool; 1920 Newcastle.

11.1892:	Completed for The Alexandra Towing Co.Ltd., Liverpool.
06.1909:	Purchased by J. Batey & Son Ltd., Newcastle.
08.1915:	Requisitioned by the Royal Navy. 03.1920: Returned to owners.
06.1920:	Transferred to Lawson-Batey Tugs Ltd.
05.1947:	Hired to Blyth Tug Co.Ltd. for 2 years.
04.1953:	Transferred, for £5,000, to Blyth Tug Co.Ltd.
08.1954:	Hired to Lawson-Batey Tugs Ltd.
11.1962:	Laid up.
05.1963:	Sold to J.J. King & Co.Ltd., Gateshead, for breaking up.

B 18 **BULLDOG** **(1912-1920)**
ON 89630 72 g. 77.0 x 16.5 x 9.3 ft. Iron Screw
B: London Drydock Co., Deptford.
E: C. 2 Cy. 15¼" & 29" x 20". 40 NHP. Windus & Dunsmore, Millwall.

PR: 1884 London; 1901 Boston; 1920 Newcastle.

06.1884:	Completed for London Steam Tug Co.Ltd.
04.1901:	Sold to Boston Steam Tug Co.
05.1912:	Purchased by J. Batey & Son Ltd., Newcastle.
06.1920:	Transferred to Lawson-Batey Tugs Ltd.
09.1925:	Sold, for £825, to Allen Brown Ltd., Newcastle, after one month's trial.
07.1934:	Sold to Risdon Beazley Marine Trading Co.Ltd., Southampton.
12.1934:	Broken up at Southampton.

B 19 **FRANCIS BATEY** **(1914-1920)**
ON 133563 151 g. 95.3 x 21.2 x 10.8 ft. Steel. Screw
B: Hepple & Co.Ltd., South Shields. (YN 637)
E: C. 2 Cy. 20" & 42" x 27". 750 IHP. 12 K. By Builder.

PR: 1914 Newcastle.

09.1914:	Completed for J. Batey & Son Ltd., Newcastle.
10.1914:	Requisitioned by Royal Navy. 05.1919: Returned to owners.
06.1920:	Transferred to Lawson-Batey Tugs Ltd.
08.1939:	Hired by Royal Navy. 11.1942: Returned to owners.
12.1946:	Sank at Bill Quay in River Tyne whilst towing REGENT TIGER. Crew of six saved.
05.1947:	Returned to service after being raised and repaired.
01.1959:	Chartered to Tyne Tugs Ltd.
11.1962:	Transferred, for £4,500, to Blyth Tug Co.Ltd.
04.1968:	Sold, for £2,576-3-6, to Hughes Bolckow Ltd., Blyth, for breaking up.

W.D.S. assisting New Londoner (1446/1894) near Newcastle Quay when owned by Tyne Tees Steam Shipping Co. Ltd. c.1911. (George Scott)

B 20 **W.D.S.** **(1919-1920)**
ON 101815 103 g. 85.2 x 19.1 x 9.5 ft. Steel Screw
B: Schlesinger Davis & Co., Wallsend. (YN 167)
E: C. 2 Cy. 17" & 32" x 24". 240 IHP. 11 K. North Eastern Marine Engineering Co.Ltd., Wallsend.

PR: 1892 Newcastle; 1946 Belfast.

12.1892:	Completed for Tyne Steam Shipping Co.Ltd.
12.1903:	Owners became Tyne-Tees Steam Shipping Co.Ltd.
09.1916:	Chartered to J. Batey & Son Ltd., Newcastle.
07.1919:	Purchased, for £5,500, by J. Batey & Son Ltd.
06.1920:	Transferred to Lawson-Batey Tugs Ltd.
1928:	Laid up.
02.1930:	Sold, for £600, to Thomas Keedy, South Shields.
12.1939:	Owners restyled Thomas Keedy & Son Ltd.
08.1941:	Sold to Berry's (Stranraer) Ltd.
1945:	Sold to J. Lee, Belfast.
1947:	Sold to Irish Bay Lines Ltd. (H.P. Lenaghan, Manager). Renamed LENASHIP.
1949:	Broken up.

• CHAPTER FOUR •

Lawson-Batey Tugs Limited

Formation and Early Years. 1920 - 1929.

Shortly after the First World War ended in 1918, tug companies, like all other companies, took some time to settle down to what they expected would be normality. Business was expected to grow as the number of ships trading to the River Tyne returned to pre-war levels. By 1920, to take advantage of the hoped-for opportunities, a number of tug companies on the north-east coast of England were thinking in terms of mergers which would reduce the large number of small companies and partnerships that existed in the ports and rivers, thus obtaining greater efficiency and competitiveness. Significant mergers at that time were the formation of United Towing Co.Ltd. at Hull, whose name speaks for itself, Tees Towing Co.Ltd. at Middlesbrough* and France, Fenwick Tyne & Wear Co. Ltd. at Sunderland and Newcastle.

On the River Tyne similar thoughts applied and as has been outlined earlier in Chapters Two and Three the two largest companies operating there decided to amalgamate their activities. These companies were The Lawson Steam Tug-boat Co.Ltd. and J. Batey & Son Ltd., incorporated respectively in 1895 and 1906 although both had been operating under different titles long before those dates. On 28th January 1920 the directors of the two companies agreed that the merger should go ahead and that the mechanism for achieving this would be to change the title of the Lawson company and to liquidate voluntarily the Batey company. The Lawson company thus became Lawson-Batey Tugs Ltd. and was officially registered as such on 17th April 1920.

The original subscribers to the combined company were Joseph Lawson, steam tug owner, Thomas Whitfield, engineer, Thomas Norne Alexander, accountant, Joseph Lawson Jr., clerk, William Alexander, master mariner, Alexander Purvis, manufacturer, and Thomas Little, clerk, and the initial directors were Matthew Coulson James (Chairman), Joseph Lawson and Frank Batey (Joint Managing Directors), Robert Jackson Thompson and John Thomas Batey whose fees were each to be £105 p.a. The secretary was G.A. Clark. The two joint managing directors were appointed as directors for their respective lives subject to not being disqualified under the rules of the company. It was stipulated that the directors should number not less than three nor more than six. At its initiation the share capital was increased from £30,000 to £150,000 and the firm was designated a 'public company'. The major shareholders were Frank Batey and John Thomas Batey. Surprisingly, the Lawsons had only minority shareholdings.

Lawson-Batey owned a combined fleet of eighteen tugs:-

Vessel Name	Gross tonnage/year	Hull construction	Means of propulsion	
Cruiser (1)	(98/1874)	Iron	Paddle	ex-Lawson
Ulysses	(118/1874)	Iron	Paddle	ex-Lawson
Criterion	(99/1876)	Iron	Paddle	ex-Lawson
Comet	(132/1876)	Iron	Paddle	ex-Lawson
Hercules	(145/1877)	Iron	Paddle	ex-Lawson
Plover	(245/1881)	Iron	Twin screw	ex-Lawson
Nestor	(122/1883)	Iron	Screw	ex-Lawson
Mentor	(141/1883)	Iron	Twin screw	ex-Lawson
Taliesin	(79/1883)	Iron	Twin screw	ex-Lawson
Royal Briton	(99/1885)	Iron	Screw	ex-Lawson
Homer	(157/1915)	Steel	Screw	ex-Lawson
Joffre	(260/1916)	Steel	Screw	ex-Lawson
Titan	(102/1865)	Iron	Paddle	ex-Batey
John Batey	(109/1884)	Iron	Paddle	ex-Batey
Bulldog	(72/1884)	Iron	Screw	ex-Batey
Langton	(157/1892)	Steel	Screw	ex-Batey
W.D.S.	(103/1892)	Steel	Screw	ex-Batey
Francis Batey	(151/1914)	Steel	Screw	ex-Batey

An impressive list no doubt, but it should be noted that they were of variable quality ranging from the 55-year-old *Titan* to the quite modern and relatively large tugs *Francis Batey*, *Homer* and *Joffre*. The average age of the fleet was thirty-four years, probably brought about by the longevity of iron hulls.

Although the companies had merged, the Lawson and Batey tugs continued to wear their old companies' funnel insignia, i.e. the former keeping their black diamond and the latter their blue Maltese cross. It was not to be until after the Second World War that the fleet achieved a uniform identity! Presumably this was done so that the original allegiance to the shipping companies to whom the tugs were contracted could be maintained, but 30 years would be a long time to persist with the idea. Perhaps it also pleased the tug crews.

At about this time, Lawson-Batey were managing the tug *St. Florence* (419/1919) for the Admiralty, but little has been discovered about this activity, which was presumably sea-going, or for how long it lasted. The vessel was sold to Canada in 1924 so the management was relatively short lived.

** See the author's history of Tees Towing Co. Ltd. 'Seahorses of the Tees' published in 1985.*

The Lawson-Batey riverside office at 15 Thrift Street, South Shields, in 1920. The entrance to Tyne Dock Engineering ship repair yard can be seen just beyond the row of shops. *(Local History Library, South Shields Central Library).*

The Lawson Steam Tug-boat Co.'s office was at 15 Thrift Street, South Shields alongside the Tyne Dock Engineering Co.'s ship-repair yard. When the merger took place the combined Lawson-Batey company used this address as their South Shields office although the registered office was at Newcastle. The rear windows of this building must have given a good view over the shipyard to the river and the tug buoy berths. The office was situated amongst several shops including: A. Ross - painter; Mrs. J. Oliver - fruiterer; T. Robinson - grocer; A. Jackson - butcher and Mrs. P. Gompertz - pawnbroker! Lawson-Batey were to use their Thrift Street premises until 1941.

An early achievement of the company was to reach an agreement with the North Eastern Railway Co. that Lawson-Batey tugs could work within the railway company's Tyne Dock. Hitherto the NER had allowed only their own tugs to work in the enclosed dock.

Problems soon began to confront the new joint company. In the spring of 1920 *Royal Briton* was in collision at Boulogne with the barge *Dr. Brooks* and subsequently sank. On being raised she was patched and taken to Grimsby for permanent repairs where about the same time *Nestor* was also being refurbished following flooding, the cause of which is not known. September saw *Taliesin* being sold for £5,000 back into the management of her original 1883 owners. This was a good price for a 37-year-old tug. However, since she lasted for another 25 years before going to the breakers in 1945 at the age of 62 years, perhaps she was worth it.

Just when things were settling down nicely the Tyne tugmen went on strike on 4th November 1920 over a claim for increased wages and a shorter working week and did not return to work until 13th January 1921. During the strike there was chaos in Shields harbour when, on 3rd December, twenty ships broke loose in a severe gale. How order was restored without tug assistance was not recorded. At the first Annual General Meeting on 15th February 1921 the directors felt that the company was now firmly established so that they could share in the benefits of an expanding economy as the effects of the war receded into the past. Like all companies they were to be severely disillusioned over the decade ahead. Also, at the beginning of 1921 there was a coal strike, which meant that there were no colliers requiring handling by the tugs. This resulted in the harbour tugs being laid-up for a while. However, sea-towage brought in some income when several French ships were towed from Ipswich to St.Nazaire. Nevertheless, the directors decided in August that no interim dividend would be paid, due to the tugboat and miners' strikes and the general depression in trade.

In October 1921 the 37-year-old paddle tug *John Batey* was sold for £3,000 to Grangemouth & Forth Towing Co.Ltd. with whom she worked for a further fifteen years before going to the breakers. This left the company with a fleet of 16 tugs.

The festive season of 1921/22 was celebrated in an unkind way for the company. They had contracted, at a price of £750, to tow the 'Majestic' class battleship *Prince George* (14,890 tons displacement, 421 ft. long x 75ft. beam) from the UK to Germany for demolition. The tugs *Joffre* and *Plover* were given the task but, when off Kamperduin on 30th December 1921, the warship foundered. As will be seen, arguments and claims regarding this loss were to go on for the next six years!

In November 1918 the Royal Navy submarine *H41* had collided with, and damaged, the tug *Ulysses*. Claims in court for this damage were not settled until early 1922 so it will be seen that litigation, both for and against the company, took a long time to resolve.

Because of the low level of trade on the Tyne, Lawson-Batey would not have been averse to selling some of their fleet if satisfactory prices could have been negotiated. However, offers of £10,500 each for *Homer* and *Francis Batey* were not felt to be acceptable. Attempts were made to find work for the tugs at ports other than the River Tyne. The possibility of working at Blyth and Hull was investigated, but shortly afterwards the decision was taken not to persist with either proposal. Economies were made when Joseph Lawson took over from G.A. Clark as company secretary in September 1922 and the size of the office staff was reduced. However, about this time, Jarrow Slake fuel-oil berth was opened, with its potential work for the tugs in handling the tanker vessels calling there.

By March 1923 thoughts again turned towards the possibility of tugs working at other ports: namely Blyth, Leith, Hull and Middlesbrough. Because of the desperate search for work no doubt these other ports were thinking about working on the Tyne! At the same time the company was looking into the prospects of acquiring a controlling interest in the Blyth Steam Tug Co. Amongst the information obtained were the profit figures, before tax, of £7,430, £5,069 and £7,382 for the years ending 31st March 1920, 1921, and 1922 respectively. Negotiations continued. Robert Bell represented the partners of the Blyth company and agreed to amend downwards their valuation of the company as a going concern to £14,000. Lawson-Batey considered this valuation to be too high.

In May 1923 the Lawson-Batey directors decided to acquire (based on a revised valuation of £10,000) a holding of £7,000 in the Blyth Steam Tug Co. and to form a new limited company named Blyth Tug Co.Ltd. Five paddle tugs were to be transferred to the new company which was formally incorporated on 9th July 1923. It was to have a capital of £20,000 in £1 shares, of which £10,000 was to be called up and fully paid. Lawson-Batey would acquire shares not exceeding £7,000 in value and would be represented on the board of the Blyth company by Matthew Coulson James, Joseph Lawson and Frank Batey. Two other directors were appointed from the partners of the Blyth Steam Tug Co.: William Thompson (grocer) and Robert Bell (gentleman). Almost immediately £1,000 of the Lawson-Batey holding was sold to others, leaving the former with £6,000. A reciprocal arrangement was made when John Thompson of Dinsdale House, Blyth was appointed a director of Lawson-Batey.

In November 1923 the London & North Eastern Railway Co.'s West Dunston coal staiths were commissioned after having been completed in December 1914. Whether this would mean more work for the tugs or simply transfer loading from other older staiths only time would tell, but, for the time being, the coal trade remained depressed.

Towage of redundant warships continued to provide

HMS Lion *arriving in the Tyne in March 1924.* *(Company archives).*

good business; provided the tow was not lost on voyage! On 26th February 1924 *Joffre* and *Plover* left Rosyth towing the 'Cressy' class armoured cruiser HMS *Sutlej* (12,000 tons displacement, 472ft. long x 70ft. beam) to Belfast. Whilst crossing the Moray Firth she broke adrift and was missing for 14 days until found 70 miles ESE of Peterhead by *Homer* which, with *Francis Batey*, continued the tow. Next came the battlecruiser HMS *Lion* (26,350 tons displacement, 700ft. long x 89ft. beam) which had been Beatty's flagship at the Battle of Jutland. She was towed from Rosyth to Jarrow by *Joffre, Plover, Francis Batey* and *Comet* where she was partly dismantled before being taken to Blyth for completion of the job. Such a dead hulk must have been quite a task for the relatively low-powered tugs of the time and indeed caused concern when she laid athwart the Tyne entrance on 30th March 1924, notwithstanding the additional assistance of *Hercules* and *Langton*.

In 1924 although few tugs were fitted with radio sets *Joffre* did have radio equipment, but this was frequently out of order and was reported by her skipper to be 'useless'. A new set was fitted by Radio Communication Co.Ltd. for £250. It is interesting to note that this had a range of only 100 miles. *Cruiser* (1) had a major engine breakdown in September 1924 when the crosshead broke in the port engine. This cost £207 to repair but, as was common with old tugs (she being 50 years old), when parts of the engine were removed it was found that repairs to the lower hull were also required.

At Christmas 1924 an offer was received from Mr. Robson of St. Anthony's to purchase *Bulldog* for £425. The company was willing to sell her but not at such a low price and tried to negotiate a higher figure with Mr. Robson, but to no avail. The paddle tug *Criterion*, which was powered with a single side-lever engine, was now 48 years old. It was proposed that she should be re-engined with a set of double-engines taken from the railway company tug *Tyne Dock No.1* (2). A price of £230 was required for the engines and £120 for boilers but since the boilers were not required, £210 was offered by Lawson-Batey for the engines. This was accepted early the following year, when work on *Criterion* cost £1,540 in addition to the engine cost - so a great deal of other refurbishment must have been necessary. At the same time *Hercules* suffered an engine breakdown but this cannot have been too serious since no details were recorded.

In the early months of 1925 dissatisfaction at the high towage rates on the River Tyne was expressed publicly by the Newcastle Chamber of Commerce following two years of discussions with the tugowners. They claimed that there was no real competition and that the tug companies were operating a price ring. There were even threats made that the shipowners would combine and provide their own tugs! As a result the Tyne Improvement Commission had a clause included in their Bill then before Parliament to allow them to control towage rates on the whole of the river.

Plover *and* Homer *taking* Ranpura *(16,688/1925) out of the Tyne for her trials after completion by Hawthorn Leslie. (Local History Library, South Shields Central Library).*

At the 5th Annual General Meeting of the company on 23rd February 1925 it was reported that the Lawson-Batey investment in Blyth Tug Co.Ltd. had produced satisfactory results; cautious optimism considering the depressed state of the business. An interest was shown in purchasing *Frisky* (185/1918) an ex-Admiralty tug from the Surplus Liquidation Department who wanted £7,500 for her, but, in the event, she was bought by others. In April *Langton* stopped working because of the need for boiler repairs. A new boiler was purchased for £1,090 from R. & W. Hawthorn Leslie & Co.Ltd. at Hebburn (one of the directors of Lawson-Batey was also managing director of Hawthorn Leslie). Other repairs were necessary whilst fitting the boiler at J.P. Rennoldson's and these cost another £390. *Langton* did not return to service until late January 1926.

A major catastrophe was now to befall the company. *Joffre* had left Falmouth in the evening of 26th May 1925 bound for Antwerp. In the early hours of the following morning she stranded near Bolt Head, Salcombe, Devon, 9 miles west of Start Point. Although all of the eleven man crew were rescued by Hope Cove Life Saving Crew from 200ft. high cliffs, Captain Ralph Hodge tragically succumbed from exposure shortly afterwards. Subsequently, the mate, William Harrison Thurlbeck, was awarded the Bronze Medal for Gallantry in Life-saving at Sea for taking the tug's boat to the shore in order to bring back a line after the Rocket Crew had failed to get a line over the vessel. This feat was extremely hazardous on a rocky shore and under very high cliffs. Members of the local life-saving team also received awards. The Liverpool and Glasgow Salvage Association was contracted to salvage *Joffre* and succeeded in refloating and beaching her at Salcombe on 30th May, where she was pumped out and her hull patched. She was towed to Falmouth on 19th June, dry-docked there the following day and stayed in the dock for the next ten days. Arrangements were made for the tug to be repaired by Tyne Dock Engineering Co.Ltd. for £1,750 to the satisfaction of Lloyds Register of Shipping and the Tyne Mutual Marine Insurance Association; the job to take 28 days. *Joffre* was towed to the Tyne in July and did not return to service until September. Although taken to TDE's dry-dock it would appear that the repairs were actually carried out by J.P. Rennoldson since, by July 1926, that company was claiming a repair cost of £4,218 from Lawson-Batey whereas the surveyor advised a payment of not more than £2,500. Two months later Rennoldson's were prepared to accept £3,500 and the surveyor thought that £3,300 was sufficient. Lawson-Batey had paid £2,750 up to this time. In March 1927 the company paid a further £400 and thus the matter ended. No doubt this all sounds like market-stall bargaining, but it must be realised that times were hard and money short, as will be seen ere long regarding financial economies. After her repair *Joffre* was soon at work again towing a pontoon from Queenborough to Scapa Flow.

Joffre *aground on the rocky Devon coast 27 May 1925.* *(Local History Library, South Shields Central Library).*

In July 1925 *Bulldog*, for which offers had already been received and rejected, was working for Allen Brown on one month's trial which, if satisfactory, would mean that the tug would be sold to that firm for £825. In fact, the sale took place in September, leaving a Lawson-Batey fleet of fifteen tugs.

The General Strike of 4th May 1926 brought business to a standstill. Although this lasted only about a week, the miners chose to continue the strike and stayed out for about seven months. Traffic on the Tyne was paralysed and only skeleton tug crews were working, with most of the men being paid off. This all came on top of the level of exports for 1925 being 4 million tons less than in 1924, as reported at the 6th Annual General Meeting. From October 1926 economies were announced; directors' fees were to be halved, the managing director's salary reduced by 20% and officials' salaries were to be reduced by between 10 and 15%.

Salvage jobs were fairly frequent in this period - a typical award being of £1,460 for services to *Crackshot* (2379/1924) by *Plover* and *Hercules* in the summer of 1926. More troubles were experienced when towing redundant warships to the breakers when the dreadnought HMS *Thunderer* (22,500 tons displacement, 581ft. long x 89ft. beam) went aground outside Blyth harbour at the end of the year when being towed from Rosyth to Blyth by *Joffre, Plover, Homer* and *Francis Batey*. She too had been at the Battle of Jutland and had been Jellicoe's flagship. The spring of 1927 saw only nine tugs working, the other six being laid up for want of jobs. Consequently, there were thoughts of disposing of some of the smaller tugs, but who would want to buy with trade at such a low ebb?

HMS Thunderer *being towed into Blyth by* Plover, Joffre *and* Homer *in early 1927.* *(Company archives).*

In May 1927 the compensation claim against the company regarding the loss of HMS *Prince George* in 1921 was still in dispute when the directors decided to make an offer of £3,000 - £5,000. Two months later they were prepared to offer £6,500 and eventually settled for £7,500. An overdraft was necessary to pay for this settlement and it would appear that actual clearance of the debt was not made until 1928. This must have placed a heavy financial strain on the company at the time and it was not until the 10th Annual General Meeting in 1930 that the directors were reporting that 'the skeleton in the cupboard the *Prince George* case has now been eliminated from the balance sheet'.

But there was another 'skeleton in the cupboard'! The Seyri Sefaine Turkish Authority had issued a writ against Lawson-Batey regarding the sale of *Francis Batey* to the Turkish Government by J. Batey & Son Ltd. in 1914. Just what this was about has not come to light, but presumably she was going to be transferred to Turkey when new in 1914 and this move was stopped because of the war. She had, of course, been hired to the Royal Navy very soon after completion so no doubt they too were involved in some way. Whatever the reason this was one more worry which the management could have done without. The claim hung around for the next eighteen months before being thrown out by the Turko-British Arbitrary Tribunal.

Although the company had received a £2,500 award for the salvage of *Ramsay* (5053/1921) by *Homer, Francis Batey* and *Hercules* they were not happy, because of several problems in recent years, about continuing with deep sea and coastal towage, so in November 1927 the board decided that they would not undertake deep sea towage. Coastal towage was to be undertaken only under special circumstances. Strong efforts were to be made to develop further the harbour towage side of the business and with this in mind *Joffre, Francis Batey* and *Homer* were surveyed for carrying passengers when involved in trial trips. Captains were instructed that for the time being they were not to carry more than twelve persons in addition to their ordinary crews until licences were obtained in 1928 for each tug to carry one hundred work people.

Difficulties or not, the company managed to pay a dividend of 2½% gross for 1927 and 4% in 1928.

In 1928 an item of incidental interest took place. Richardson & Brock's wooden floating dock was sent away to be broken up. It had been in commission for over seventy years, having been built in 1854. Over two thousand vessels, including many tugs, had been docked in it. About 1860 the dock was stationed at the north side of the Tyne when, during a gale, it broke loose and drifted to the south side with a vessel in it. There it remained for the rest of its life!

Salvage activities which needed a large number of tugs resulted in the local companies, who were normally in competition, working together and sharing the benefits. One such occurrence in January 1928 saw *Joffre, Homer, Titan, Ulysses* and *Francis Batey* of Lawson-Batey, *Warrior* of Joseph Crosthwaite (Tugs) Ltd., and *Hendon* and *Cullercoats* of

Homer *and* Francis Batey *towing the Anglo-American Oil Co.'s oil-barge* Navahoe *(7,718/1908) in the 1920s.* *(Company archives).*

France, Fenwick Tyne & Wear Co.Ltd. combining their efforts. The same Lawson-Batey tugs combined on another salvage job with *President* and *Contest* of the Brown Tug Co.Ltd.

The company had been wanting to sell *Criterion* for almost two years and they eventually found a buyer for her, but as scrap! Surprisingly, they received only £290 for her, having spent £1,750 on a new engine only four years before. Costly incidents still arose when on the night of 19th November 1929 *Homer, Nestor, Ulysses* and *Cruiser* (1) were broken adrift from their moorings by the collier *Flashlight* (934/1920). Nearly a month later, in the early hours, a fire occurred on *Ulysses* which caused severe damage; her engine house sides were destroyed and part of the bridge deck and deck around the engine house were badly affected. The cause was unknown.

The Depression in Trade Continues. 1930 - 1939

The tug *W.D.S.* was sold for £600 to Thomas Keedy of South Shields in 1930 after having been laid up for two years. She had cost the Batey company no less than £5,500 in 1919 and was to survive for another nineteen years, ending up working in Ireland. This sale reduced the Lawson-Batey fleet to thirteen tugs (5 paddle and 8 screw), with the average age now increased to 42.5 years. Although the board had decided three years previously not to seek work outside the harbour, it is interesting to note that a 1930 advertisement showing *Joffre* offered continental and deep sea work as well as service on the Tyne!

Changes to the board of directors took place in August 1930, when Joseph Lawson, who had been seriously ill for three years, was disqualified from holding office as a director and therefore ceased to be joint managing director and secretary. Frank Batey took over as secretary temporarily and was eventually replaced by John Robert Lawson from 1st December. A new director, Robert Ingo James, was appointed on 25th September 1930. For both 1929 and 1930 the company's financial position had improved considerably, and for these years they were able to pay a gross dividend of 6%.

Advertisement of 1930. (Company archives).

Mercantile Chambers, Quayside, Newcastle upon Tyne in the 1930s. The entrance is just to the right of the left-hand car. The Lawson-Batey offices are to the right of the door on the ground floor, and France, Fenwick Tyne & Wear occupied the whole of the first floor. Redhead & Dry's Tugs had their offices in the first floor of the building to the right. (Harrington collection).

Homer *towing the whaling factory ship* Southern Princess *(12,156/1915) in the 1930s. (Company archives).*

On 7th January 1931 disaster struck again, this time within the River Tyne. The paddle tug *Hercules* was attending the Blue Star Line *Napier Star* (10,583/1927) which was proceeding to sea from Palmer's Hebburn Quay. The *Napier Star* collided with and sank the tug, involving the loss of three lives - John B. Jago, engineer; Frank Watts, fireman and his son Frank Watts Junior. The latter had been on board without the knowledge of the master Peter Newton. The tug sank in midstream abreast of Northumberland Dock and lay bottom-up with her bows buried in the mud. She had a large

Hercules *lying upside-down on Jarrow Slake after the accident of 7 January 1931. It is interesting to note the very small area of her rudder. (Company archives/Frank & Sons).*

Joffre *and* Francis Batey *towing the tanker* Cadillac *(12,074/1917) in 1930. (Company archives).*

Another view of Hercules *after her collision with* Napier Star. *(Company archives/Frank & Sons).*

gash in her port side extending through the keel. With considerable difficulty divers were successful in getting lifting chains underneath the hull, by means of which, on 14th January, the floating crane *Titan* was able to lift *Hercules,* still bottom up, on to the edge of Jarrow Slake. Only in early February was it possible to move the tug further up on to the Slake so that she was out of the water at low tide. The accompanying photograph shows her in this position in a very sad state. She was broken up where she lay. Her insured value was £4,000. *Hercules* had a life of 54 years and had seen service at Hull before coming to the Tyne in 1893 when purchased by Joseph Lawson. At the subsequent enquiry *Napier Star* and *Hercules* were adjudged to be two-thirds and one-third respectively at fault. £3,000 was received from the Tyne Mutual Marine Insurance Association.

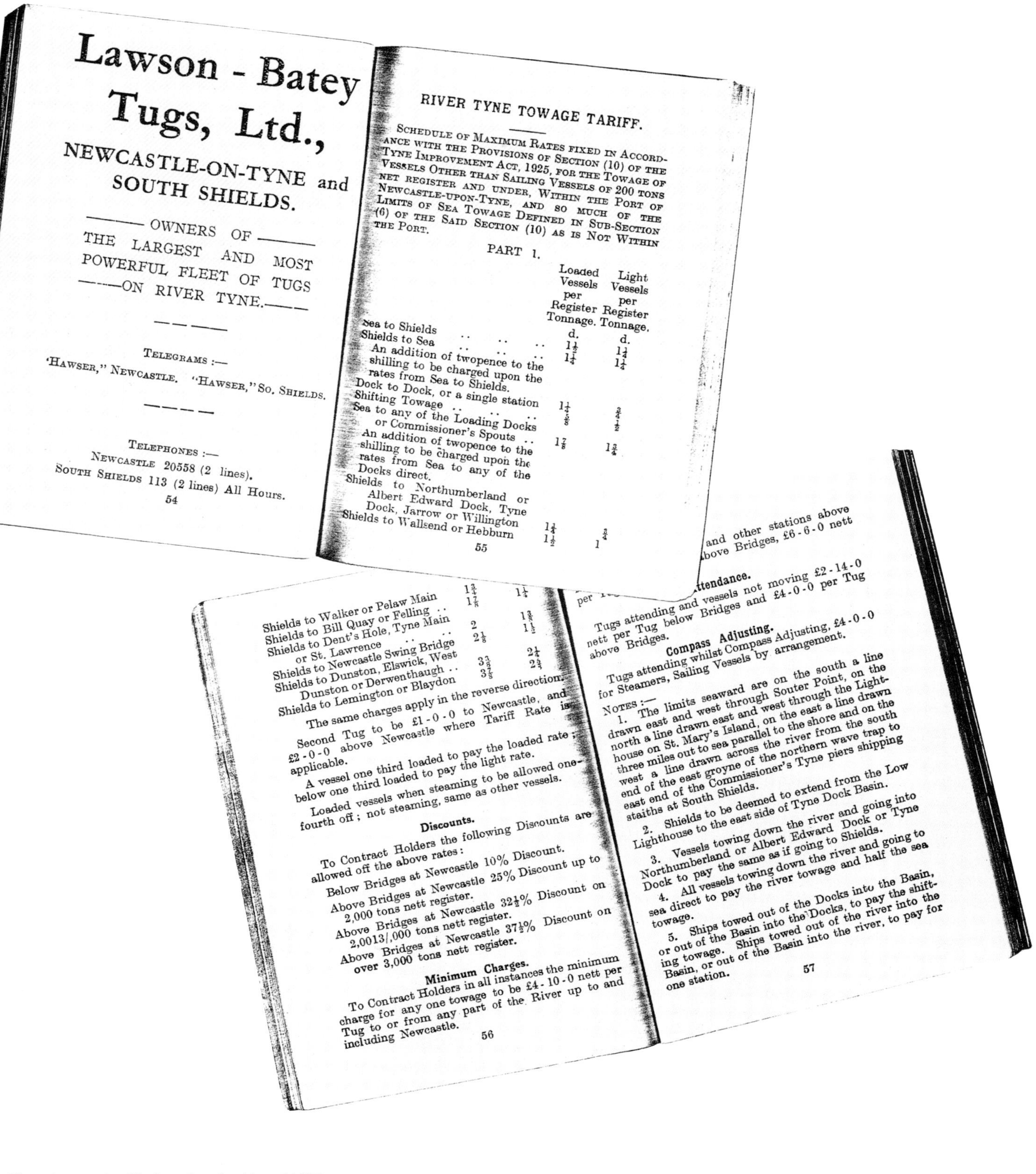

Lawson - Batey
Tugs, Ltd.,
NEWCASTLE-ON-TYNE and
SOUTH SHIELDS.

OWNERS OF
THE LARGEST AND MOST
POWERFUL FLEET OF TUGS
ON RIVER TYNE.

TELEGRAMS :—
"HAWSER," NEWCASTLE. "HAWSER," SO. SHIELDS.

TELEPHONES :—
NEWCASTLE 20558 (2 lines).
SOUTH SHIELDS 113 (2 lines) All Hours.

54

RIVER TYNE TOWAGE TARIFF.

SCHEDULE OF MAXIMUM RATES FIXED IN ACCORDANCE WITH THE PROVISIONS OF SECTION (10) OF THE TYNE IMPROVEMENT ACT, 1925, FOR THE TOWAGE OF VESSELS OTHER THAN SAILING VESSELS OF 200 TONS NET REGISTER AND UNDER, WITHIN THE PORT OF NEWCASTLE-UPON-TYNE, AND SO MUCH OF THE LIMITS OF SEA TOWAGE DEFINED IN SUB-SECTION (6) OF THE SAID SECTION (10) AS IS NOT WITHIN THE PORT.

PART 1.

Loaded Vessels per Register Tonnage. d. — Light Vessels per Register Tonnage. d.

Sea to Shields .. .. .. $1\frac{1}{2}$ $1\frac{1}{4}$
Shields to Sea .. .. .. $1\frac{1}{4}$ $1\frac{1}{4}$
An addition of twopence to the shilling to be charged upon the rates from Sea to Shields.
Dock to Dock, or a single station
Shifting Towage .. .. .. $1\frac{1}{4}$ $\frac{3}{4}$
$\frac{5}{8}$ $\frac{1}{2}$
Sea to any of the Loading Docks or Commissioner's Spouts .. $1\frac{7}{8}$ $1\frac{3}{4}$
An addition of twopence to the shilling to be charged upon the rates from Sea to any of the Docks direct.
Shields to Northumberland or Albert Edward Dock, Tyne Dock, Jarrow or Willington $1\frac{1}{4}$
Shields to Wallsend or Hebburn $1\frac{1}{2}$ $1^{\frac{3}{4}}$

55

$1\frac{3}{4}$ $1\frac{1}{4}$
Shields to Walker or Pelaw Main $1\frac{7}{8}$
Shields to Bill Quay or Felling .. $1\frac{3}{8}$
Shields to Dent's Hole, Tyne Main or St. Lawrence .. .. 2 $1\frac{1}{2}$
$2\frac{1}{8}$
Shields to Newcastle Swing Bridge $2\frac{1}{4}$
Shields to Dunston, Elswick, West Dunston or Derwenthaugh .. $3\frac{3}{8}$ $2\frac{3}{4}$
Shields to Lemington or Blaydon $3\frac{7}{8}$

The same charges apply in the reverse direction.

Second Tug to be £1 - 0 - 0 to Newcastle, and £2 - 0 - 0 above Newcastle where Tariff Rate is applicable.

A vessel one third loaded to pay the loaded rate; below one third loaded to pay the light rate.

Loaded vessels when steaming to be allowed one-fourth off; not steaming, same as other vessels.

Discounts.

To Contract Holders the following Discounts are allowed off the above rates:

Below Bridges at Newcastle 10% Discount.

Above Bridges at Newcastle 25% Discount up to 2,000 tons nett register.

Above Bridges at Newcastle $32\frac{1}{2}$% Discount on 2,0013/,000 tons nett register.

Above Bridges at Newcastle $37\frac{1}{2}$% Discount on over 3,000 tons nett register.

Minimum Charges.

To Contract Holders in all instances the minimum charge for any one towage to be £4 - 10 - 0 nett per Tug to or from any part of the River up to and including Newcastle.

56

and other stations above
above Bridges, £6 - 6 - 0 nett
per

Attendance.

Tugs attending and vessels not moving £2 - 14 - 0 nett per Tug below Bridges and £4 - 0 - 0 per Tug above Bridges.

Compass Adjusting.

Tugs attending whilst Compass Adjusting, £4 - 0 - 0 for Steamers, Sailing Vessels by arrangement.

NOTES :—

1. The limits seaward are on the south a line drawn east and west through Souter Point, on the north a line drawn east and west through the Lighthouse on St. Mary's Island, on the east a line drawn three miles out to sea parallel to the shore and on the west a line drawn across the river from the south end of the east groyne of the northern wave trap to east end of the Commissioner's Tyne piers shipping staiths at South Shields.

2. Shields to be deemed to extend from the Low Lighthouse to the east side of Tyne Dock Basin.

3. Vessels towing down the river and going into Northumberland or Albert Edward Dock or Tyne Dock to pay the same as if going to Shields.

4. All vessels towing down the river and going to sea direct to pay the river towage and half the sea towage.

5. Ships towed out of the Docks into the Basin, or out of the Basin into the Docks, to pay the shifting towage. Ships towed out of the river into the Basin, or out of the Basin into the river, to pay for one station.

57

The River Tyne towage tariff given in a booklet of 1932.
(Company archives).

In late January 1931 Lawson-Batey made an offer to purchase the shareholdings of the late Joseph Crosthwaite Jr. and the late Nicholas Brown in Joseph Crosthwaite (Tugs) Ltd. They offered 12/- (60p.) per share for a total of 16,500 shares which, if accepted, would have meant the transfer of the company to Lawson-Batey. Captain F.B. West and Lewis Crosthwaite declined to accept the offer.

During 1931, *Homer* was chartered to the Gamecock Steam Towing Co.Ltd. on the Thames for a short period, in an attempt to obtain as much work as possible for the tugs. The level of trade was very low indeed and it was reported that 70% of steamers were laid up the world over. This situation was reflected in the fall in the company's gross dividend which was 5% in 1931, 4% in 1932 and would drop further to 3% in 1933. In 1931 one of the directors, R.J. Thompson, who had been a director of Lawson-Batey since its inception in 1920, died, reducing the size of the board to five. The company's fleet of twelve tugs was valued during 1932 at £68,847. *Royal Briton* or *Nestor* were offered to the Halal Shipping Co.Ltd. for £3,000 and £2,750 respectively, but neither offer was taken up. In 1933 the company had contracts to handle the ships of no less than thirty-four U.K., and twenty-nine foreign shipping companies; no mean feat in this difficult period.

The company's office (with the flagpole) at 15 Thrift Street, South Shields in 1932. It will be noted that the area is looking quite run-down as compared with the earlier photograph of 1920. Between Thrift Street and the river is the repair yard of Tyne Dock Engineering Co. In the view can be seen one of the company's tugs, probably Plover. *(Local History Library, South Shields Central Library).*

In July 1933 Lawson-Batey was approached by France, Fenwick Tyne & Wear Co.Ltd. and Redhead & Dry's Tugs Ltd. jointly to ask whether Lawson-Batey would be prepared to enter into a joint scheme with themselves to purchase Joseph Crosthwaite (Tugs) Ltd. on an equal division of shares. The capital of the Crosthwaite company was £17,000 and France Fenwick already had an option to purchase at 10/- (50p.) per share. Lawson-Batey agreed with the scheme but thought that the price was too high. It should be noted that Lawson-Batey had offered 12/- (60p.) for these same shares in 1931 and had been turned down. However, a week later, Lawson-Batey agreed the price and in September 1933 they purchased 5,666 £1 shares in Joseph Crosthwaite, allocating 500 each to their directors Frank Batey and R.I. James, who became members of the Crosthwaite board. Two members of the France Fenwick board, J.E. Dawson and F. Dawson became chairman and secretary respectively, of Joseph Crosthwaite (Tugs) Ltd. Thus Lawson-Batey became one-third owners of the Crosthwaite company which continued to operate under its own name until 1949 with initially three tugs *Champion* (3), *Joseph Crosthwaite* and *Lewis Crosthwaite*.

As 1933 drew to a close Lawson-Batey were still trying to sell two of their tugs so that the size of their fleet could match the business available, but to no avail because prices being offered were very low.

At the 14th Annual General Meeting of the company on 19th February 1934 it was reported that 1933 had been one of the worst years experienced by the shipping industry. The company had obtained no salvage work and there had been no income from their majority holding in Blyth Tug Co.Ltd. Sadly, on 5th May 1934, the company chairman, M.C. James, died. He had joined the board of the Lawson Steam Tug-boat Co.Ltd. in May 1909, had been elected chairman of that company on the death of T.N. Alexander in March 1918, and had also, of course, been chairman of Lawson-Batey since its formation in 1920. In June 1934 J.T. Batey was appointed to succeed him and R.I. James was appointed to represent Lawson-Batey on the Blyth board in place of his father.

In June 1934 the 49-year-old *Royal Briton* was sold to Jarrow Dredging & Salvage Co.Ltd. for only £500, making a loss against book value of no less than £3,628. This sale reduced the tug fleet to eleven. Further financial problems arose due to the fact that no income was received from the one-third share in the Crosthwaite company and *Joffre* needed a new boiler which, with installation expenses, was to cost £1,945 by early 1935 when she was re-commissioned.

In the local newspaper *'Shields Gazette'* of 25th February 1935 there appeared a story recounted by a woman reporter who took a trip on a Lawson-Batey tug. This article gives such a good insight into the times of the 1930s that it is worth repeating here almost in its entirety:

"The Day's Work on a Tyne Craft - Tugboatmen's Life through the Eyes of a Girl."

"Last week I went out on a tug that was taking a ship up river to Dunston Staiths. She was due about 3pm. and was coming from Middlesbrough. I was rowed over to the *Comet;* a yellow funnel with a black diamond showed that she belonged to the Lawson-Batey company. I found the men on the tug waiting for their ship to appear, ready at any moment to set out to sea. They had been aboard since 8am. that morning.

Joffre, Francis Batey *and* Homer *towing the Brocklebank liner* Manipur *(8,652/1920) downstream past Newcastle Quay in 1935.* *(Company archives).*

"When the ship's smoke was sighted beyond the piers the tugboat's engines began to throb, the paddles to thrash the water and we moved out into the river as our ship, a Spaniard, entered the harbour. Another black diamond tug, the *Homer,* screw driven, waited to help. Before we took up our position for towing we had to deliver a letter to the ship.

"As we drew nearer we could make out her name, *Urkiola-Mendi* (3,552/1902), registered in Bilbao. The tug seemed to me quite a large vessel but the approaching ship, though her masts were lowered, stood high above us like a wall that flew along. We raced along by her side ready to pass up the letter. I was on the bridge beside Captain Elsdon whilst the mate Septimus Cummings was going to pass the letter up. He fixed it in a slit at the end of a long bamboo pole then stepped out on top of the paddle box. Wavering, the pole reached up the ship's side until an officer leant over the rail and took it.

Plover *towing the Spanish* Aritz-Mendi *(5,754/1920)*
(Local History Library, South Shields Central Library).

"*Homer* had taken up the bow tow whilst at the stern a Spaniard in a beret passed a rope down to us which was quickly fastened to our bow samson post. Everything seemed to be done with so little fuss. The foyboatmen hooked on to the *Comet* and then came aboard.

"The first thing you have to do on board a boat is to take a piece of cotton-waste to wipe your hands on and to clean the brasses and to get hold of very oily things. Peter Newton, the engineman, gave me a piece. He wore a seaman's peaked cap and brown overalls. The fireman has mighty muscles, and the boy who does all the washing and polishing wore a natty little jersey with short sleeves.

"We were passing steadily up the river now, guiding *Urkiola-Mendi* round difficult bends on the way. Tyne Dock, Northumberland Dock, Jarrow and Howdon were behind us. I was employing myself polishing the galley door handle and found that it takes a lot of effort to remove the salt-spray stains from brass. Then the men led me down an iron ladder into the stokehold and let me fling coal into the furnaces. Once we passed a half-built ship on the stocks and the battleships lying at Hebburn were alive and hope-giving. Opposite the quays where the Portuguese sloops lay berthed was a grey battleship for India. By her side the shipbuilders worked on rafts.

"We had tea in the men's quarters on a clean scrubbed table. The captain gave me some home-baked bread and someone else offered me a kipper. My next job was to light the lamps that hang to port and starboard. I carried them up from the hold, great heavy things shielded by thick glass. Then Newcastle Quay came in sight; lights twinkled in the windows of the shipping offices along the waterfront making them look as romantic as a painting. The swing bridge opened and we sailed through then under the High Level bridge looking very grim.

"So we reached the quiet of Dunston Staiths and the ship was made fast to the buoys, then home by moonlight as we sped down the river. The last quarter of an hour of my trip was spent removing half the funnel and other black parts of the tug from my face. This I had the privilege of doing in the captain's cabin with the oil-cloth pushed back from his table for my dish to stand on and a mirror hanging in front of the brass-bound lamp swinging from the skylight. Reluctantly I combed my hair and powdered my nose and shortly stepped ashore at the South Shields ferry landing. Darkness engulfed the *Comet*, her skipper and the crew."

A nice piece of journalism of times long since past.

Cover of a booklet about the company published in 1935.
(Company archives).

Joffre *and* Plover *about to pick up the tow of the Dutch Ocean Steamship Co.'s vessel* Phrontis *(6,181/1925) as she enters the Tyne.*
The arrangement of the superstructure, funnel and masts of Phrontis *is very unusual.* *(Company archives).*

The White Star liner Olympic *arriving at Jarrow for breaking up on 13 October 1935.*
Towage was shared between the Lawson-Batey, France Fenwick and Redhead & Dry's companies. *(Company archives).*

Now to record a few miscellaneous events from 1935. Henry Jobling, master of *Langton,* was fined £1 for mooring his tug to the private buoys of the British Mexican Petroleum Co.Ltd. after being told not to do so by the Harbour Master. In defence, Henry claimed that the tug mooring facilities were inadequate. John Francis Batey, son of Frank Batey, joined the company at nineteen years of age. He was later to become managing director. The chairman, J.T. Batey, was awarded an honorary doctorate (D.Sc.) of Durham University. The Cunard-White Star liner *Olympic* (46,439/1911) arrived in the Tyne on 13th October 1935 for breaking up at Jarrow and was attended by the Lawson-Batey tugs *Joffre* and *Plover, Hendon* (2) and *Wearmouth* of France Fenwick, and *George V* (2) and *Great Emperor* (2) of Redhead & Dry's. A sharing of the work in the current hard times. After being cut down to a hulk, *Olympic* was towed again to Inverkeithing in September 1937 for completion of her demolition.

The 16th Annual General Meeting on 9th March 1936 still reported difficulties due to lack of trade. In particular, there had been large reductions in coal shipments to Italy from the railway company staiths at Dunston. Dr. J.T. Batey was appointed to represent Lawson-Batey on the Blyth Tug board. He was shortly to retire from his position as managing director of R. & W. Hawthorn Leslie & Co.Ltd. where he had started as an apprentice and been a director since 1907. *Titan,* now 71 years old, was sold for breaking up at Dunston and realised £205. Further reductions in the fleet occurred in 1937 when *Mentor* was sold for £570 to Cappon of Dundee for whom she continued to work for another nine years. *Nestor,* 54 years old, and *Ulysses,* 63 years old, were both sold for demolition at Dunston for about £450 each. This left the Lawson-Batey fleet with only seven tugs - a severe reduction from the eighteen they had owned when established in 1920.

Company letterhead of c.1936.

Thus by mid-1937 the fleet consisted of:-

Cruiser (1)	(98/1874)	Iron	Paddle
Comet	(132/1876)	Iron	Paddle
Plover	(245/1881)	Iron	Twin screw
Homer	(157/1915)	Steel	Screw
Joffre	(260/1916)	Steel	Screw
Langton	(157/1892)	Steel	Screw
Francis Batey	(151/1914)	Steel	Screw

- with an average age of 41.5 years.

The company had offices at Mercantile Chambers on the Quayside, Newcastle and 15 Thrift Street, South Shields.

In 1938 business on the Tyne was fluctuating severely, making forward planning very difficult. Also, the Crosthwaite company had not been able to pay a dividend for the year ended 5th April 1938. In December another large passenger liner arrived for breaking up at Jarrow. This time the vessel was the Cunard-White Star *Berengaria* (52,101/1913) and the same mixed company of tugs which had handled the *Olympic* in 1935 brought her in. She was to stay in the river until 1946 when sufficiently cut down to be towed to Rosyth.

In May 1939 holidays with pay for the tug crews were granted for the first time. The tugboatmen were to be allowed seven days off per year, but whether they were able to take advantage of this for some time due to forthcoming events, is doubtful. When tug crews were needed during the night, in the days before widespread use of the telephone, they had to be roused from their beds by 'Night Callers' who, in 1939, were paid 6d. (2.5p) per week per man.

The Second World War Period. 1939 - 1945.

The Second World War began on 3rd September 1939, but a few days before that, on 28th August, the tugs *Francis Batey* and *Homer* had been requisitioned under an agreement with the Director of Sea Transport. They left the Tyne the following day and arrived at Kirkwall early on 30th August. The former was to be away for over three years, serving in the Irish Sea as well as Orkney. *Homer* was away for nineteen months. This left Lawson-Batey with only five tugs for their work on and around the Tyne. John Francis Batey, who by now was assistant manager, was called up quickly to join the Royal Field Artillery, for he was a Lieutenant in the Territorial Army.

Life for the tugboatmen on the Tyne, and elsewhere, was onerous, the crews having to work very long hours in difficult and dangerous conditions. This was particularly so at night when there were no lights to give them the usual guidance. Ships arriving in convoy presented problems for the tugs in having to deal with a large number of ships in a concentrated period of time. There was also the constant stream of new vessels being launched and undergoing trial trips some of which were battleships, cruisers, destroyers and escort vessels for the Royal Navy. During the war, of course,the number of vessels requiring repair work increased enormously.

On the afternoon of 14th December 1939 news was received that the tanker *Atheltemplar* (8,939/1930) had struck a mine and was lying powerless eleven miles from the Tyne. Notwithstanding the presence of mines off the harbour *Joffre* and *Langton,* which were then at Blyth, set off to give assistance to the tanker along with Redhead & Dry's *Great Emperor* (2). They were met by the destroyers HMS *Mohawk* and HMS *Kelly* to act as escort because of the risk of enemy interference from air or sea. Visibility was poor and it was almost dark when they reached *Atheltemplar* at 4.30pm. The casualty's stern was high out of the water, her forecastle was awash and she was rolling very badly. A short distance away was another tanker *Inverlane* (9,141/1938), derelict and on fire. Whilst deciding how best to tackle these two vessels HMS *Kelly* struck another mine and was left without power. The tug *Great Emperor* (2) went to the assistance of *Kelly* whilst *Inverlane* had to be left to her fate, subsequently drifting ashore. The story of *Great Emperor* (2) and *Kelly* will be completed in the chapter relating to the history of Redhead & Dry's Tugs Ltd.

Meanwhile, *Joffre* and *Langton* were having great difficulty in making fast to *Atheltemplar* but were finally able to do so when members of *Mohawk's* crew boarded the tanker. The tanker took violent sheers, increasing the danger of the tugs' towing hawsers fouling one another, but they managed to arrive off the Tyne entrance by 11pm., only to be told that they could not enter. They then proceeded to tow the stricken vessel between the Tyne entrance and Souter Point, a distance of 2 miles, throughout the night - an exasperating job which, with safety so near at hand, was made worse still by dense fog which descended upon them. By mid-day on the 15th it was decided that the tanker was in such dire straits that it was too risky to bring her into the Tyne and she would have to be beached south of the river entrance. With assistance from *Great Emperor* (2) and *George V* (2), *Joffre* and *Langton* completed the beaching operation by 7.15pm. and then remained standing-by throughout the night in case the *Atheltemplar* refloated. The tug crews had thus been at work for about 43 hours, presumably with no chance of any sleep in those conditions - a typical example of how unpredictable are the hours of tugboatmen. It has always been and can still be so.

Joffre and *Langton* were not, however, finished with the *Atheltemplar,* for after the tanker was lightened they towed her off the beach into the Tyne where her remaining cargo was recovered and later she was towed to a repair yard. This rescue of the damaged *Atheltemplar* goes down as one of the greatest services rendered by the Tyne tugs during the war.

At the 20th Annual General Meeting on 27th February 1940 it was reported that the financial state of the company was now quite good and it is notable that the gross dividend had risen gradually to 12.5% from the 3% of 1935. All five directors were present at a board meeting on 29th July 1940, but very soon afterwards John Thompson, who had been a director since 1923, died. It was decided that John Francis Batey would be appointed a director as from 1st January 1941 as a replacement. Although serving in the army he attended the board meeting on 17th December 1940, whilst on leave.

On 1st December 1940 the Lawson-Batey tugs were involved in another major salvage operation. Again it was a tanker that was in trouble and again it was due to enemy mines. This time the *British Officer* (6,990/1922) was seriously damaged and without steam off the mouth of the Tyne. She was being towed into the river by *Joffre* and *Langton* together with Ridley's *Wonder* and *King Edward VII* and Redhead & Dry's *Great Emperor* (2), when her stern grounded, her bows being high out of the water. Ropes were passed ashore to the south pier to prevent the tanker blocking the channel. *British Officer* now took on a serious list, whilst *Joffre* and *Langton* continued towing ahead, later assisted by *Plover.* Gale force winds did not help matters at all. The tugs had been taking turn and turn about, trying to keep the ship's head parallel to the channel, when, during *Plover's* stint, just after midnight on 13th December, the tug suddenly found that she was moving slowly ahead and that she was towing only the fore-part of the tanker up-river, whilst the after part sank. The fore-part was still fastened to the south pier and it took until 5am. to get her released. For twelve days the tugs had struggled to save this ship and had only been partly successful, but they had been wholly successful in preventing the blocking of the river entrance which would have been disastrous.

The offices of John Cowie & Co. Ltd. in Ferry Street, South Shields, to which Lawson-Batey moved in 1941 after being bombed out of 15 Thrift Street.

The company was finding difficulty in carrying out repairs to its tugs due to the high work-load, which is not surprising since of their seven tugs only five were available on the River Tyne. However, the situation was eased slightly when *Homer* returned from Kirkwall and Lawson-Batey took over the management of the railway company tug *NER No.4* which normally worked at Hartlepool, on behalf of the Ministry of War Transport. Problems of a different kind arose when the old Lawson Tug-Boat office at 15 Thrift Street, South Shields was damaged by a bomb, which exploded alongside one of the Tyne Dock Engineering Co.'s drydocks and lifted a ship off the keel blocks, laying it on its side. Fortunately there were no casualties. Lawson-Batey moved temporarily to John Cowie & Co. Ltd., ship store merchants in Ferry Street. The company also occupied a new office at Exchange Buildings, Quayside, Newcastle upon Tyne.

In October 1941 £600 was awarded for the salvage of the tanker *British Reliance* (7,000/1928), £200 of which was passed on to the France Fenwick company for the assistance which their tug *Wearmouth* had given.

In July 1942 news was received that John Francis Batey, now a Captain in the Royal Artillery (and soon to be promoted major), had been wounded in his eyes at Gazala. Thankfully this was not too serious and he was back in action again before long. However, one did not necessarily have to be in the forces to suffer at the hands of the enemy in those times; the skipper of *Plover,* A. Berry, was killed by enemy bombing at his home in South Shields.

After being in temporary premises for a while, Lawson-Batey set up office at 15 Wright's Buildings, Dean Street, South Shields. Near the end of 1942 *Francis Batey* was returned from her period of requisition and the management of *NER No.4* ceased when she returned to Hartlepool. The company fleet was now back to full strength on the Tyne, with *Comet, Cruiser* (1), *Joffre, Homer, Plover, Francis Batey* and *Langton* together as they had been since 1937.

Even during the war the company advertised its existence and one such publicity document referred to Lawson-Batey as owners of the 'Black Diamond' and 'Blue Cross' Line of Tugs. Thus, through their funnel markings they were perpetuating the identity of the separate Lawson and Batey companies even though the two had been completely amalgamated since 1920!

A large number of tugs had been built during the war for the Ministry of War Transport, often with the standard naming system *'Empire————'*, and these were placed under the management of many of the UK tug companies. One such tug was *Empire Sybil* (276/1943), newly-built by Cochranes at Selby. At the beginning of September 1943 she was being modified for service on the Tyne although only completed on 24th August! Just what the special requirements were for the Tyne has not been discovered. She remained under Lawson-Batey management until put up for sale in 1946 when the company unsuccessfully tendered £19,750 for her. She went to Liverpool in May of that year.

In February 1944 *Comet* assisted the France Fenwick tug *Hendon* (2) to fight a fire in *Coity Castle* (2,767/1919). The ship's magazine was in danger of exploding and although it was difficult to gain access because she was moored between two other vessels, the fire was successfully extinguished, preventing a very serious incident.

A company 'Black Diamond' and 'Blue Cross' advertisement which still, in the 1940s, differentiated between the Lawson and Batey companies which had been combined over twenty years earlier.

Langton *displaying the wartime identification T.391 on her bow. These numbers are discussed in Chapter Eight.* *(Company archives).*

The company secretary, John Robert Lawson, resigned in March 1944. He had held that post since 1930 and was one of the principal shareholders in the company. His service with the Lawson and Lawson-Batey companies had lasted over fifty years and he had attended fifty consecutive Annual General Meetings: quite a record! Arthur Scott took over as secretary from 1st May 1944. In December, R.I. James died; he had been a director of the company since 1930.

The France, Fenwick Tyne & Wear Co.Ltd., in June 1944, purchased Redhead & Dry's Tugs Ltd. including their one-third holding in Joseph Crosthwaite (Tugs) Ltd. Lawson-Batey negotiated with France Fenwick to purchase half the Crosthwaite shares formerly held by Redhead & Dry, namely 2,833 fully paid up of £1 each at 15/- (75p.) per share. By July this had been agreed, so that now France Fenwick and Lawson-Batey were joint owners of Joseph Crosthwaite (Tugs) Ltd. on a 50/50 basis. However, the latter company continued to trade under its own name for a further five years.

In 1945 Lawson-Batey celebrated their Silver Jubilee at their Annual General Meeting on 20th March. The company capital was reported as being £50,000 in 10/- (50p.) shares and their seven tugs were valued at £20,400. This latter figure is quite low, but it should be remembered that the average age of the tugs was almost fifty years, the youngest being *Joffre* at twenty-nine years old. It is really quite amazing that they were kept operating during the difficulties and shortages with which they had to contend during the war. The war in Europe ended on 8th May 1945, and that in the Far East on 15th August 1945, without Lawson-Batey having lost any tugs to enemy action.

When Germany was defeated many of her merchant ships were confiscated by the British Government as reparations. Amongst these was *Vegesack* (142/1939), a steam screw tug owned by Unterweser Reederei A.G. of Bremen. She was renamed *Exemplar* and her management was entrusted to Lawson-Batey by the Ministry of War Transport. After two months undergoing modifications on the Tyne she worked there for only four months, until early 1946, before being sent to a French port and allocated to the USA by the tripartite commission. Although known to exist c.1952, her ultimate fate is not known. At the end of the war *Cruiser* (1) was chartered to Blyth Tug Co.Ltd. and worked there for the rest of her days.

Post-War Expansion and the Change to Diesel Propulsion. 1945 - 1957

At the board meeting on 13th September 1945 Major J.F. Batey was nominated to fill the vacancies on the boards of the Blyth and Crosthwaite companies occasioned by the recent death of R.I. James. The board discussed the possibility of acquiring an interest in The Ridley Steam Tug Co.Ltd. They proposed to purchase up to nine hundred £5 shares at £20 per share, the number required to give Lawson-Batey control of a majority of the shares. Frank Batey, the company chairman, was requested to discuss the matter with Francis Christian Batey, managing director of Ridley. In the event, Lawson-Batey purchased 873 shares, Frank Batey became the Ridley chairman and J.F. Batey was appointed to the Ridley board. The latter was demobilised from the army as Major 74th Field Regiment, Royal Artillery (TA), after six-and-a-half years' active service and recommenced office duties on 25th March 1946.

As previously recounted, Lawson-Batey, having failed to purchase *Empire Sybil*, tendered slightly more at £20,400 for *Empire Cherub*, which was accepted. She was then taken over at Belfast on 17th June 1946 and arrived in the Tyne on 21st June thence going to Smiths Dock Co. at North Shields for repairs. In October, after at first considering the name *Cherub* for her, she was renamed *Tynesider* (1) (262/1942) and became the first tug to be purchased in the Lawson-Batey history and the first to carry the '*——sider*' suffix. She had been built by Hall, Russell & Co.Ltd. of Aberdeen and was powered by a triple expansion steam engine giving 1000 IHP and a speed of 10 knots. She was to become a very useful member of the fleet.

A company advertisement of 1946. *(Company archives)*

Francis Batey *being raised by floating crane and pontoons after her sinking in December 1946 near Bill Quay.*
(Local History Library, South Shields Central Library).

Elation at the acquisition of a new powerful tug was soon dampened when *Francis Batey* sank off Bill Quay on 7th December 1946 whilst towing the tanker *Regent Tiger* (10,125/1938). Fortunately all of the crew of six were saved. She was raised on 11th January 1947 by the floating crane *Titan II* (1,042/1922) and towed to Smiths Dock Co. for repairs. It was to be over five months before she returned to service. The owners of *Regent Tiger* admitted liability and the company received £3,500 in compensation.

During 1946 Stanley Spencer Woodbridge Dalgliesh was appointed to the board of directors and Frank Batey was elected chairman of the British Tugowners Association for two years. Early in 1947 the Lawson-Batey directors were discussing the threat of nationalisation of shipping services, including towage, on the Tyne, Blyth and Wear rivers. One reason for the government's decision to leave things alone was that there had not been a strike for over twenty-five years. All of the local tug owners formed a joint meeting with the Northern Group of Members of Parliament to discuss the situation.

In May *Langton* was hired on time charter to Blyth Tug Co.Ltd. at a rate of £30 per week. Towards the end of the year an offer of £5,125 was made, and soon accepted, for the Admiralty tug *West Hyde,* after an inspection at Chatham. She had been built by Crabtree & Co.Ltd., Great Yarmouth and had a two-cylinder compound steam engine developing 430 IHP. She was renamed *Seasider* (1) (154/1919) and commenced operations, after alterations, on 27th February 1948, bringing the fleet strength up to nine. At this time Lawson-Batey had a contract with the John Readhead shipyard to handle all of the new ships built by that yard. On 3rd November 1947 Frank Batey, the managing director, then aged 67, celebrated 50 years on Newcastle Quayside and in recognition of this he was presented with the parchment of the Newcastle Stock Exchange by Sir Arthur Sutherland. Joseph Lawson, who had resigned as joint managing director in 1930, died on 5th November 1947, aged 73.

The whole company was saddened when, on 13th March 1948, Frank Batey died. He had been managing director since 1920 and was the third generation of the Batey family, who had been tugowners on the Tyne for over 100 years. He was also a director of Joseph Crosthwaite (Tugs) Ltd. and chairman of Blyth Tug Co.Ltd. and The Ridley Steam Tug Co.Ltd. He had been educated at the Royal Grammar School, Newcastle and had served an apprenticeship as a shipbroker with Cairns, Young & Noble. As has been related, he joined the family firm in 1904. He was a JP, a past chairman of the Tyne Tugowners' Association, vice-chairman of the Tyne Mutual Marine Insurance Association, a director of Whitley Bay pleasure gardens, and had been a special constable during the Second World War. Quite clearly he had led a full and useful active life. He left a widow, a son, John Francis Batey, and a daughter, Mrs. G.L. Redhead.

Other officials of the company at this time were Fred G. Hutton - Manager; Albert E. Dutton - Riverside Manager, and A.S. Barnard - Marine Superintendent. Radio telephone had now been fitted into some of the tugs, allowing immediate communication with the Newcastle and South Shields offices. This facility must have been very welcome for all concerned since it permitted more efficient use of the tugs when orders were being given; a far cry from the days when information was transmitted by bicycle, bus, hand signals and whistle.

In mid-1948 John Francis Batey was appointed managing director and Frank Sewell Batey joined the board of

Lawson-Batey. *Cruiser* (1) was in dire trouble. Two of the furnace crowns in the starboard boiler had collapsed, two cylinders showed fractures repaired with patches, two cast-iron bedplates were defective, the air pump gave frequent trouble, the boiler was 42 years old and patched in many places, and a new funnel was required. Not surprisingly, since she was 74 years old, it was decided that she should be scrapped. She was sold to the British Iron & Steel Corporation for £275, and allocated to J.J. King & Co.Ltd. at Gateshead for demolition, temporarily reducing the Lawson-Batey fleet to eight vessels. With *Langton* being chartered to Blyth Tug Co.Ltd. this meant that only seven tugs were available for the Tyne, whereas eight were needed.

A decision had to be made on whether to recall *Langton* or obtain another tug. The latter action was chosen and *West Creek* (138/1918) was inspected whilst in the Tyne for bunkers. Her owners were asking £8,500 for her, but in the event she was found to be unsuitable. Another tug, *La Morinie*, soon came on the market and was inspected at Boulogne in November 1948. She was owned by the Boulogne Chamber of Commerce and had been built in Holland. Lawson-Batey paid a deposit of £725 to secure her. Early in 1949 the balance of £6,156 was paid for the vessel and the directors knew that a further £5,495 would be needed for repairs, for the tug was already 30 years old.

Match-book covers were a useful and attractive way to publicise a company. Here Lawson-Batey and Blyth Tug Co. Ltd. together make the most of the small space available.

La Morinie left Boulogne in tow of United Towing's *Serviceman* (325/1945) and arrived in the Tyne on 28th March 1949. The following month she was towed to Sunderland by *Joffre* to be repaired by John Crown & Sons Ltd. It was anticipated that the tug would be back in service by August. In the event, the repairs cost no less than £12,275 and she was not ready until the end of September, by which time she had been renamed *Waysider* (166/1919). She had perhaps proved to be an expensive buy for a tug of her age, but nevertheless, she gave over fourteen years' service to the company.

Towards the end of 1948 S. Fisher, who was a Lawson-Batey nominee on the Blyth Tug Co.Ltd. board of directors, resigned and his place was taken by S.S.W. Dalgliesh and F.S. Batey early in 1949.

Post-war reconditioning of large passenger vessels was still proceeding. Two such vessels were in the River Tyne for lengthy periods during 1948/49: *Queen of Bermuda* (22,501/1933) of Furness-Withy, and *Strathmore* (23,580/1935) of P. & O., the former being handled by *Francis Batey* and *Joffre*, and the latter by *Joffre* and *Tynesider* (1). In May 1949 *Langton* completed her time-charter to the Blyth company but was then out of action for five months whilst a new condenser was fitted by Smiths Dock Co. *Comet* was also laid-up for repairs for three months about the same time.

Discussions had been taking place with the France Fenwick company, the joint owners, with Lawson-Batey, of Joseph Crosthwaite (Tugs) Ltd., whether that company should now be wound up and the tugs, existing contracts and business divided between the two shareholders.

Joffre *and* Francis Batey *taking the Furness-Withy liner* Queen of Bermuda *to sea on 23 January 1949 after her post-war reconditioning. Note that* Joffre *now carries the Lawson-Batey funnel colours.* *(Company archives).*

Ships monthly
The magazine for shiplovers ashore and afloat

Kottingham House
Dale Street
Burton-on-Trent DE14 3TD
England
Telephones 0283 - 64290
0283 - 42721
Fax 0283 - 61077

Our ref:
Your ref:

13 January 1994

Captain Sandy Kinghorn
15 Kendal Avenue
Cullercoats
North Shields
TYNE & WEAR
NE30 3AQ

Dear Sandy

'SHIPS LIBRARY'

Book Title: 150 YEARS OF THE MALTESE CROSS 1840-1990

Author: JOHN H. PROUD

Publisher: TYNE & WEAR TUGS LTD Price: £27.00

I am enclosing a copy of the above book, which I hope you will be able to review for 'SHIPS MONTHLY'.

We offer the alternative of the retention of the book or payment of a review fee of £10.00. Perhaps you will let me know which you prefer.

It would be helpful if you could adhere strictly to the style published in 'Ships Library' when listing the book's title, author, publisher, etc., in your review. Please also type your review in double spacing on one side of the paper only.

Please telephone me if you have any queries. Otherwise I look forward to receiving your review as soon as you are able.

Yours sincerely

(as discussed)
R

Robert

ROBERT SHOPLAND
EDITOR

Enc

Joffre *and* Tynesider *(1) taking the P & O liner* Strathmore *to sea on 8 October 1949 after her re-conversion to peace-time use.* *(Company archives).*

Agreement was reached and consequently *Lewis Crosthwaite* went to France Fenwick and *Joseph Crosthwaite* to Lawson-Batey where she was renamed *Northsider* (1) (149/1896). The Crosthwaite funnel colour of all black thus disappeared from the Tyne and the two tugs took up the colour scheme of their respective new owners late in 1949. It should be noted that *Northsider* (1) was already fifty-three years old and had been on the Tyne since 1911! Before going into service she needed a new funnel and boiler lagging.

In September 1949 *Seasider* (1) was transferred to the associated Blyth Tug Co.Ltd. at a valuation of £10,450, and, apart from a short period on the Tyne in August 1953, she stayed at Blyth until broken up in 1969. Because of the increasing number of large ships that needed to be handled, the directors were giving consideration to acquiring additional tugs. This resulted in an offer of £20,000 being made in December 1950 for the Hamburg tug *Heros,* but at the time her owners did not consider this a sufficient price.

Tynesider *(1) with foyboat in tow awaiting an incoming ship at the mouth of the Tyne.* *(Local History Library, South Shields Central Library).*

The stern section of the Norwegian tanker Rondefjell *(15,067/1951) leaving Sunderland for the Tyne after being built by John Crown & Sons Ltd.* Waysider *has the task of steering her unwieldy charge.* *(Company archives).*

The stern section of Rondefjell *arriving in the Tyne from Sunderland in 1951 with* Joffre *and* Homer *as the head (stern?) tugs.* *(Company archives).*

It may be useful to summarise the Lawson-Batey fleet working on the Tyne at the end of 1950:-

Comet	(132/1876)	Iron	Paddle
Plover	(245/1881)	Iron	Twin screw
Homer	(157/1915)	Steel	Screw
Joffre	(260/1916)	Steel	Screw
Langton	(157/1892)	Steel	Screw
Francis Batey	(151/1914)	Steel	Screw
Tynesider (1)	(262/1942)	Steel	Screw
Waysider	(166/1919)	Steel	Screw
Northsider (1)	(149/1896)	Steel	Screw

- with an average age of forty-four years. All were steam-driven and only one, *Tynesider* (1), could be described as modern.

An increased offer of £22,000 was made for the German *Heros* in February 1951. This price was accepted and was to include delivery, under her own power, to the Tyne by a German crew. She had been built during the war in Holland and had been operated by the Bugsier company since new. On arrival in the Tyne she went to Smiths Dock Co. for repairs and had her name changed to *Southsider* (239/1940). In March she was on trials when it was found that the crankshaft of her triple-expansion engine was fractured and, as a result, she did not come into effective Lawson-Batey service until June 1951; not an auspicious start, but nevertheless she was to serve for almost seventeen years on the Tyne.

On 23rd July 1951 Dr. J.T. Batey died at the grand age of eighty-eight having held the position of company chairman since 1934. He was succeeded by Frank Sewell Batey who had been a director since 1948. Later in the year George Harold Readhead Towers, J.P., who was managing director of the shipbuilders John Readhead & Sons, was appointed to the board and also as a Lawson-Batey nominee to the board of the Blyth Tug Co.Ltd.

The year 1951 was celebrated as the 'Festival of Britain' and one of the events was the visit to several ports in the U.K. of the Royal Navy escort carrier *Campania* (12,450 displ./1943) fitted out as an exhibition ship so that people in the provinces could see this travelling display if they could not see the permanent exhibitions in London. Whilst on the Tyne *Campania* was handled by *Joffre, Tynesider* (1) , *Waysider* and *Francis Batey*.

The crew of the paddle tug Comet *in June 1951. Left to right: Stratford, Unknown, skipper Alfred Softley, Unknown, Bill Callender.*
(The Softley family).

Tragedy struck suddenly in February 1952 when there was a fatal accident on the old paddle tug *Comet*. Ernest Durrant, 17, fell into the engine when the tug was on passage to Dunston. Whilst waiting to pass through the Newcastle swing bridge it was necessary to disconnect the engines and it was when attempting to carry out this operation that he slipped off the platform and was killed.

Two months later *Comet* was sold to the Blyth company for £1,650, but less than a year later she was scrapped having given no less than seventy-seven years of service at Great Yarmouth, Liverpool, Sunderland and the last fifty-six of them on the Tyne.

Homer suffered a fractured engine bedplate and had other faults too, requiring extensive repairs for her special survey and renewal of her passenger certificate. *Southsider* had got over her initial troubles and was noted as assisting at the launch of *Temple Hall* (8,003/1954) at the Caledon yard at Dundee. Also *Waysider* was employed at the launch of *Eumaeus* (7,841/1953) at the same yard for a charge of £212-10-0 (£212.50). These jobs had been obtained because the local towage company at Dundee had ceased to operate. Nearer to home the Norwegian passenger vessel *Blenheim* (4,766/1951), in dense fog, struck *Joffre* and *Plover* at Stanhope Buoys. Damage to masts and bulwarks cost £1,090 to repair.

At the end of 1952 it became known that the Danish tug *Pluto* was for sale at Copenhagen where she was inspected by Lawson-Batey management on 7th January 1953. They offered £20,000 for her which the owners Em. Z. Svitzer accepted, and agreed to deliver her to the Tyne for that price. She was built in Holland and had spent all her twenty-nine years in the employ of Svitzers. During the Second World War this tug, amongst many others, smuggled Danes to Sweden to escape the occupying forces. *Pluto* arrived on the Tyne on 26th February and was refitted at Brigham & Cowans Ltd. at South Shields where she was renamed *Eastsider* (162/1924) and commenced service on 23rd March.

Up to this time all the Lawson-Batey tugs had been steam-powered. Diesel tugs had already appeared at Sunderland and were proving very successful and more economical to run. The management were looking into the possibility of buying a new diesel tug when the chairman, F.S. Batey, and Francis Christian Batey, managing director of The Ridley Steam Tug Co.Ltd., went to talk to prospective builders at Hamburg. It will be remembered that since 1945 Lawson-Batey had a controlling interest in the Ridley company so it was not surprising that the two were working together in evaluating the possibilities of obtaining diesel tugs. The reasons for going to Germany were that new tugs were cheaper there, the delivery time was less and they could be bought at a fixed price, none of which considerations could be met by any UK shipyard. Such a vessel would cost c.£50,000, an enormous sum when compared with the values of steam tugs of the recent past. In the event it was the Ridley company which first ventured into the diesel market when they placed an order in Germany in September 1953 with a predicted launch date of April 1954 and delivery in June 1954. This delivery date was less than one month late.

In April *Langton* was transferred to Blyth at a value of £5,000 but stayed there for only about sixteen months before being hired back to Lawson-Batey on the Tyne. *Plover* had a fractured crankshaft in August 1953, possibly the result of some mis-alignment caused by being run into by *Blenheim* seven months earlier. At the end of 1953 more potential business became available for the Tyne tug companies when the Tyne Improvement Commission decided to cease operating tugs in their Tyne Dock due to shortage of work and the high cost of repairs.

Tynesider *(1) with her anchor down, is having difficulty in controlling* Georgidore *(8,063/1954) following her launch at John Readhead's yard at South Shields.* *(Company archives).*

At the 34th Annual General Meeting of the company on 29th March 1954 the directors reported that the experiment of ordering a diesel tug for the Ridley company would be a guide as to future policy in that respect. Comment was made that more shiprepairing work was going to the continent than previously, but on the other hand the new riverside Iron Ore Quay had opened, which should give the tugs more work. *Tynesider* (1) was to be fitted with Decca radar equipment.

In May, Lawson-Batey were issuing enquiries for a diesel tug similar to that being built in Germany for Ridley but with a more powerful engine. The cost of a UK-built tug was between £70,000 and £95,000 with delivery ranging from 16 to 33 months, whilst Mützelfeldwerft at Cuxhaven quoted £57,500 with a delivery of 7/8 months. The difference between the UK and the German quotations was so great that, even had they wished to favour UK yards, Lawson-Batey had no alternative but to buy from abroad. By the time that Lawson-Batey came to order their first diesel tug Ridley had taken delivery of theirs, named *Impetus,* in July 1954.

It was obvious that diesel tugs, size for size, required a smaller crew to operate them since there was no need for a fireman to stoke the boilers and the engines would be under direct control from the bridge. The tugmen became anxious for their jobs, seeing that the introduction of diesel power was likely to take place gradually throughout the whole Lawson-Batey fleet and throughout the fleets of other owners too. The management wanted a manning level of four men and a boy on diesel tugs. A meeting was held between the management and the crews' representatives of the companies involved, but after two full days' discussion no agreement could be reached. In September 1954 the crews stopped work but later returned on condition that five men and a boy be employed on Ridley's *Impetus* and the recently-converted France Fenwick's *George V.* (2) and that the matter be put to arbitration. The crews threatened an overtime ban and the owners threatened a lock-out of all 140 tugboatmen unless they did work overtime. All twenty-eight tugs on the Tyne were affected. In December the Industrial Disputes Tribunal met to consider the disagreement and decided that for the manning of diesel tugs of over 700 BHP the manning should be five men and a boy. The crews were obviously delighted at the outcome of the hearing and felt that their future would be more secure. The owners were disappointed, particularly as the Tribunal did not give any reasons for its decision.

At the end of 1954 Albert E. Dutton retired as office manager at South Shields after 47 years' service. At a presentation made by the chairman it was remarked that he had in earlier times chased about the river to give orders to the tugs when communication with them was far from easy. At the time of his retirement he had two sons and a grandson working for Lawson-Batey. The skipper of *Joffre,* Harry Jobling, had been a tugboatman for 40 years. His father and grandfather had also been tugboatmen before him and for over 100 years a Jobling had been in Tyne tugs. A remarkable record! Harry was soon to be appointed skipper of the new diesel tug being built at that time.

The name for the new tug was chosen as *Quaysider* - an appropriate choice since the Batey and Lawson-Batey companies had had offices on the Quayside at Newcastle for a very long time. Whilst there was, no doubt, eager anticipation of the arrival of the new vessel, there were still problems to be dealt with amongst the old steam tugs; *Northsider* (1) had recently needed a new condenser; *Eastsider's* boiler was in a poor state and when a new one was fitted in 1956 it was to cost £5,455 plus a further £2,480 for fitting at Smiths Dock Co., North Shields. *Joffre* required repairs estimated to cost between £10,000 and £12,000. An expensive time for the company was imminent.

Quaysider was christened and launched at Cuxhaven by Mrs. Joan Batey, the wife of J.F. Batey, on 1st March 1955. At the beginning of July the new tug was ready, but the company declined to accept her because of a serious vibration in the gear-box between 170 and 250 RPM engine speed. By September the faults had been rectified by fitting an outer rim to the engine fly-wheel, and she was commissioned on the Tyne on 16th September, having in fact cost a total of £55,316-2-0 (£55,316.10.). She had the typical appearance of a German tug of the time and looked very powerful and purposeful when running at speed. Built by Mützelfeldwerft Gmbh and engined by Klöckner-Humboldt-Deutz AG of Cologne with an 8-cylinder engine of 1200 BHP, she was fitted with a Seebeck towing hook and had a Kort nozzle around her propeller. The purpose of this nozzle was to concentrate and increase the available thrust from the propeller. A month after her delivery *Quaysider* (137/1955) was experiencing gearbox troubles again, and these were corrected when engineers from the builders came over to the Tyne to make adjustments. However, as will be seen later, this was not the end of her gearbox problems.

Launch of Quaysider *into an ice-bound harbour at Cuxhaven on 1 March 1955. (Company archives).*

In April 1955 the mate of *Tynesider* (1) was taken to court and fined £2 for stealing 1/- (5p.) worth of coal. France Fenwick and Ridley men were similarly fined as an example. Reading between the lines, it would appear that this was a common occurrence and was treated as a perk of the job, until perhaps it got out of hand. The tugboatmen requested an increase in wages of 10% and a reduction in the standard weekly hours from fifty to forty-five. The claim was referred to an Industrial Disputes Tribunal where it was revealed that the average hours actually worked over the past twelve months were eighty-and-a-half per week! A rise of 7.5% was granted but the standard hours were to remain at fifty.

January 1956 found the shipyard fitters on a strike which was to last several weeks and severely affected the work available for the tugs. This resulted in *Joffre, Waysider* and *Langton* being laid idle. Discussions regarding working conditions continued and agreement was reached on a rota system of one rest day per week. To achieve this the company would need to recruit another twenty-seven men and because they would have difficulty in finding sufficient relief skippers the rota system could not be implemented until the autumn of 1956.

Photographic evidence shows *Joffre,* in 1956, wearing the Batey funnel colours of buff with a blue Maltese cross, which proves that after the Second World War the Lawson black diamond funnel disappeared from the scene, though no documentary evidence has been found to confirm this. Also, when *Quaysider* was delivered in 1955 she was flying the Lawson-Batey house flag in colours corresponding to the Batey funnel marking, this being the first photographic evidence of this flag being used. One therefore presumes that the flag with similarities to the original pre-1920 Batey flag had been used up to this time.

Another court case was reported in January 1957 but this time it was the company who were the defendants. They were fined £4 for failing to show riding lights on *Quaysider* and *Joffre.* However, crew members also appeared as offenders when one from *Southsider* was fined £2 for throwing two bucketsful of ash into the river. Again, one assumes that this was a common practice for it to be necessary to take the man to court for a seemingly trivial offence. Due to shipyard workers being on strike in March 1957 five tugs out of the Lawson-Batey fleet were laid-up for want of work.

In April the iron twin-screw tug *Plover* was sold to the British Iron & Steel Corporation for £4,500 and was then allocated to C.W. Dorkin & Co.Ltd. at Gateshead for demolition. She was 76 years old and had seen service at Fleetwood and Liverpool before coming to the Tyne in 1909. In August 1957 *Charing Cross* was purchased for £6,500 from Tees Towing Co.Ltd. at Middlesbrough, notwithstanding the fact that the management had decided, in 1952, not to purchase her because she was too old! She was a compound-engined tug which had spent all her life on the River Tees, her original owners being the Tees Conservancy Commissioners. She was soon renamed *Westsider* (1) (153/1913) but was to work for the company for only two-and-a-half years.

Discussions had been going on since June 1957 between the tugboatmen and the various managements regarding pay and conditions without them being able to come to any agreement. Consequently by October the companies withdrew their towage services, making 150 men and 26 tugs idle for almost a month. It was reported that this was the first stoppage for over thirty years apart from a one-day three years earlier. Work was resumed early in November 1957 after it had been agreed that there would be an extra man in each crew so that each could have one day off per week. The men were also awarded a 7.5% pay rise. One can perhaps sympathise with the crew members since this rota-system had been promised to be in operation by the autumn of 1956.

France, Fenwick Tyne & Wear Co.Ltd. and Lawson-Batey Ltd. had been competitors on the River Tyne since 1920, mostly by division of the work between them due to the fact that they each had their own established contracts with individual shipping companies. In December 1957 proposals were made by K.R. Pelly, chairman of France Fenwick, that the two companies should work more closely together with the possibilities of future amalgamation, thus giving a better service and reducing costs. At this time the Lawson-Batey fleet consisted of ten tugs and that of France Fenwick twenty-four, fourteen of which worked on the Tyne. Ridley Tugs Ltd., who had been retitled in 1955, was an associated company which would come within the scope of any working arrangement. At the time they were operating five tugs. .

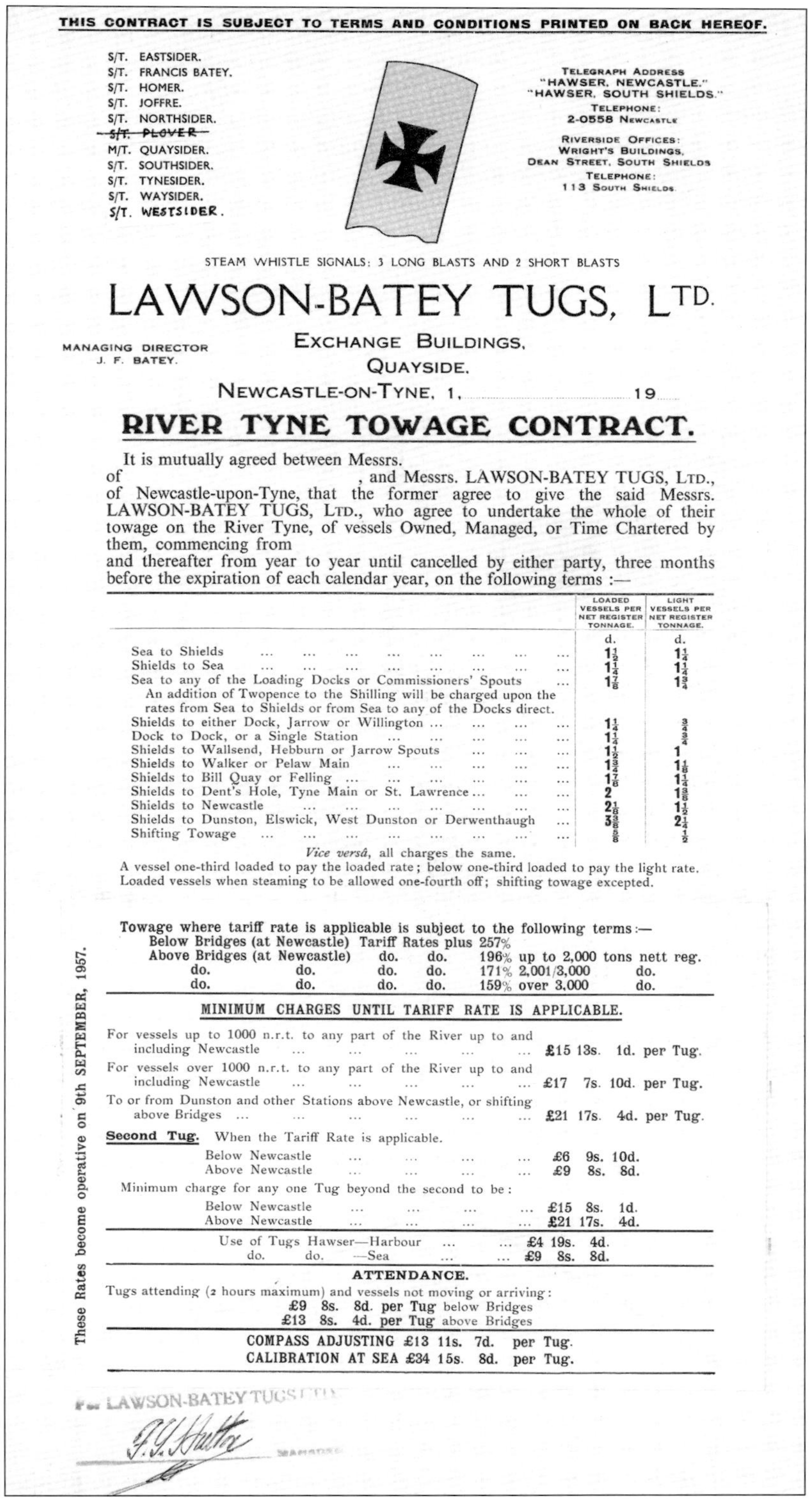

THIS CONTRACT IS SUBJECT TO TERMS AND CONDITIONS PRINTED ON BACK HEREOF.

S/T. EASTSIDER.
S/T. FRANCIS BATEY.
S/T. HOMER.
S/T. JOFFRE.
S/T. NORTHSIDER.
~~S/T. PLOVER~~
M/T. QUAYSIDER.
S/T. SOUTHSIDER.
S/T. TYNESIDER.
S/T. WAYSIDER.
S/T. WESTSIDER.

TELEGRAPH ADDRESS
"HAWSER, NEWCASTLE."
"HAWSER, SOUTH SHIELDS."
TELEPHONE:
2-0558 NEWCASTLE

RIVERSIDE OFFICES:
WRIGHT'S BUILDINGS,
DEAN STREET, SOUTH SHIELDS
TELEPHONE:
113 SOUTH SHIELDS

STEAM WHISTLE SIGNALS: 3 LONG BLASTS AND 2 SHORT BLASTS

LAWSON-BATEY TUGS, LTD.

MANAGING DIRECTOR
J. F. BATEY.

EXCHANGE BUILDINGS,
QUAYSIDE,
NEWCASTLE-ON-TYNE, 1, ... 19......

RIVER TYNE TOWAGE CONTRACT.

It is mutually agreed between Messrs.
of , and Messrs. LAWSON-BATEY TUGS, LTD., of Newcastle-upon-Tyne, that the former agree to give the said Messrs. LAWSON-BATEY TUGS, LTD., who agree to undertake the whole of their towage on the River Tyne, of vessels Owned, Managed, or Time Chartered by them, commencing from
and thereafter from year to year until cancelled by either party, three months before the expiration of each calendar year, on the following terms :—

	LOADED VESSELS PER NET REGISTER TONNAGE.	LIGHT VESSELS PER NET REGISTER TONNAGE.
	d.	d.
Sea to Shields	$1\frac{1}{2}$	$1\frac{1}{4}$
Shields to Sea	$1\frac{1}{4}$	$1\frac{1}{4}$
Sea to any of the Loading Docks or Commissioners' Spouts ...	$1\frac{7}{8}$	$1\frac{3}{4}$
An addition of Twopence to the Shilling will be charged upon the rates from Sea to Shields or from Sea to any of the Docks direct.		
Shields to either Dock, Jarrow or Willington	$1\frac{1}{4}$	$\frac{3}{4}$
Dock to Dock, or a Single Station	$1\frac{1}{4}$	$\frac{3}{4}$
Shields to Wallsend, Hebburn or Jarrow Spouts	$1\frac{1}{2}$	1
Shields to Walker or Pelaw Main	$1\frac{3}{4}$	$1\frac{1}{8}$
Shields to Bill Quay or Felling	$1\frac{7}{8}$	$1\frac{1}{4}$
Shields to Dent's Hole, Tyne Main or St. Lawrence	2	$1\frac{3}{8}$
Shields to Newcastle	$2\frac{1}{8}$	$1\frac{1}{2}$
Shields to Dunston, Elswick, West Dunston or Derwenthaugh ...	$3\frac{3}{8}$	$2\frac{1}{4}$
Shifting Towage	$\frac{5}{8}$	$\frac{1}{2}$

Vice versâ, all charges the same.
A vessel one-third loaded to pay the loaded rate; below one-third loaded to pay the light rate.
Loaded vessels when steaming to be allowed one-fourth off; shifting towage excepted.

Towage where tariff rate is applicable is subject to the following terms:—

Below Bridges (at Newcastle)	Tariff Rates plus 257%					
Above Bridges (at Newcastle)		do.	do.	196%	up to 2,000 tons nett reg.	
do.	do.	do.	do.	171%	2,001/3,000	do.
do.	do.	do.	do.	159%	over 3,000	do.

MINIMUM CHARGES UNTIL TARIFF RATE IS APPLICABLE.

For vessels up to 1000 n.r.t. to any part of the River up to and including Newcastle ... ... £15 13s. 1d. per Tug.
For vessels over 1000 n.r.t. to any part of the River up to and including Newcastle ... ... £17 7s. 10d. per Tug.
To or from Dunston and other Stations above Newcastle, or shifting above Bridges ... ... £21 17s. 4d. per Tug.

Second Tug. When the Tariff Rate is applicable.
Below Newcastle ... ... £6 9s. 10d.
Above Newcastle ... ... £9 8s. 8d.
Minimum charge for any one Tug beyond the second to be:
Below Newcastle ... ... £15 8s. 1d.
Above Newcastle ... ... £21 17s. 4d.

Use of Tugs Hawser—Harbour ... ... £4 19s. 4d.
do. do. —Sea ... ... £9 8s. 8d.

ATTENDANCE.

Tugs attending (2 hours maximum) and vessels not moving or arriving:
£9 8s. 8d. per Tug below Bridges
£13 8s. 4d. per Tug above Bridges

COMPASS ADJUSTING £13 11s. 7d. per Tug.
CALIBRATION AT SEA £34 15s. 8d. per Tug.

These Rates become operative on 9th SEPTEMBER, 1957.

For LAWSON-BATEY TUGS LTD.

A Lawson-Batey towage contract of 1957.

Amalgamation and Modernisation. 1958-1976.

In January 1958 further discussions took place on the amalgamation proposals, beginning with a review of the evolution of the companies operating on the Tyne. After the amalgamation of The Lawson Steam Tug-boat Co.Ltd. and J. Batey & Son Ltd. in 1920 there were six individual companies, namely:-

Lawson-Batey Tugs Ltd.	
France, Fenwick Tyne & Wear Co.Ltd.	
Robert Redhead & Son Ltd.	Later Redhead
The John Dry Steam Tugs Ltd.	& Dry's Tugs Ltd.
The Ridley Steam Tug Co.Ltd.	Later Ridley Tugs Ltd.
Joseph Crosthwaite	Later Joseph Crosthwaite (Tugs) Ltd.

Redhead and Dry had joined forces in 1929 and were subsequently taken over by France Fenwick in 1944. Crosthwaite was taken over jointly by Lawson-Batey, France Fenwick, and Redhead & Dry's in 1933 and liquidated in 1950. In 1945, Lawson-Batey had obtained a controlling interest in the Ridley company.

By March 1958 the amalgamation discussions had progressed to a point where it had been decided to form a new organisation with a small capital which would be a tug operating company and not a holding company. The tugs on the Tyne were to be chartered, on a bare-boat basis, to the new company for a fixed period of ten years with Lawson-Batey and France Fenwick receiving charter hire on a 50/50 basis.

The new company was to be known as Tyne Tugs Ltd. and was to be incorporated on 28th November 1958 with the intention of operating, at the latest, from 1st January 1959. The initial directors were F.S. Batey, J.F. Batey and F.C. Batey representing the Lawson-Batey and Ridley companies, and three representatives of France Fenwick. The shareholders had been informed of the proposals at the 38th Annual General Meeting on 1st April and were officially notified of the intentions on 31st October. The capital of Tyne Tugs Ltd. was to be £2,400 in £1 shares subscribed as follows: £1,000 by Lawson-Batey who would contribute 10 tugs; £200 by Ridley (2 tugs), and £1,200 by France Fenwick (12 tugs).

Throughout 1958, whilst all the important discussions with France Fenwick were going on, Lawson-Batey were also negotiating the purchase of Ridley Tugs Ltd. in which they had had a controlling interest since 1945. They already owned 883 Ridley £5 shares thus they would have to purchase a further 732 shares at approximately £50 each to achieve wholly-owned status. Ridley considered that the shares were worth £60 each and initially Lawson-Batey felt that they could not afford to purchase all the outsiders' shares. Later, in September 1958, the three Batey family shareholders in the Ridley company said that their capital in Ridley would be left in the future parent company as a loan. This offer, therefore, considerably reduced the money which Lawson-Batey would have to find at the time of the buy-out, and they agreed to purchase the outsiders' shares and to take over the company officially on 28th October 1958. In the event the £5 shares did change hands at £50 each, putting a value of £80,750 on Ridley Tugs Ltd.

Whilst these commercial negotiations were proceeding the activities of the tugs, of course, were continuing as normal. A notable ship movement which occurred in April 1958 was the undocking of the tanker *Zaphon* (24,802/1957) from Swan Hunter's No.4 drydock which required the assistance of six tugs, *Tynesider* (1), *Northsider* (1), *Joffre, Southsider, Quaysider* and *Francis Batey*. This operation showed how the

The undocking of the Shell tanker Zaphon *with six of the company's tugs in attendance, 29 April 1958.* *(The Shields Gazette).*

The whale-factory ship Southern Venturer *(14,066/1945) in tow of* Joffre *and* Eastsider *c.1955.* *(Company archives).*

increasing size of vessels was now demanding more tugs, or alternatively more power available per tug. The National Coal Board was to carry out drilling work off the north-east coast to determine their future resources of coal. To this end they had constructed at Hartlepool a tall drilling tower which would be floated out and then lowered to sit on the sea bed. *Quaysider* and *Tynesider* (1) were contracted to tow this unusual structure to its working location. This was to be the first of a great many movements of strange towers and platforms, both floating and on barges, that were handled by the Tyne tugs. At about this time the engineer from *Tynesider* (1), who had worked for the company for 30 years, was in court for stealing coal, the property of Lawson-Batey. He had obviously not heeded the warning of the similar 'crime' in 1955 and was ordered to pay costs of 15/- (75p.).

Lawson-Batey Tugs Ltd., and soon also Tyne Tugs Ltd., set up office in the Market Place, at South Shields when they signed the lease for 21 years of the 2nd and 3rd floors of a wing of the new National Assistance Board Office. The rental was £175 per annum. At the end of 1958 T.G. Hedley was Office Manager, F.G. Hutton was Towage Manager and the Marine Superintendent was Edgar Heal.

As the time drew near for the chartering of the Lawson-Batey fleet to Tyne Tugs Ltd. a certain amount of shuffling around of tugs took place so that the correct number would be available to meet the charter agreement. *Roughsider* (143/1958), which had been completed in May for Blyth Tug Co.Ltd., was, the following month, transferred to work on the Tyne and in November was sold to Lawson-Batey for £71,000. She was a 750 BHP diesel tug built by Richard Dunston Ltd. at Thorne, but with a German-built engine from the same source as that for *Quaysider*. The steam tug *Battleaxe* was transferred to Lawson-Batey from the Ridley company at a book value of £8,000 and was immediately renamed *Hillsider* (1) (163/1911). To balance these moves *Northsider* (1), now 62 years old and valued at only £1,500, and *Homer*, valued at £8,000, were transferred to Blyth Tug Co.Ltd. France Fenwick were also making similar adjustments to their fleet.

To finalise the arrangements for the commencement of the operations by Tyne Tugs Ltd. from 1st January 1959 all of the Lawson-Batey and Ridley tug crews and office staff had their employment with those companies terminated and were immediately re-employed by Tyne Tugs Ltd.

In the continuing history of Lawson-Batey Tugs Ltd. we will deal with the ownership of the fleet, whereas the story of the operation of the fleet will appear in Chapter Twenty concerning Tyne Tugs Ltd., later additionally Wear Tugs Ltd., and finally Tyne & Wear Tugs Ltd.

On 1st January 1959 all of the Lawson-Batey tugs came under the operational control of Tyne Tugs Ltd. They were, in order of acquisition by Lawson-Batey:-

Joffre	(260/1916)	Steam
Francis Batey	(151/1914)	Steam
Tynesider (1)	(262/1942)	Steam
Waysider	(166/1919)	Steam
Southsider	(239/1940)	Steam
Eastsider	(162/1924)	Steam
Quaysider	(137/1955)	Diesel
Westsider (1)	(153/1913)	Steam
Roughsider	(143/1958)	Diesel
Hillsider (1)	(163/1911)	Steam

Ten tugs with an average age of 29 years.

Ridley Tugs Ltd. contributed two diesel tugs, *Impetus* (141/1954) and *Maximus* (141/1956), which were in effect owned by Lawson-Batey, making a total contribution of 12 tugs. France Fenwick's contribution was also 12 tugs thus making up the 50/50 agreement which had been reached the previous year. Ridley had, in fact, a third tug which they chartered to France Fenwick for a while to replace one of the latter's tugs which had unfortunately sunk in October 1958. Thus the balance was very neatly achieved.

The familiar Lawson-Batey, Ridley and France Fenwick funnel colours were soon to disappear, much to the disapproval of many onlookers, when Tyne Tugs Ltd. opted for a red funnel with black top after, it is believed, experimenting with various reds, brown and blue. The choice was perhaps inevitable since the colour would have to be different from any colours the chartering companies already used.

The tugs were demise-chartered for a period of ten years at a minimum hire rate of £2,500 per annum for a steam tug and £3,000 for a diesel tug to be paid in calendar monthly instalments. There would, of course, be hope of additional charter hire income based on actual results from Tyne Tugs Ltd. operations, since Lawson-Batey were 50% owners of the charter company. Adjustments soon had to be made to match the number of tugs required on the Tyne and so, by September 1959, *Westsider* (1) and *Hillsider* (1) were laid up. In February 1959 *Roughsider* suffered a damaged gearbox due to a wire being caught in the propeller when towing a Stephenson Clarke collier. Lawson-Batey were to lose her hire income for six months as a result.

So we come to the end of the first year of the combined operation of tugs on the Tyne with a utilisation rate of an average of 2.18 jobs per day per tug amounting to a total of 15,099 jobs per year on the River Tyne. This was not quite the peak figure reached but we will follow the steady decline in jobs which was soon to follow, in the chapter on the Charter Companies.

At this juncture, it may be pertinent to quote a few statistics of the Lawson-Batey company. Its registered office in 1960 was at Exchange Buildings, Quayside, Newcastle, but at the beginning of the following year this was changed to Watergate Buildings, (2nd Floor), Quayside, Newcastle. The directors were F.S. Batey (chairman), J.F. Batey (managing), S.S.W. Dalgliesh and G.H.R. Towers. The secretary was A. Scott. The paid-up capital was £50,000 in 100,000 ordinary shares of 10/- (50p.) each. The company did not have any of its tugs mortgaged. The principal share-holders were, in alphabetical order, Mrs. A.M. Batey; F.S. Batey; J.F. Batey; J.R. Lawson; P.D. Nickirk, S. Hawdon and N. Stockdale jointly; Mrs. E. Readhead; Mrs. I.M. Readhead; Miss A. Rewcastle and Mrs. C.A. Whitfield, who together owned almost 55,000 of the shares.

In January 1960 *Westsider* (1) was sold by the company for breaking up at Gateshead, for £1,375, after paying £6,500 for her only two-and-a-half years before. *Hillsider* (1) also went to breakers at Blyth in October for the same price, Ridley having bought her in 1957 for £8,550. These two must have been expensive purchases since the above figures represent an average depreciation of about 30% per annum.

Enquiries had been made to various shipyards for the building of a diesel tug of 1000 BHP. Quotations had been received from Germany at £73,500; Hall Russell of Aberdeen - £96,000 and Richard Dunston, Thorne - £85,000. Clelands on the Tyne, Scots of Bowling and Cochranes, Selby were unable to quote because delivery would be c.3 years. An order was placed on 29th December 1960 with Richard Dunston with delivery expected to be in March 1962. She was to be built at Dunston's Hessle yard which had, until recently, operated under the name of Henry Scarr Ltd., a subsidiary company of Dunston. The new tug was to have an 8-cylinder diesel engine of German manufacture and her propulsion system included a Kort nozzle.

At the company's 41st Annual General Meeting on 5th May 1961 it was announced that a dividend of 20% was to be paid. This, together with the substantial amount of money being spent now and in the near future on new tugs, suggests that the company was on a firm financial basis.

A serious problem arose in September 1961 when *Quaysider* suffered a major gearbox breakdown. It will be remembered that she had had gearbox vibrations when newly built and again shortly after delivery in 1955. This time the gearbox had to be sent back to Germany to be repaired, resulting in the tug being out of use until November 1962. Why it should have taken so long to repair has not been revealed.

The new tug was christened and launched at Hessle on 4th April 1962 by Mrs. Ida Redhead, sister of J.F. Batey. The vessel was named *Appelsider* (173/1962). Why or how this enigmatic name was chosen for her has not been discovered with certainty, but Sir Edward Elgar would have approved of the choice. Was it a deliberate play on the words 'Apple Cider'? If so, was the spelling deliberate or a mistake not noticed until it was too late? Was a painter at the Dunston yard sacked about this time?! Whatever the intention nobody now, thirty years later, knows the reason nor will admit to

Mrs. Ida Redhead ready to christen Appelsider *on 4 April 1962. (Walter Fussey & Son).*

A detailed view of the Kort rudder of Appelsider *just before launching. (Walter Fussey & Son).*

Appelsider *on trials in the Humber estuary, July 1962.* *(Walter Fussey & Son).*

knowing. It seems most likely that it was a deliberate play on the words 'Apple Cider' following on from the previous naming within the group of *Roughsider* itself a distorted reference to 'Rough Cider'. No wonder that the newspapers of 1962 were confused, for in one article they referred to her as *'Apple Cider'* and *'Cider Apple'! Appelsider* entered service in July on charter to Tyne Tugs Ltd.

In mid-summer Lawson-Batey had six steam and three diesel tugs chartered out and their subsidiary company Ridley Tugs Ltd. had two which were diesel-powered. By November work was slack on the Tyne and so *Joffre* was laid-up and *Francis Batey* was sold for £4,500 to Blyth Tugs Ltd. where she was to work for a further five-and-a-half years before being broken up.

Continuing with their modernisation plan of replacing steam propulsion with diesel Lawson-Batey had received tenders for a further 1000 BHP tug which was to be a near-sister to *Appelsider* but with a more straightforward name. The prices quoted by the firms who tendered were: Richards Ironworks Ltd., £94,000; Richard Dunston Ltd., £97,500; Clelands, £98,000; John Lewis, £102,000; Alexander Hall, £104,000 and Cochrane, £107,000, all promising delivery in less than twelve months. This tug was to have a six-cylinder engine, though giving the same power output as that obtained from *Appelsider's* eight cylinders. She was ordered from Richard Dunston at a revised price of £91,000 in February 1963 and was launched on 16th December 1963 by Mrs. F.G. Hutton, wife of the towage manager of Tyne Tugs Ltd., who christened her *Westsider* (2) (151/1964). The tug was handed over at Hessle on 2nd March 1964 and soon entered service under Tyne Tugs' colours.

Mrs. F. G. Hutton about to christen Westsider *(2) on 16 December 1963.* *(Walter Fussey & Son).*

Westsider *(2) ready for launching in darkness and with a strong wind obviously blowing.* *(Walter Fussey & Son).*

John Francis Batey, right, when chairman of the British Tugowners' Association in 1966-68. *(Company archives).*

Whilst *Westsider* (2) was being completed *Waysider* was sold for £1,710 for demolition by Hughes Bolckow Ltd. at Blyth at 45 years of age - one more step on the way to an all-diesel-powered fleet.

About this time, the British Tugowners' Association had been awarded a grant of arms and their choice of motto was 'Parvis potestas', translated as 'in small things there is power'. A very appropriate choice one would think. Lawson-Batey's managing director, John Francis Batey, was elected vice-chairman of that body in October 1964 and two years later he was to become its chairman for two years; the first Tyne tugowner in this post for eleven years. John Robert Lawson, who had been superintendent engineer and company secretary until his retirement in 1944 died on 2nd May 1965. He was the son of Benjamin Dain Lawson.

Further new diesel tugs were required to replace the ageing and under-powered steam tugs still in the fleet. In March 1966 an order was placed with Richard Dunston for two tugs similar to *Westsider* (2) but with 1420 BHP eight-cylinder engines. Delivery was planned for April 1967 and January 1968. The price was to be £106,500. At the same time *Eastsider* and *Southsider* were laid up and the fifty-year-old *Joffre* was sold for £4,075 for demolition at Bo'ness.

At the 46th Annual General Meeting on 23rd May 1966 the business outlook was gloomy since the country was in the midst of a British seamen's strike which was to last about six weeks. Some tugmen on the Tyne had to be paid off as a result. By early 1967 the general situation was no better, with comments being made about the dying trade on the Tyne. Jarrow Oil Berth was closing, with the terminal to be supplied by rail, and Tyne Dock coal staiths had closed. On top of all this the Government were proposing to nationalise the ports by 1970, which did not particularly please the Lawson-Batey management.

The first of the new tugs being built at Richard Dunston was christened *Northsider* (2) (156/1967) on 26th February 1967, by Mrs. Frances Heal, wife of Tyne Tugs' superintendent engineer but was not launched until the following day because of high winds. *Northsider* (2) ran trials on 20th April and was handed over to her owners on the 22nd, exactly on time. The second vessel, *Ironsider* (156/1967), was christened by Mrs. J.F. Batey on 25th July 1967, and, after trials in the Humber, arrived in the Tyne on 26th September, some three months ahead of schedule. Surprisingly, *Northsider* (2) and *Ironsider* were not fitted with the Kort nozzles which had been a feature of the three previous new buildings. Both tugs were chartered to Tyne Tugs on delivery. *Southsider* and *Eastsider* were laid up and in early 1968 were disposed of for breaking up by Hughes Bolckow at Blyth, for £2,050 and £1,850 respectively.

Mrs. Joan Batey about to christen Ironsider *on 25th July 1967. (Walter Fussey & Son).*

Ironsider *entering the waters of the Humber.* *(Walter Fussey & Son).*

The company secretary, Arthur Scott, retired on 31st December 1967, and was replaced from the beginning of 1968 by Douglas Frank Hamilton, who had joined Tyne Tugs Ltd. in 1964 as a chartered accountant after serving articles with Peat, Marwick, Mitchell & Co. in Newcastle.

On the river, shipyards, which had regularly employed Lawson-Batey tugs, amalgamated to form a new company named Swan Hunter and Tyne Shipbuilders Ltd. The constituent companies were Hawthorn Leslie, John Readhead, The Swan Hunter Group and Vickers Ltd.

Tynesider (1) worked at Blyth and later in 1968 was sold to the Blyth Tug Co. for £5,000. *Quaysider* was laid-up at this time. At the company's 48th Annual General Meeting it was reported that there were no steam tugs working for the company and that the change-over to diesel propulsion had taken thirteen years. The Tyne Tugs Ltd. arrangement came to the end of its ten-year agreed term at the end of 1968. All parties had been highly satisfied with the way things had worked out so a further agreement was signed on 24th May 1968 to continue with the arrangement for a further ten years from 1st January 1969. As will be seen later, however, important changes were to take place before that period was completed. The all-diesel fleet of Lawson-Batey consisted of six tugs, all chartered to Tyne Tugs Ltd., namely:-

Quaysider	(137/1955)	1,200 BHP	15 BP
Roughsider	(143/1958)	750 BHP	11 BP
Appelsider	(173/1962)	1,000 BHP	18 BP
Westsider (2)	(151/1964)	1,000 BHP	18 BP
Northsider (2)	(156/1967)	1,420 BHP	18.5 BP
Ironsider	156/1967)	1,420 BHP	18.5 BP

- with the low average age of just under six years - a remarkable change from the days when the fleet had an average age of around forty years! Ridley Tugs Ltd. also had two diesel tugs in the Tyne Tugs' fleet, and, of course, they were in effect owned by Lawson-Batey making a total input to the scheme of eight tugs. France, Fenwick Tyne & Wear Co.Ltd. at this time had seven tugs chartered to Tyne Tugs and a further seven working on the River Wear.

In April 1968 William France, Fenwick & Co.Ltd., the collier owners of Sunderland, had been taken over by Jessel Securities Ltd., a finance company of London. France, Fenwick Tyne & Wear Co.Ltd. was a subsidiary company of William France Fenwick and thus came under the control of Jessel Securities. In January 1969 Jessel Securities had put France, Fenwick Tyne & Wear on the market and the tug firms of William Cory & Son Ltd., United Towing Co.Ltd. and Tees Towing Co.Ltd. all showed an interest. The Lawson-Batey management were seriously concerned about these developments since they could have affected the joint operation of Tyne Tugs Ltd. Additional worry was occasioned by the falling level of business on the Tyne; out of twenty-four drydocks only twelve were occupied and there were discussions going on about the closure of the Iron Ore Quay. The number of ships being built or repaired had fallen dramatically over recent years and in 1970 Palmer's Shipyard at Hebburn was to close completely.

The year 1970 was the 50th anniversary of the formation of the Lawson-Batey company and at the Annual General Meeting on 14th May not a very happy situation could be presented, since the charter hire income from Tyne Tugs had continued to fall. The only happy note came from the information that 'the Government is not intending to nationalise the towage industry for the moment'. The company chairman, Frank Sewell Batey, was now aged over seventy years and by the rules had to stand down and be re-confirmed as a director. This was done on 30th June 1970 when he was also re-elected chairman. At the Annual General Meeting in 1971 it was reported that Lawson-Batey now had more than a 75% holding in Blyth Tug Co.Ltd. and were gradually hoping to increase this with a view to the Blyth company becoming a wholly-owned subsidiary. Nevertheless, prospects there were reported as 'very grave'.

Appelsider, which had been chartered away during part of 1971, was sold for £110,000 to the Dublin Docks & Harbour Board on 28th January 1972, thus reducing the Lawson-Batey fleet to five tugs. She was renamed *Coliemore* and was still going strong in Dublin after a further twenty years.

For many years, the Lawson-Batey tugs had been insured with the Tugowners' Mutual Insurance Association. This organisation closed in 1972 and after that the company's tugs were insured with C.T. Bowring & Co.Ltd. The company's income had been affected by the strike of tug crews in October/November 1971 and, to counteract the low level of work on the river, Tyne Tugs were attempting to find other charter work and coastal towage. Although, since April 1971, Ridley Tugs Ltd. had not owned any tugs, no decision had been reached at the end of 1972 as to whether or not this company should be wound up. In fact, though under a different name, it remained in being until 1989.

In April 1973 a quotation was received from Richard Dunston for a new tug of 3,180 BHP at a price of £380,000, and with delivery planned for April 1975. A tender from Richards (Shipbuilders) Ltd. was £5,000 more, but delivery was to be six months sooner. This tug was, by a long way, much more powerful than anything that had been owned on the Tyne up to that time. With a bollard pull of 40 tons and a free running speed of 13 knots she was designed for handling the very large tankers then being built on the Tyne and elsewhere. Richard Dunston's quotation was accepted and the order was placed in November 1973 by which time the contract price had risen to £425,000. France Fenwick were to purchase a similar tug and both would work under the operational control of Tyne Tugs Ltd. In this case, notwithstanding the excellent delivery performances on *Northsider* (2) and *Ironsider,* the delivery dates were to go seriously awry.

Lawson-Batey, already having a controlling interest of 81% in Blyth Tug Co.Ltd., was seeking to persuade the minority shareholders to sell their shares so that the Blyth company would be a wholly-owned subsidiary. During 1973 iron ore imports ceased at the Riverside Quay, Tyne Dock, and were transferred to the Redcar Terminal on the River Tees - one more step in the relentless reduction in trade on the Tyne. Nevertheless, at the 54th Annual General Meeting held on 7th June 1974, the directors were reporting 'a very successful year'.

As noted above, very serious delays were being reported in the construction of the new tug by Richard Dunston. There was a twelve-months' delay in delivery, due to the three-day working week being imposed by the government, and to shortages of steel and labour. Because it had been difficult to get information from the builders the Lawson-Batey directors decided to look into the financial stability of the company and its American-owned parent. Discussions between the two companies took place in March 1975 when it came to light that Dunston were set to lose no less than

£110,000 on the fixed price contract. To give an idea of the size of the problems facing the builders, it was revealed that a similar tug being ordered then would be priced at £785,000 compared with the price of £425,000 to which they were committed. No conclusions were reached, but the Dunston management insisted that the financial stability of the company was not in doubt.

Jessel Securities Ltd., owners of France, Fenwick Tyne & Wear Co.Ltd., who, it will be remembered, had tried to sell France Fenwick in 1969, again put the company on the market at the beginning of 1975 and received bids from various shipowners and the Rothschild Organisation. If a sale was to take place then, under the terms of the Tyne Tugs Ltd. agreement, Lawson-Batey would have to give their consent. An offer by Jessel to sell France Fenwick to Lawson-Batey was declined at this time due to the latter company's existing financial commitments.

Lawson-Batey had suffered several misfortunes over their fifty-five years' history but these were mainly of a material or financial nature. They were now to experience a human tragedy, so unexpected and so awful as to be almost beyond belief. John Francis Batey, the managing director, was in London on business for the company, and in the late evening of 13th November was having dinner at Scott's Restaurant, Mayfair, when he was severely injured by a bomb thrown through a window by the Irish Republican Army. He died soon afterwards in St. George's Hospital, Hyde Park. His death was a terrible loss to his wife, Joan, and to the company; J.F. Batey was the fourth generation of tugowners spanning 135 years, he had been a director since 1941 and managing director of the company since 1948. He was 59 years of age. It, no doubt, took all the employees of the company a long time to get over the shock of this dreadful episode. The respect in which he was held by the crews was shown by the fact that fifty tugmen attended his funeral at the packed Corbridge Parish Church.

In January 1976 the forecast regarding the progress of the new tug was that she would now be launched in March and completed in August. Even at this stage, over two years into the contract, these promises were to slip back even further.

Thomas Anthony Greenwell and D.F. Hamilton, the company secretary, were appointed to the board of directors, the former becoming managing director in place of J.F. Batey. T.A. Greenwell had, until 1973, been joint managing director of Doxford & Sunderland Shipbuilding Group.

Jessel Securities Ltd., who had, for some time, been in financial difficulties, had gone into voluntary liquidation in January 1976. Jessel owned over 90% of the France Fenwick shares but again, in March 1976, Lawson-Batey declined to purchase that company. They did, however, consider making a joint offer, with the Letzer Tug Co. of Antwerp, of £630,000 to £680,000 against a valuation of France Fenwick's assets of £610,000. By May 1976 they had revised the valuation of the shares held by Jessel as £495,000 and the bid was now to be £454,454. It was being proposed that the Lawson-Batey/Letzer bid should purchase all the France Fenwick shares in the relative proportion 51%/49%. Letzer was now being referred to as the Nassau Group and they and Lawson-Batey were talking of forming a new company, probably in Bermuda, to which the two sister tugs being built at Richard Dunston would be chartered. Lawson-Batey would share in the Drumochty Towing Co. project at Peterhead. All of this came to nought, since the liquidators of Jessel turned down the joint offer, considering that values of over £700,000 would have been more appropriate.

The new tug was at last launched and christened *Cragsider* (266/1976) by Mrs. Elizabeth (Bet) Batey on 25th August 1976 after further postponements due to labour troubles. She was finally delivered on 9th December 1976, twenty months late but by then was said to be worth about £1,000,000 after having been bought for £425,000. Maybe it is true that every cloud has a silver lining!

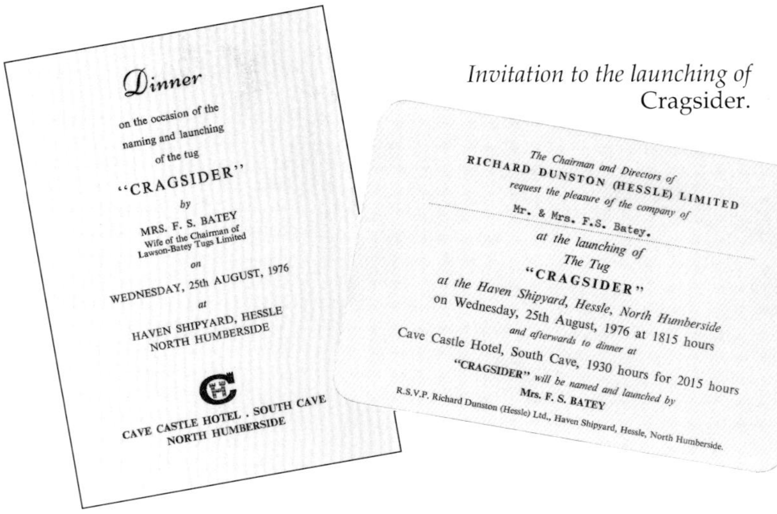
Dinner
on the occasion of the
naming and launching
of the tug
"CRAGSIDER"
by
MRS. F. S. BATEY
Wife of the Chairman of
Lawson-Batey Tugs Limited
on
WEDNESDAY, 25th AUGUST, 1976
at
HAVEN SHIPYARD, HESSLE
NORTH HUMBERSIDE
CAVE CASTLE HOTEL . SOUTH CAVE
NORTH HUMBERSIDE

The Chairman and Directors of
RICHARD DUNSTON (HESSLE) LIMITED
request the pleasure of the company of
Mr. & Mrs. F.S. Batey.
at the launching of
The Tug
"CRAGSIDER"
at the Haven Shipyard, Hessle, North Humberside
on Wednesday, 25th August, 1976 at 1815 hours
and afterwards to dinner at
Cave Castle Hotel, South Cave, 1930 hours for 2015 hours
"CRAGSIDER" will be named and launched by
Mrs. F. S. BATEY
R.S.V.P. Richard Dunston (Hessle) Ltd., Haven Shipyard, Hessle, North Humberside.

Invitation to the launching of Cragsider.

Cover of dinner menu following launch of Cragsider.

Cragsider *entering the water on 25 August 1976.*
(Walter Fussey & Son).

The last minority shareholders in Blyth Tug Co.Ltd. eventually sold out to Lawson-Batey so that Blyth Tug Co.Ltd. became a wholly-owned subsidiary.

Having now arrived at the end of another phase of the Lawson-Batey story in which the fleet had been fully modernised and integrated, from an operational point of view, on the Tyne, with the France Fenwick company, there was still a state of flux, due to the liquidation of Jessel Securities. The outcome of further financial discussions and the eventual acquisition of the France Fenwick fleet will be continued in the next section of the history. At this juncture, it might be useful to remind ourselves of the six tugs owned by Lawson-Batey at the end of 1976. They were:-

Quaysider	(137/1955)	1,200 BHP	15 BP
Roughsider	(143/1958)	750 BHP	11 BP
Westsider (2)	(151/1964)	1,000 BHP	18 BP
Northsider (2)	(156/1967)	1,420 BHP	18.5 BP
Ironsider	(156/1967)	1,420 BHP	18.5 BP
Cragsider	(266/1976)	3,180 BHP	46 BP

- all were chartered to Tyne Tugs Ltd. It will be noted how, since 1959 when Tyne Tugs was first formed, the number of tugs had gradually reduced to match the ever-falling level of work available on the River Tyne.

In Control on the Three Rivers. 1977 - 1990.

Having at first decided not to make a bid for France, Fenwick Tyne & Wear Co.Ltd. themselves, and later having had their joint offer with Letzer turned down, Lawson-Batey, in 1977, felt they were now in a position to reconsider the matter. As previously, there were also other companies interested, notably Cory Ship Towage Co.Ltd., but nothing further had been heard from the Nassau Towage Group. By April 1977 the London & Yorkshire Trust Ltd., representing a consortium of companies, was considering buying Jessel Securities' holding of France Fenwick shares. If this went ahead then Lawson-Batey said that they would be willing to buy certain assets and the tug business of France Fenwick. In May, agreements were drawn up between London & Yorkshire and Lawson-Batey and between France Fenwick and Lawson-Batey regarding the purchase, by Lawson-Batey, of ten tugs together with other assets of France Fenwick. More precisely, an agreement was signed on 5th May 1977 relating to the acquisition, by Lawson-Batey, of:-

1) The goodwill of the business of France, Fenwick Tyne & Wear Co.Ltd. as a going concern for £1.
2) Stock of fuel and oil aboard the Sunderland tugs, together with spares, for £29,472.
3) Plant and machinery, motor vehicles and office furniture for £7,880.
4) 1200 'B' shares of £1 each in Tyne Tugs Ltd. for £1,200.
5) Ten tugs together with spares and stores aboard for £517,000.

making a total purchase price of £555,553.

The prices paid for the individual tugs were:-

George V (2)	£7,000	*Whitburn* (2)	£18,000
Cornhill	£18,000	*Prestwick*	£40,000
Ashbrooke	£40,000	*Alnwick*	£40,000
Bamburgh	£45,000	*Marsden* (4)	£45,000
Alnmouth	£134,000	*Dunelm* (2)	£130,000

An immediate payment of £300,000 was made, with further instalments being due in November 1977 and November 1978. The agreement did not include the purchase of *Cragsider's* sister tug *Hendon* (3), built for France Fenwick by Richard Dunston in May 1977. *George V* (2), *Ashbrooke, Alnwick, Bamburgh* and *Alnmouth* were already on charter to Tyne Tugs Ltd. and thus working on the Tyne. The other five were working at Sunderland.

With these acquisitions Lawson-Batey became the sole owner of Tyne Tugs Ltd. and the owner of sixteen tugs - more than they had possessed since 1924. This number was too large, and *George V* (2) was soon sold to Coastal Marine Services of Plymouth for £11,500, but the actual transfer was to be delayed until January 1978. *Whitburn* (2) was sold to Egypt for £36,000, a rather surprising price considering that she was 34 years old and had been purchased for far less from France Fenwick only five months earlier. However, the Egyptians were probably quite satisfied since this war-time ex-US Army tug is still in service with them fifteen years later.

At the 57th Annual General Meeting the directors reported 'a most satisfactory year' and said that they had paid £100,000 off the price of *Cragsider* which had cost £425,000. The balance, being a bank loan guaranteed by the Secretary of State for Trade and Industry under Section 10 of the Industry Act 1970, was to be repaid over seven years. Lawson-Batey now controlled all towage at Blyth, and on the Tyne and Wear. It was pointed out that R.B. Harrison's shipyard at Bill Quay and the offices at Mercantile Chambers on the Quayside at Newcastle had not been included in the deal with France Fenwick.

At the end of 1977 the Lawson-Batey fleet consisted of fifteen tugs with an average age of 21 years. They were, in descending order of age:-

George V (2)	(217/1915)	1,086 BHP		Awaiting completion of sale.
Cornhill	(176/1943)	700 BHP		Working on the Wear.
Prestwick	(119/1955)	1,080 BHP	14 BP	Working on the Wear.
Ashbrooke	(119/1955)	1,080 BHP	14 BP	Chartered to Tyne Tugs Ltd.
Quaysider	(137/1955)	1.200 BHP	15 BP	Chartered to Tyne Tugs Ltd.
Alnwick	(119/1955)	1,080 BHP	14 BP	Chartered to Tyne Tugs Ltd.
Bamburgh	(122/1956)	1,080 BHP	14 BP	Chartered to Tyne Tugs Ltd.
Marsden (4)	(122/1956)	1,080 BHP	14 BP	Working on the Wear.
Roughsider	(143/1958)	750 BHP	11 BP	Chartered to Tyne Tugs Ltd.
Alnmouth	(173/1962)	1,080 BHP	16.5 BP	Chartered to Tyne Tugs Ltd.
Westsider (2)	(151/1964)	1,000 BHP	18 BP	Chartered to Tyne Tugs Ltd.
Dunelm (2)	(132/1964)	970 BHP	14 BP	Working on the Wear.
Northsider (2)	(156/1967)	1,420 BHP	18.5 BP	Chartered to Tyne Tugs Ltd.
Ironsider	(156/1967)	1,420 BHP	18.5 BP	Chartered to Tyne Tugs Ltd.
Cragsider	(266/1976)	3,180 BHP	46 BP	Chartered to Tyne Tugs Ltd.

In December 1977, the Lawson-Batey subsidiary company Ridley Tugs Ltd., which had remained in existence since 1971 though not owning any tugs, was retitled Wear Tugs Ltd. and set up as a charter company to operate the tugs at Sunderland. Thus the four Lawson-Batey tugs listed above as 'working on the Wear' were chartered to Wear Tugs Ltd. early in 1978. From 1st January 1978 all the tugs in the three fleets were insured with Sunderland Marine Mutual Insurance Co. The Tyne Tugowners' Association was wound up, since there was now only one tugowner operating on the Tyne. The tug *Maximus* (141/1956) was transferred from the Blyth Tug Co.Ltd. to Lawson-Batey for £22,500 and immediately chartered back to the Blyth company. Thus the ownership of all tugs was now formally in the hands of Lawson-Batey.

Richard Dunston was asked for an indicative price for a tug of 1,000 BHP and 18 tons bollard pull, similar to *Westsider* (2). In March 1978 they quoted £700,000 with a delivery of 12/14 months. An enquiry for the purchase of *Cragsider* for £800,000 had been received but this was turned down by the Lawson-Batey directors. From 1st January 1979 a new demise charter with Tyne Tugs Ltd. became effective. The minimum hire rates were £600 per month for *Cragsider*, £500 per month for *Northsider* (2) and *Ironsider* and £400 per month for the others. This was a renewal of the charter arrangements which had been established in 1959 and renewed in 1969 each for a ten-year period.

In June 1979 definitive quotations were received for two new 1,125 BHP tugs from Richard Dunston, Richards Shipbuilders, Cochrane Shipbuilders and Clelands. The prices varied between £685,000 and £875,000 with delivery periods of between eleven and sixteen months for the first, and between thirteen and twenty months for the second. Richard Dunston was again favoured with the contract which was placed in July at a price of £701,000 for the first, and £721,000 for the second vessel, delivery being promised for October 1980 and February 1981.

At the close of 1979 the company's valuation of their fifteen tugs was as follows:-

Cornhill	£30,000	*Prestwick*	£48,000
Ashbrooke	£48,000	*Quaysider*	£75,000
Alnwick	£48,000	*Maximus*	£52,000
Bamburgh	£50,000	*Marsden* (4)	£50,000
Roughsider	£75,000	*Alnmouth*	£134,000
Westsider (2)	£130,000	*Dunelm* (2)	£130,000
Northsider (2)	£250,000	*Ironsider*	£250,000
Cragsider	£1,000,000.		

The total valuation was thus £2,370,000 to be depreciated over 30 years. The first of the new tugs was christened *Wearsider* (150/1980) and launched by Mrs. Muriel Greenwell, the wife of the company managing director, on 14th July 1980. *Wearsider* was completed in November, arriving on the Tyne on the 19th and was immediately chartered to Tyne Tugs Ltd. She made an inauspicious debut when her towing hook fell off due to a loose pin in the towing arm!

A stern view of Wearsider *just prior to her launch. (Walter Fussey & Son).*

Mrs. Muriel Greenwell ready to christen Wearsider *on 14 July 1980. (Walter Fussey & Son).*

Wearsider *ready for launching. (Walter Fussey & Son).*

Tynesider *(2) on charter to Tyne Tugs Ltd.* *(Swan Hunter Ltd.).*

The second of the newbuildings *Tynesider* (2) (150/1981) was named and launched by Mrs. Sheila M. Hamilton, the wife of D.F. Hamilton, in the same month with delivery expected in March 1981. *Cornhill* was sold, for £25,000, to W.G.S. Crouch & Sons Ltd., Greenhithe, on 11th July 1980 and was replaced at Sunderland by *Roughsider* whose charter to Tyne Tugs Ltd. was transferred to Wear Tugs Ltd.

Michael Francis Pyman, whose family had run Mercantile Dry Dock Co. Ltd. on the Tyne, was appointed a director of Lawson-Batey Tugs Ltd. in September 1980, bringing the strength of the board to six. At the 61st Annual General Meeting on 23rd April 1981 the directors reported on the very difficult trading position on the Tyne, in particular noting that all of the repair yards were very short of work. As a result *Quaysider* and *Alnwick* were laid-up, and *Ashbrooke* and *Bamburgh* were put up for sale. In September 1981 it was decided that Lawson-Batey Tugs Ltd. should become a Public Limited Company and was so registered on 8th March 1982. This re-registration did not apply to the subsidiary companies.

Whilst laid-up at Sunderland *Alnwick* sank at her moorings and was soon afterwards sold to her salvors, Eurosalve Ltd., Folkestone. In May 1982 *Ashbrooke* and *Bamburgh*, which had been laid up for more than 3 years and 2 years respectively, were sold to Orchid Maritime Co.Ltd., Colombo, Sri Lanka for £15,000 each, thus reducing the fleet to thirteen.

S.S.W. Dalgliesh retired from the board during 1982 and further changes took place on 1st January 1983 when T.A. Greenwell retired as managing director and was appointed vice-chairman, his place being taken by D.F. Hamilton, who had deputised in that post since 1979. The latter relinquished his position of company secretary, which he had held since 1968, and was followed in that post by Harold Reginald Gilbert.

Late in 1982, an offer for the purchase of Lawson-Batey PLC was received from Clyde Shipping Co.Ltd. who made an offer of £4,089,000 for the whole of the share capital of the company, based on the 1982 Lawson-Batey profit, the value of the thirteen tugs and the goodwill of the business. This offer equated to £13.63 per ordinary share and was to be paid as 60% cash plus 40% guaranteed loan stock 1984-86. Lawson Batey became a subsidiary of Clyde Shipping Tugs Ltd., itself a wholly-owned subsidiary of Clyde Shipping Co.Ltd., when the take-over was agreed in April 1983. A preliminary announcement had been made to the employees of Lawson-Batey and its subsidiaries as to what was being arranged so that they were kept fully informed while negotiations proceeded.

As a result of the take-over two new directors were appointed to the Lawson-Batey board to represent the Clyde Shipping Tugs Ltd. interests: William Moncrieff Cuthbert and Dr. Leonard John Paterson. Because Lawson-Batey Tugs PLC was now a subsidiary of a private limited company it ceased to be designated as a public limited company and its title reverted to Lawson-Batey Tugs Ltd. In November *Westsider* (2) was transferred to Blyth Tug Co.Ltd. and *Maximus* was laid-up at South Dock, Sunderland. No offers had been received for *Quaysider* or *Maximus* when put on the market.

In September 1983 a contract was signed for the building of an additional tug by Richard Dunston at Hessle. She was to be of 1,125 BHP and would have a controllable pitch propeller and a 'Clyde' hook. The price was £900,000.

On 1st January 1984 T.A. Greenwell was appointed chairman of the board of directors of Lawson-Batey and its subsidiary companies, since F.S. Batey had retired from this position. The latter had been chairman for 32 years and was now 84 years of age. He was made Honorary Life President in recognition of his long service as a director of the company.

Changes in the Lawson-Batey fleet continued apace. In January 1984 the 28-year-old *Maximus* was sold to M. Tugs Ltd., Lowestoft, for £25,000, and in July *Quaysider* went to T.S.A. Tugs Ltd. for £24,000 where she was renamed *Towing Wizard*. At the end of the year *Westsider* (2) was transferred back to the Lawson-Batey company from the Blyth Tug Co. She had been idle at Blyth for 35 weeks as a result of the miners' strike, and was the last company tug to be stationed there. As a consequence the Blyth office closed at that time. *Roughsider* and *Marsden* (4) were offered for sale at £55,000 for the pair.

The latest tug building at Hessle was christened *Holmsider* (164/1984) by Mrs. Ethel Rutherford, the wife of William Rutherford, manager of Tyne Tugs Ltd., on 15th May 1984 and delivered to Lawson-Batey on 27th June. She was chartered to Wear Tugs Ltd. and began work at Sunderland in place of *Roughsider*. In August 1984, at the Hessle yard, the keel was laid of a sister tug to *Holmsider*, but on this occasion she was to be owned by Clyde Shipping Co.Ltd. She was christened by Mrs. Kay Towers, the wife of G.H.R. Towers, a Lawson-Batey director, on 10th December 1984 and named *Seasider* (2) (164/1985) and was to enter service with Tyne Tugs Ltd. in March 1985.

The long-lasting miners' strike was continuing to depress the company's results and this stoppage was not to end until February 1985. The shipping brokers, H.E. Moss, who were trying to sell the surplus tugs for Lawson-Batey, had sent out more than three hundred and fifty circulars to possible purchasers without success. However, *Roughsider* had to be withdrawn from the sale list when she suffered severe engine damage. Whilst laid-up the oil had emulsified due to rain entering via funnel vents! *Wearsider's* charter to Tyne Tugs Ltd. was cancelled and she was re-chartered to Wear Tugs Ltd. for operation at Sunderland. *Westsider* (2), which had been laid-up since transfer from Blyth was re-activated in December 1985.

In February 1986 *Roughsider* was sold for £4,700 to J.A. White, Inverkeithing, for breaking up. She was towed there by *Cragsider* whilst on her way to Dundee. *Prestwick* and *Dunelm* (2) were laid-up at Sunderland at that time.

Ironsider *on charter to Tyne Tugs Ltd.* *(Swan Hunter Ltd.).*

It was felt that there was now no need for two separate chartering companies for Lawson-Batey's tugs. So, in April 1986, Wear Tugs Ltd. transferred its business and assets to Tyne Tugs Ltd. which in turn changed its name to Tyne and Wear Tugs Ltd. This company now had all of the Lawson-Batey tugs on charter, viz: *Prestwick, Marsden* (4), *Alnmouth, Westsider* (2), *Dunelm* (2), *Northsider* (2), *Ironsider, Cragsider, Wearsider, Tynesider* (2), and *Holmsider* plus the Clyde-owned *Seasider* (2).

In July 1986 H.R. Gilbert retired as company secretary and was succeeded by John Gordon Mucklow, who was also appointed a director, representing Clyde Shipping Tugs Ltd. At the end of the year G.H.R. Towers retired from the board, having been a director since 1951. D.F. Hamilton had recently been appointed to the board of Clyde Shipping Tugs Ltd.

Shipments of coal from Sunderland ceased in December 1986 after 600 years. In the previous year the new Tyne Coal Terminal had been opened by HM Queen Elizabeth the Queen Mother, on land reclaimed from the eastern end of Jarrow Slake. It was intended that this terminal would handle all coal shipments from the area, utilising much larger bulk-carriers than had been possible in the past.

The company's Honorary Life President, F.S. Batey, died on 16th February 1987 at the grand old age of eighty-seven, after a very active life. He had been chairman of Lawson-Batey until 1983 and was the son of Dr. J.T. Batey who had also been chairman before him, so together they were at the helm for a total of almost fifty years. In the same month, the chairman, T.A.Greenwell retired as a working director, and sadly he too died, aged 65, on 23rd June 1987. He was succeeded by Dr. L.J. Paterson who had been a director since 1983 as a nominee of Clyde Shipping Tugs Ltd. of which company he was managing director. In November 1987 D.F. Hamilton was appointed a director of Sunderland Marine Mutual Insurance Association, the organisation which insured all of the company's tugs.

Early in 1987 *Alnmouth* had been sold, for £32,500, to Greek owners and renamed *Thiela*. In December *Prestwick* and *Marsden* (4) were sold to Medway Port Services Ltd., Strood, without change of name. These sales reduced the fleet strength to nine. *Dunelm* (2) was sold in March 1988 for £55,000 to Fowey Harbour Commissioners and renamed *Pendennick. Westsider* (2) was also sold to an Anguillan company in 1989.

W.M. Cuthbert, who was another of the Clyde Shipping Tugs Ltd. nominees on the board of directors, retired in May 1988 to be succeeded in February 1989 by Peter Henry P. Southern of M.H. Southern & Company Limited. W.M. Cuthbert died almost exactly one year after the date of his retirement.

Wearsider had been laid-up since June 1990 due to shortage of work, when an opportunity arose to charter her in December of that year to North Sea Tugs Ltd., based at Sunderland for a contract which entailed working on the River Medway.

This then brings to a close the story of Lawson-Batey Tugs Ltd. up to the end of 1990. The company had survived through good times and bad for a period of seventy years from the time of the merger of the two family companies of Lawson and Batey. Lawson-Batey itself had been involved with mergers and the take-over of other firms in its locality only to find in 1983 that it too was taken over by Clyde Shipping Co.Ltd. This whole book is a portrayal of how the many tug companies of Tyne, Blyth and Wear were gradually absorbed until only one company, Lawson-Batey, exists today. One's mind wonders what will be the ultimate result of these take-overs and mergers, in view of the unremitting course of events. Might there be only one tug company in the whole of the United Kingdom, or perhaps even a 'Eurotug' concern, with its headquarters in Brussels or Bonn, not even within sight of the sea? One hopes not. Ever since the formation of Tyne Tugs Ltd. in 1959 there has been a steady reduction of work available for the tugs, and this had been all too clearly reflected in the tug numbers in the fleet and, of course, in the dwindling number of employees. All of this has been caused not only by the increase in size and reduction in numbers of vessels trading to the area but also by the smaller numbers of ships being launched or repaired. Also, technical improvements, like the increasing use of bow thrusters on ships, have meant that, whereas a ship might previously have needed two tugs for berthing it, it now needs only one or none at all.

The period of the Lawson-Batey company has seen the demise of the paddle tug and the gradual, post-Second World War changeover from steam to diesel propulsion. As ships grew larger there was a need to provide tugs with the capability of exerting an ever increasing pull at the towing hook. Lawson-Batey, having started in 1920 with eighteen tugs, never again reached this size of fleet, though they did own sixteen just after their take-over of the France Fenwick business in 1977. All told they have owned forty-seven different tugs throughout their existence. Their fleet, at the end of 1990, having an average age of 14 years, was:-

Northsider (2)	(156/1967)	1,420 BHP	18.5 BP	
Ironsider	(156/1967)	1,420 BHP	18.5 BP	
Cragsider	(266/1976)	3,180 BHP	46 BP	
Wearsider	(150/1980)	1,125 BHP	17 BP	Chartered away.
Tynesider (2)	(150/1981)	1,125 BHP	17 BP	
Holmsider	(164/1984)	1,125 BHP	18 BP	
Seasider (2)	(164/1985)	1,125 BHP	18 BP	Owned by Clyde Shipping Co.Ltd.

These tugs are normally based on the Tyne and service the needs of Blyth and Sunderland from there as required.

Since 1920 there have been only eighteen different directors of the company. Those serving at the end of 1990 were:-

Chairman	Dr. L.J. Paterson
Managing Director	D.F. Hamilton
Directors	M.F. Pyman
	J.G. Mucklow
	P.H.P. Southern
Secretary	J.G. Mucklow

Who can predict what the future will bring for this company? It must be admitted that sometimes prospects do not look good, because of the very low level of trade at the three ports, but circumstances can, as we have seen, change dramatically, and one definite fact about the current period is that, when change occurs, it does so very quickly indeed. So, 150 years after John R. Lawson and John Batey started separately in the tug business, the present Lawson-Batey company will carry on, overcoming difficulties as they are encountered, just as their forebears did.

APPENDIX A TO CHAPTER FOUR

Lawson-Batey Tugs Ltd. 1920 - 1982
Lawson-Batey Tugs Plc. 1982 - 1983
Lawson-Batey Tugs Ltd. 1983 -

Board of Directors

Chairmen	M.C. James	1920-1934	Died
	Dr. J.T. Batey	1934-1951	Died
	F.S. Batey	1951-1983	Appt.Hon.Life President
	T.A. Greenwell	1984-1987	Died
	Dr. L.J. Paterson	1987-	
Managing Directors	J. Lawson)	1920-1930	Disqualified
	F. Batey)Joint		Became sole M.D.
	F. Batey	1930-1948	Died
	J.F. Batey	1948-1975	Died
	T.A. Greenwell	1976-1982	Vice-Chairman in 1983
	D.F. Hamilton	1976-1982	Deputy M.D.
	D.F. Hamilton	1983-	
Directors	R.J. Thompson	1920-1931	Died
	J.T. Batey	1920-1934	Became Chairman
	J. Thompson	1923-1940	Died
	R.I. James	1930-1944	Died
	J.F. Batey	1941-1948	Became M.D.
	S.S.W. Dalgliesh	1946-1982	Resigned
	F.S. Batey	1948-1951	Became Chairman
	G.H.R. Towers	1951-1986	Retired
	D.F. Hamilton	1976-1978	Became Deputy M.D.
	M.F. Pyman	1980-	
	W.M. Cuthbert	1983-1988	Retired
	Dr. L.J. Paterson	1983-1987	Became Chairman
	J.G. Mucklow	1986-	
	P.H.P. Southern	1989-	
Secretaries	G.A. Clark	1920-1922	
	J. Lawson	1922-1930	Resigned
	F. Batey	1930	
	J.R. Lawson	1930-1944	Resigned
	A. Scott	1944-1967	Retired
	D.F. Hamilton	1968-1982	Became M.D.
	H.R. Gilbert	1983-1986	Retired
	J.G. Mucklow	1986-	

APPENDIX B TO CHAPTER FOUR

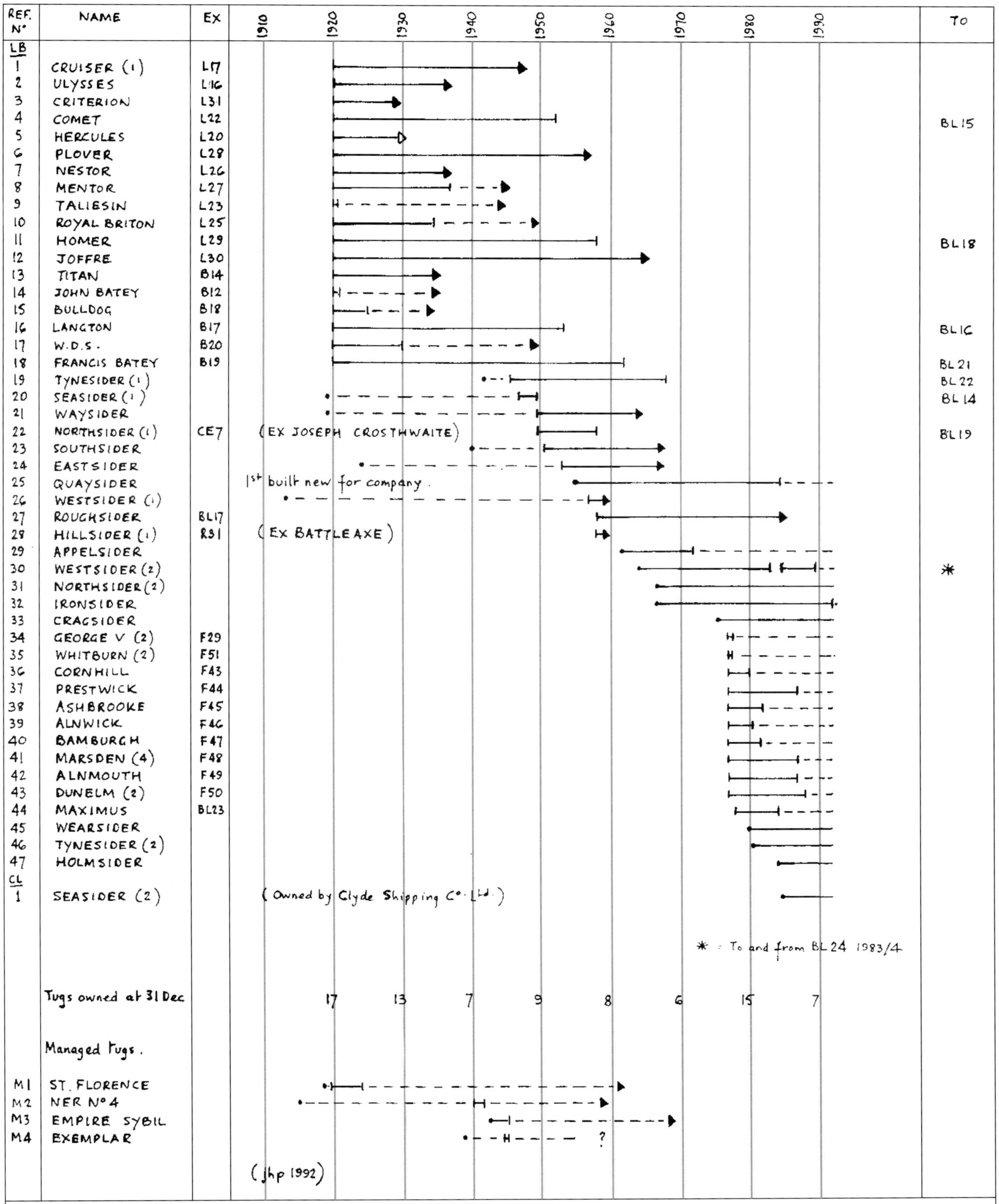

Tug Ownership Chronology.

LAWSON-BATEY TUGS LTD. 1920–1992

APPENDIX C TO CHAPTER FOUR

Lawson-Batey Tugs Ltd. Fleet List

LB 1 **CRUISER (1)** **(1920-1948)**
See Lawson fleet list, L 17

Ulysses *towing* Pear Branch *(4,786/1921). (Local History Library, South Shields Central Library).*

LB 2 **ULYSSES** **(1920-1937)**
See Lawson fleet list, L 16

LB 3 **CRITERION** **(1920-1929)**
See Lawson fleet list, L 31

Comet. *(Courtesy Paul Hood).*

LB 4 **COMET** **(1920-1952)**
See Lawson fleet list, L 22

LB 5 **HERCULES** **(1920-1931)**
See Lawson fleet list, L 20

Plover *passing St. Anthony's. (Les Turner).*

LB 6 **PLOVER** **(1920-1957)**
See Lawson fleet list, L 28

LB 7 **NESTOR** **(1920-1937)**
See Lawson fleet list, L 26

LB 8 **MENTOR** **(1920-1937)**
See Lawson fleet list, L 27

LB 9 **TALIESIN** **(1920 only)**
See Lawson fleet list, L 23

Royal Briton. *(John Clarkson, Longton).*

LB10 **ROYAL BRITON** **(1920-1934)**
See Lawson fleet list, L 25

Homer *with main-mast removed. Post-Second World War. (John Clarkson, Longton).*

LB11 **HOMER** **(1920-1958)**
See Lawson fleet list, L 29

Joffre. *(Local History Library, South Shields Central Library).*

LB12 **JOFFRE** **(1920-1966)**
See Lawson fleet list, L 30

LB13 **TITAN** **(1920-1936)**
See Batey fleet list, B 14

John Batey *when owned by Grangemouth & Forth Towing Co. Ltd. (Local History Library, South Shields Central Library).*

LB14 **JOHN BATEY** **(1920-1921)**
See Batey fleet list, B 12

LB15 **BULLDOG** **(1920-1925)**
See Batey fleet list, B 18

Langton *with shorter funnel. (Company archives).*

LB16 **LANGTON** **(1920-1953)**
See Batey fleet list, B 17

LB17 **W.D.S.** **(1920-1930)**
See Batey fleet list, B 20

Francis Batey. *(John Clarkson, Longton).*

LB18 **FRANCIS BATEY** **(1920-1962)**
See Batey fleet list, B 19

Tynesider *(1). (A. Duncan).*

LB19 **TYNESIDER (1)** **(1946-1968)**
ON 165256 262 g 107.8 x 26.2 x 12.5 ft Steel Screw
B: Hall, Russell & Co.Ltd., Aberdeen. (YN767).
E: T.3Cy. 16", 26" & 43" x 30". 1000 IHP. 10 K. Ailsa Shipbuilding Co.Ltd., Troon.

PR: 1942 Aberdeen; 1946 Newcastle.

04.1942:	Completed, as EMPIRE CHERUB, for Ministry of War Transport. (J. Cooper, Belfast, Mgrs.)
07.1946:	Purchased, for £20,400, by Lawson-Batey Tugs Ltd. and renamed TYNESIDER
01.1959:	Chartered to Tyne Tugs Ltd. until:-
11.1968:	Transferred, for £5,000, to Blyth Tug Co.Ltd.
03.1970:	Sold, for £4,150, to Hughes Bolckow Ltd., Blyth, for breaking up.

LB20 **SEASIDER (1)** **(1947-1949)**
ON 143296 154 g 88.5 x 21.2 x 10.6 ft. Steel Screw
B: Crabtree & Co.Ltd., Great Yarmouth. (YN174).
E: C.2Cy. 15" & 32" x 24". 430 IHP. 10 K. By builders.

PR: 1919 London; 1948 Newcastle.

07.1919:	Completed, as WEST HYDE, for Secretary of the Admiralty, London. Based at Chatham Dockyard.
12.1947:	Purchased, for £5,125, by Lawson-Batey Tugs Ltd. Arrived River Tyne in tow of GUARDSMAN.
02.1948:	Commenced service after alterations. Renamed SEASIDER.
09.1949:	Transferred, for £10,450, to Blyth Tug Co.Ltd.
08.1953:	Worked temporarily on Tyne for short period.
03.1969:	Sold, for £2,100, to Hughes Bolckow Ltd., Blyth, for breaking up.

Waysider. *(Maritime Photo Library).*

LB21 **WAYSIDER** **(1949-1964)**
ON 169221 166 g 101.2 x 22.5 x 10.4 ft. Steel Screw
B: Wed.C. Boele & Zonen, Slikkerveer.
E: T.3Cy. 14", 23" & 39" x 27". 600 IHP. 11 K. Alblasserdam Machinefabriek, Alblasserdam.

PR: 1919 Boulogne; 1949 Newcastle.

1919:	Completed, as LA MORINIE, for Boulogne Chamber of Commerce
02.1949:	Purchased for £6,881 by Lawson-Batey Tugs Ltd.
03.1949:	Arrived River Tyne in tow of SERVICEMAN
04.1949:	Towed to Sunderland by JOFFRE for repair at John Crown & Sons Ltd. Renamed WAYSIDER.
09.1949:	Commenced service on River Tyne
01.1959:	Chartered to Tyne Tugs Ltd.
01.1964:	Sold for £1,710 to Hughes Bolckow Ltd., Blyth, for breaking up.

Northsider *(1). (A. Duncan).*

LB22 **NORTHSIDER (1)** **(1949-1958)**
See Crosthwaite fleet list, CE 7, as JOSEPH CROSTHWAITE

Southsider. *(Maritime Photo Library).*

LB23 **SOUTHSIDER** **(1951-1968)**
ON 169228 239g 103.7 x 25.5 x 12.6 ft. Steel Screw
B: T. van Duijvendijk's Scheepswerf, N.V., Lekkerkerk.
E: T.3Cy. 370, 610, & 1000 x 610mm. 650IHP. 10 K. By Builders.

PR: 1940 Hamburg; 1951 Newcastle

1940:	Completed, as HEROS, for Bugsier Reederei und Bergungs A.G., Hamburg
02.1951:	Purchased, for £22,000, by Lawson-Batey Tugs Ltd., renamed SOUTHSIDER
03.1951:	On trials, after repairs, found to have a fractured crankshaft.
06.1951:	Began work on River Tyne
01.1959:	Chartered to Tyne Tugs Ltd.
03.1966:	Laid up until 12.1966 and from 11.1967.
02.1968	Sold, for £2,050, to Hughes Bolckow Ltd., Blyth, for breaking up.

Eastsider *with mainmast on funnel. (John Clarkson, Longton).*

LB24 **EASTSIDER** **(1953-1968)**
ON 169251 162g 99.1 x 21.3 x 11.1 ft. Steel Screw
B: Gebr. G.& H.Bodewes, Scheepswerven, N.V., Martenshoek.
E: T. 3 Cy. 335, 540 & 886 x 580mm. 600IHP. 10 K. Gebr. Päglow, Hamburg. 10K

PR: 1924 Copenhagen; 1953 Newcastle

1924:	Completed, as PLUTO, for Em.Z.Svitzer's Bjergnings Entreprise, Copenhagen
02.1953:	Purchased, for £20,000 delivered Tyne, by Lawson-Batey Tugs Ltd. and renamed EASTSIDER
01.1959:	Chartered to Tyne Tugs Ltd.
1966:	Laid up
02.1968:	Sold, for £1,850, to Hughes Bolckow Ltd., Blyth, for breaking up.

Westsider *(1). (A. Duncan).*

LB26 **WESTSIDER (1)** **(1957-1960)**
ON 128826 153g 106'0" x 22'1" x 13'5" Steel Screw
B: J.P. Rennoldson & Sons, South Shields. (YN 282)
E: C. 2 Cy. 20" & 42" x 27". 700 IHP. by builders.

PR: 1913 Middlesbrough; 1957 Newcastle.

03.1913:	Completed, as JOSEPH CONSTANTINE, for Tees Conservancy Commissioners
05.1917:	Hired by Royal Navy. 02.1919: Returned to owners
05.1934:	Sold, for £1,500, to Tees Towing Co.Ltd., renamed CHARING CROSS
08.1957:	Purchased, for £6,500, by Lawson-Batey Tugs Ltd., renamed WESTSIDER
01.1959:	Chartered to Tyne Tugs Ltd.
01.1960:	Sold, for £1,375, to British Iron & Steel Corporation. Allocated to J.J.King & Co.Ltd. Gateshead, for breaking up.

Quaysider *when new.* *(Company archives/Turners).*

LB25 **QUAYSIDER** **(1955-1984)**
ON 186843 137g 97'2" x 23'5" x 10'5.75" Steel Screw Kort rudder
B: Mu"tzelfeldwerft Gmbh, Cuxhaven. (YN 139)
E: Oil. 4SA. 8Cy. 320 x 450mm. 1200 BHP. 15 BP Klöckner-Humboldt-Deutz AG, Cologne.

PR: 1955 Newcastle; 1984 Jersey; 1987 Maldon

09.1955:	Completed, for £55,316, for Lawson-Batey Tugs Ltd.
01.1959:	Chartered to Tyne Tugs Ltd.
07.1984:	Sold, for £24,000, to T.S.A. Tugs Ltd., Leigh-on-Sea, renamed TOWING WIZARD.
1992:	Still in service

Roughsider *when on charter to Tyne Tugs Ltd. and with gaff on mainmast. (World Ship Photo Library).*

LB27 **ROUGHSIDER** **(1958-1986)**
See Blyth fleet list, BL 17

Hillsider *(1) in Ridley colours. (Maritime Photo Library).*

LB28 **HILLSIDER (1)** **(1958-1960)**
See Ridley fleet list, R 31, as BATTLEAXE

Appelsider *on trials in Tyne Tugs Ltd. funnel colours. (Company archives).*

LB29 **APPELSIDER** **(1962-1972)**
ON 186930 173g 99'6" x 25'3" x 10'8.25" Steel Screw Kort rudder
B: Richard Dunston (Hessle) Ltd., Hessle. (YN S.783)
E: Oil. 4SA. 8 Cy. 320 x 450mm. 1000 BHP. 18 BP. Klockner-Humboldt-Deutz, Cologne

PR: 1962 Newcastle; 1972 Dublin

07.1962:	Completed, for £85,887, for Lawson-Batey Tugs Ltd., Chartered to Tyne Tugs Ltd.
01.1972:	Sold, for £110,000, to Dublin Port & Docks Board; renamed COLIEMORE
1992:	Still in servce

Westsider *(2) on trials with Tyne Tugs Ltd. funnel colours and flying the Lawson-Batey houseflag. (Company archives).*

LB30 **WESTSIDER (2)** **(1964-1983; 1984-1989)**
ON 305422 151 g. 99'6" x 25'3" x 11'51/4" Steel Screw Kort rudder
B: Richard Dunston (Hessle) Ltd., Hessle. (YN S.804)
E: Oil. 4SA. 6 Cy. 320 x 450mm. 1000 BHP. 18 BP. Klöckner-Humboldt-Deutz, Cologne.

PR: 1964 Newcastle.

01.1964:	Completed, for £91,000, for Lawson-Batey Tugs Ltd., chartered to Tyne Tugs Ltd.
11.1983:	Transferred to Blyth Tug Co.Ltd.
12.1984:	Transferred back to Lawson-Batey Tugs Ltd. and laid up at Sunderland.
11.1985:	Chartered to Tyne Tugs Ltd.
04.1986:	Chartered to Tyne & Wear Tugs Ltd. 07.1987: Laid up at Sunderland.
05.1989:	Sold, for £46,000, to Clipper Cargoes Ltd., Anguilla.
1992:	Still in service.

Northsider *(2) when on charter to Tyne Tugs Ltd. (Author's collection).*

LB31 **NORTHSIDER (2)** **(1967-)**
ON 305447 156g 98'8" x 26'3" x 10'6.5" Steel Screw
B: Richard Dunston (Hessle) Ltd., Hessle. (YN S.848)
E:Oil. 4 SA. 8 Cy. 320 x 450mm. 1420 BHP. 18.5 BP. 11 K. Klöckner-Humboldt-Deutz, Cologne.

PR: 1967 Newcastle

04.1967:	Completed, for £106,500, for Lawson-Batey Tugs Ltd., Chartered to Tyne Tugs Ltd.
04.1986:	Chartered to Tyne & Wear Tugs Ltd.
1992:	Still in service.

Ironsider *when on charter to Tyne Tugs Ltd. (John Clarkson, Longton).*

LB32 **IRONSIDER** **(1967-1992)**
ON 305451 156g 99'8" x 26'3" x 9'9" Steel Screw
B: Richard Dunston (Hessle) Ltd., Hessle. (YN S.849)
E: Oil. 4 SA. 8 Cy. 320 x 450 mm. 1420 BHP. 18.5 BP. 11 K. Klöckner-Humboldt-Deutz. Cologne.

PR: 1967 Newcastle

09.1967:	Completed, for £106,500, for Lawson-Batey Tugs Ltd., Chartered to Tyne Tugs Ltd.
04.1986:	Chartered to Tyne & Wear Tugs Ltd.
06.1992:	Sold to Proodos Maritime Co., Piraeus.

Cragsider *leaving the Tyne at speed when on charter to Tyne Tugs Ltd. (Company archives/Turners).*

LB33 **CRAGSIDER** **(1976-)**
ON 363069 266g 32.72 x 9.66 x 4.152 mtrs Steel Screw Kort rudder
B: Richard Dunston (Hessle) Ltd., Hessle. (YN H.906)
E: Oil. 4 SA. 6 Cy. 380 x 457mm. 3180 BHP. 46BP. 13 K. Mirrlees Blackstone Ltd., Stockport.

PR: 1976 Newcastle

12.1976:	Completed for Lawson-Batey Tugs Ltd. Chartered to Tyne Tugs Ltd.
04.1986:	Chartered to Tyne & Wear Tugs Ltd.
1992:	Still in service

LB34 **GEORGE V (2)** **(1977-1978)**
See Redhead fleet list, RH 18.

LB35 **WHITBURN (2)** **(1977 only)**
See France Fenwick fleet list, F 51

LB36 **CORNHILL** **(1977-1980)**
See France Fenwick fleet list, F 43

LB37 **PRESTWICK** **(1977-1987)**
See France Fenwick fleet list, F 44

LB38 **ASHBROOKE** **(1977-1982)**
See France Fenwick fleet list, F 45

LB39 **ALNWICK** **(1977-1981)**
See France Fenwick fleet list, F 46

LB40 **BAMBURGH** **(1977-1982)**
See France Fenwick fleet list, F 47

LB41 **MARSDEN (4)** **(1977-1987)**
See France Fenwick fleet list, F 48

LB42 **ALNMOUTH** **(1977-1987)**
See France Fenwick fleet list, F 49

LB43 **DUNELM (2)** **(1977-1988)**
See France Fenwick fleet list, F 50

LB44 **MAXIMUS** **(1978-1984)**
See Ridley fleet list, R 30

LB45 **WEARSIDER** **(1980-)**
ON 363093 150g 28.02 x 7.83 x 2.81 mtrs Steel Screw Kort rudder
B: Richard Dunston (Hessle) Ltd., Hessle (YN H.929)
E: Oil 4SA. 8 Cy. 222 x 292mm. 1125 BHP. 17 BP. 11.5 K. Mirrlees Blackstone Ltd., Stamford.

PR: 1980 Newcastle

11.1980:	Completed, for £750,000, for Lawson-Batey Tugs Ltd. Chartered to Tyne Tugs Ltd.
03.1985:	Chartered to Wear Tugs Ltd.
04.1986:	Chartered to Tyne & Wear Tugs Ltd. 06.1990: Laid up
12.1990:	Chartered to North Sea Tugs Ltd. for approx. 6 weeks for work on River Medway.
05.1991:	Chartered to North Sea Tugs Ltd. for work on River Tees, renamed SAMURAI SWORDSMAN.
06.1992:	Charter completed. Returned to River Tyne. Renamed WEARSIDER. Still in service.

Tynsider *(2) when on charter to Tyne & Wear Tugs Ltd. (World Ship Photo Library).*

LB46 **TYNESIDER (2)** **(1981-)**
ON 363096 150g 28.02 x 7.83 x 2.81 mtrs Steel Screw Kort rudder
B: Richard Dunston (Hessle) Ltd., Hessle. (YN H.930)
E: Oil. 4SA. 8 Cy. 222 x 292 mm. 1125 BHP. 17 BP. 11 K. Mirrlees Blackstone Ltd., Stamford.

PR: 1981 Newcastle

03.1981:	Completed, for £750,000, for Lawson-Batey Tugs Ltd. Chartered to Tyne Tugs Ltd.
04.1986:	Chartered to Tyne & Wear Tugs Ltd.
1992:	Still in service

Wearsider *on trials in the colours of Tyne Tugs Ltd.* *(Company archives).*

Holmsider *when new in the charter colours of Wear Tugs Ltd. (Walter Fussey & Son).*

LB47 HOLMSIDER (1984-)
ON 398265 164 g 28.02 x 7.83 x 2.891 mtrs Steel Screw(CP) Kort rudder
B:Richard Dunston (Hessle) Ltd. Hessle. (YN H.943)
E: Oil. 4SA. 8 Cy. 222 x 292mm. 1125 BHP. 18 BP. 12.25 K. Mirrlees Blackstone Ltd., Stamford.

PR: 1984 Newcastle

06.1984:	Completed for Lawson-Batey Tugs Ltd. Chartered to Wear Tugs Ltd.
04.1986:	Chartered to Tyne & Wear Tugs Ltd.
1992:	Still in service

Seasider *(2) when on charter to Tyne & Wear Tugs Ltd. (World Ship Photo Library).*

CL 1 SEASIDER (2) (1985-)
ON 398266 164 g 28.22 x 7.83 x 2.891 mtrs Steel Screw (CP) Kort rudder
B:Richard Dunston (Hessle) Ltd., Hessle. (YN 949)
E: Oil. 4SA. 8 Cy. 222 x 292mm. 1125 BHP. 18 BP. 12.25 K. Mirrlees Blackstone Ltd., Stamford.

PR: 1985 Newcastle

03.1985:	Completed for Clyde Shipping Co.Ltd. Chartered to Tyne Tugs Ltd.
04.1986:	Chartered to Tyne & Wear Tugs Ltd.
1992:	Still in service

LB/M.1 ST. FLORENCE (c.1920-1924)
ON 143307 419 g 135.2 x 29.1 x 13.6 ft. Steel Screw
B: J. Chrichton & Co.Ltd., Saltney, Chester. (YN 286)
E: T. 3 Cy. 181/4", 281/2" & 481/2" x 28". 1200 IHP. Fawcett, Preston & Co.Ltd., Liverpool.

PR: 1919 London; 1925 Victoria, B.C.

03.1919:	Completed for The Secretary of the Admiralty.
c.1920:	Managed by Lawson-Batey Tugs Ltd.
11.1924:	Sold to Canadian Pacific Railway Cp., renamed KYUQUOT.
01.1925:	Left Leith for Victoria, B.C. where arrived 03.1925.
11.1931:	Ashore, salvaged 02.1932.
07.1962:	Sold to General Shipbreaking Co., Vanvouver, for breaking up.

LB/M.2 N.E.R. No.4 (1941-1942)
ON 135920 177 g 95.0 x 24.1 x 11.4 ft. Steel Twin screw
B: J.P. Rennoldson & Sons Ltd., South Shields. (YN 298)
E: Two-C. 2 Cy. 17" & 36" x 24". 800 IHP. By builder

PR: 1916 West Hartlepool.

03.1916:	Completed, for £9,985, for North Eastern Railway Co.
01.1923:	Transferred to London & North Eastern Railway Co.
11.1941:	Managed by Lawson-Batey Tugs Ltd. for Minister of War Transport
11.1942:	Returned to owners
01.1948:	Transferred to British Transport Commission, afterwards Docks & Inland Waterways Executive and later British Transport Docks Board.
01.1960:	Sold to British Iron & Steel Corporation. Allocated to C.W. Dorkin & Co.Ltd., Gateshead for breaking up.

Empire *Sybil as* Assistant. *(World Ship Photo Library).*

LB/M.3 EMPIRE SYBIL (1943-1946)
ON 169313 276 g. 105.2 x 26.6 x 12.2 ft. Steel Screw
B: Cochrane & Sons Ltd., Selby. (YN 1268)
E: T. 3 Cy. 15", 25" & 42" x 27". 850 IHP. Amos & Smith Ltd., Hull.

PR: 1943 Hull; 1946 Liverpool.

08.1943:	Completed for Ministry of War Transport, Lawson-Batey Tugs Ltd. appointed managers.
05.1946:	Sold, for £21,000, to Mersey Docks & Harbour Board, Liverpool, renamed ASSISTANT
1962:	Sold to Alexandra Towing Co.Ltd., Liverpool, renamed CASWELL.
1969:	Sold to Haulbowline Industries Ltd., Passage West. Arrived in tow 03.1969 for breaking up.

LB/M.4 EXEMPLAR (1945-1946)
ON 169204 142 g 87.9 x 23.1 x 11.1 ft. Steel Screw
B: Deutsche Schiff-und Maschinebau A.G.Werk, Seebeck. Wesermu"nde. (YN 624)
E: T. 3 Cy. 310, 510 & 850 x 650 mm. 800 IHP. By builders.

PR: 1939 Bremen; 1945 Newcastle; 1946 ?

03.1939:	Completed, as VEGESACK, for Unterweser Reederei A.G.
1945:	Taken over by Ministry of War Transport as war reparations, renamed EXEMPLAR
08.1945:	Arrived River Tyne for management by Lawson-Batey Tugs Ltd.
02.1946:	Transferred to French port. Allocated to the U.S.A. by the Tripartite Commission.
1946:	Transferred to United States War Shipping Administration
c.1947:	Transferred to United States Maritime Commission.
c.1952:	Transferred to United States Department of Commerce.
1955:	Not in Lloyds Register. Fate unknown.

Alnwick *almost submerged after sinking at Sunderland during a period of lay-up in May 1981.* *(Company archives).*

Alnwick *looking in a very sorry state after being raised.* *(Company archives).*

• CHAPTER FIVE •

The Blyth Companies

Blyth Steam Tug Co. 1887 - 1923

The early ownership of tugs at the port of Blyth is difficult to determine since most were owned by individuals holding a number of sixty-fourth shares rather than by a clearly defined partnership or company. What is known, however, is that Captain George Dinsdale in 1887, after having retired from the sea, became involved in tug ownership. Along with others he purchased the wooden paddle tug *Earl of Beaconsfield* (1) (75/1877) from Cardiff owners and operated her under the style of the Blyth Steam Tug Co. That George Dinsdale held the major interest in this company is suggested by the fact that the black funnels of the tugs carried a large white letter 'D'. On board *Earl of Beaconsfield* (1) on her delivery voyage from Cardiff was the 'boy' Samuel Fisher. He was in fact a boy in the true sense of the word as he was aged only fourteen and was just a few weeks out of school. As will be seen he was to give the company and its successor over sixty-one years' service.

In 1889 the company purchased its second paddle tug, *Ruby* (98/1862) from Shields owners. She was considered something of a novelty at Blyth because of her double side-lever engine. About that time George Dinsdale's nephew, William Dinsdale, was introduced into the company as a partner. The wooden paddle tug *General Gordon* (97/1885) was purchased from Anthony Irwing of Sunderland in April 1890. She was built in the year that General Gordon was killed at Khartoum so that is the likely reason why she was so named. Two years later the relatively large double-engined iron paddle tug *Flying Squadron* was purchased from the Clyde Shipping Co. and renamed *Bothal* (149/1868). In her twenty-four years she had served first on the Thames as *Fergus MacIvor,* then at Queenstown, Ireland, before being given her second name when arriving on the Clyde. When at Queenstown she had often been used as a tender to trans-Atlantic mail ships. On her arrival at Blyth, Samuel Fisher was appointed as her skipper. A contemporary reference to *Bothal* described her as 'the pride of the fleet, having graceful flowing lines, spacious upper deck, and tall, slim twin funnels, every inch a lady and the envy of tugboat owners up and down the coast.' In the days before towage contracts were usual and tugs had to go to sea 'seeking' jobs, her speed placed her in a class by herself.

In September 1894 another quite large double-engined iron-hulled paddle tug was purchased from owners at Llanelli. She was *George Peabody* (115/1867) and cost her new owners £2,400. She had started life on the Thames where she was owned by a partnership of river pilots who kept her for eighteen years before her sale to South Wales owners. Her record includes a mention of a collision with *General Gordon* at Blyth in November 1900. *Ruby* was disposed of for breaking-up in 1895, leaving the Blyth Steam Tug Co. with a fleet of four tugs at the turn of the century:-

Name	Gross Tonnage/Year	Hull construction	Means of Propulsion
Earl of Beaconsfield (1)	(75/1877)	Wood	Paddle
General Gordon	(97/1885)	Wood	Paddle
Bothal	(149/1868)	Iron	Paddle
George Peabody	(115/1867)	Iron	Paddle

They had an average age of almost twenty-six years.

George Peabody was sold in 1907 for a further thirteen years' service on the River Tees. In 1911 the well-known shipbreaking company Hughes Bolckow & Co.Ltd. was established at North Blyth. This enterprise, over many years, was to provide a very welcome addition to the jobs available for the tugs at Blyth. In 1913 Samuel Fisher came ashore and took over as manager from William Dinsdale.

Until 1897 the port of Blyth had come under the jurisdiction of the North Shields Custom House. In that year Blyth acquired its own Custom House and the first vessel to be registered at Blyth had been the company's tug *George Peabody.* Only in 1913, however, did the other tugs have their registry transferred from North Shields to Blyth. At that time, although trading as Blyth Steam Tug Co., the tugs were owned by Alexander Watt, John Thompson, William Thompson, Mary Ann Thompson and Robert Bell, with William Dinsdale (and soon Samuel Fisher) as the designated manager.

The company received delivery, in March 1913, of the steel-hulled tug *George Dinsdale* (105/1913) from Scott & Sons of Bowling. She was notable in four respects: being, for the company, the first built new, the first screw propelled, the first steel-hulled, and the first to have a triple-expansion steam engine. In the same year *Earl of Beaconsfield* (1) was sold to the recently established shipbreakers Hughes Bolckow & Co.Ltd. at Blyth. *George Dinsdale* had quite a short life with the company, for, in 1914, she was hired by the Royal Navy, and whilst under their control, ownership passed to Grangemouth & Forth Towing Co.Ltd., who, in the 1920s, sold her to owners in Adelaide, Australia. Why such a fine powerful tug should stay at Blyth for such a short time is not known, but it could be guessed that she may not have been sufficiently manoeuvrable for the confines of Blyth Harbour.

In 1919 the Blyth Steam Tug Co. took over the towage business of Octavius Wilson Jewels at Blyth. Thus the single-engined iron paddle tug *Jupiter* (102/1876) came into the company's ownership. *Corsair* (2) (85/1877) was acquired in a similar manner in August 1921 from Richard Lough Manners of Blyth for £1,400, and in 1922 the North Eastern Railway Company's double-engined steel paddle tug *Salt* was purchased. This latter vessel had spent all of her thirty-three years' life working in the docks at Hartlepool and West Hartlepool. On arrival at Blyth she was renamed *Earl of Beaconsfield* (2) (114/1889).

Lawson-Batey Tugs Ltd., of Newcastle and South Shields, had for a while been looking at the possibilities of operating tugs at Blyth. One way of doing so, which was under consideration, was for them to purchase the Blyth Steam Tug Co. Enquiries showed that in the years ended March 1920, 1921 and 1922 the Blyth company had made profits, before tax, of £7,430, £5,069 and £7,382 respectively and that the asking price for an outright purchase was £14,000. Lawson-Batey considered this price too high. In May 1923, Lawson-Batey offered to take a majority holding if the valuation was reduced to £10,000. This new valuation was agreed on 29th May and Lawson-Batey undertook to take a holding of £7,000 in a new company to be formed in which they would have the controlling interest.

Blyth Tug Co. Ltd. 1923 - 1988

The Blyth Tug Co.Ltd. was formally incorporated on 9th July 1923 with a share capital of £20,000 in £1 shares, of which £10,000 was to be fully paid up. The objectives of the new company were 'to acquire and take over as a going concern, and carry on the business of, steam tug owners now carried on by Alexander Watt, John Thompson, William Thompson, Mary Ann Thompson, Robert Bell and William Dinsdale under the style or firm of Blyth Steam Tug Co.' The directors were to be:-

Matthew Coulson James	(chairman)	Representing Lawson-Batey Tugs Ltd.
Frank Batey	joint managing	
Joseph Lawson		
William Thompson		
Robert Bell		Ex Blyth Steam Tug Co.

The fleet of five tugs transferred formally on 16th July 1923 were:-

General Gordon	(97/1885)	Wood	Paddle
Bothal	(149/1868)	Iron	Paddle
Jupiter	(102/1876)	Iron	Paddle
Corsair (2)	(85/1877)	Wood	Paddle
Earl of Beaconsfield (2)	(114/1889)	Steel	Paddle

- with an average age of forty-four years.

The initial subscribers were Robert Bell and William Thompson, described respectively as 'Gentleman' and 'Grocer'! Samuel Fisher remained as manager, and the funnel markings of the tugs retained the 'Dinsdale D' as described previously. One of the earliest actions of the new board was to dispose of the tugs *Corsair* (2) and *Jupiter*. These were sold for £200 and £275 respectively for further service on the Tyne. This quick sale of two very old tugs (though not the oldest) is not surprising, since, early in 1924, shipping activity was reported as being at a very low ebb. This situation was, of course, not just at Blyth, but at all ports throughout the country.

When, in 1925, the Coulson Tug Co.Ltd. went into liquidation and was acquired by France, Fenwick Tyne & Wear Co.Ltd., discussions took place between France Fenwick and the Blyth Tug Co.Ltd. regarding Coulson's operations at Blyth. In 1926 the outcome was that the Coulson presence at Blyth would cease and that the Blyth company would purchase their two tugs which worked there. The wooden paddle tug *Harwich* (125/1877) and the iron screw tug *Dunera* (91/1892) were purchased for £1,500 and £2,000 respectively on 1st March 1926. Interestingly, when *Harwich* came into the Blyth Tug Co.Ltd. ownership, this was the fifth company which had owned her and whose history is included in this volume. At the time of her purchase *Dunera* was in need of

General Gordon, Earl of Beaconsfield *(2)*, William Fallows *and* Harwich *moored in Blyth Harbour, c.1928.* *(George Scott).*

Walmer Castle *(12,546/1902) entering Blyth Harbour on her way to the breakers in 1932. The head tugs are* Francis Batey *and* Joffre.
(Company archives).

damage repairs so this no doubt would have affected the price offered. During the miners' strike of 1926 only two of the fleet of five tugs were in use, and at the third Annual General Meeting of the company on 14th September it was reported that there had been no employment for the tugs at Blyth since 1st May. One thing that irked the management at this difficult time was that Blyth Harbour Commissioners continued to charge dues on the tugs in the harbour. In fact it was December before the tugs got back to work, when there was a good volume of traffic to handle for about three months. Thereafter, the level of business in 1927 fell away badly again.

At the AGM in 1928 steady work over the previous year was reported, the company having made a 'good' profit of £802. In July 1929 an offer of £878-10-0 (£878.50) was made for the paddle tug *Troon* then owned by the London, Midland & Scottish Railway Co., but this offer proved to be too low. She did, however, come to the north-east coast, being bought by Middlesbrough owners for £1,000 the following year. If *Troon* had been purchased then it had been intended to sell *Bothal* for a minimum of £400. In February 1930 *General Gordon* was the tug to go when she was sold for £140 to Sunderland shipbreakers.

In July 1930 William Thompson resigned his directorship of the company, and in September Joseph Lawson was disqualified from being a director. John Thompson and Samuel Fisher respectively replaced them on the board from September 1930. Also, in July, at the Seventh Annual General Meeting, it was reported that Alexander Watt, one of the early shareholders, had died. Business outlook at that time was said to be 'not good'. In an agreement with The 'Ridley' Steam Tug Co.Ltd. in 1930 it was arranged that the Blyth company would take over all of the Ridley contracts at Blyth. What the consideration was for this arrangement has not been discovered. However, the Ridley tug *Granville* was chartered to the Blyth Tug Co.Ltd. for £10 per week from October. When the Canadian Pacific liner *Empress of Scotland* (24,581/1905) caught fire at Hughes Bolckow's shipbreaking yard in December 1930 *Harwich* and *Earl of Beaconsfield* (2) had given valuable assistance in fighting the fire.

In 1932 the sixty-four-year-old *Bothal* and the wooden-hulled *Harwich* were sold to Clayton & Davie Ltd. at Dunston for breaking up, for the princely sums of £75 and £41-10-0 (£42.50) respectively. It had been hoped that they would have been worth £100 and £75. To offset the loss of these two paddle tugs a further vessel was acquired. She was *Steel* (118/1889) belonging to the London & North Eastern Railway Co. and was a sister to *Earl of Beaconsfield* (2), ex *Salt*, having a steel hull and being fitted with a powerful salvage pump and fire engine. After survey at Tyne Dock the company made an offer of £286 for her which was accepted by the railway company. The board minutes record that 'a good tug was obtained at a very low price'. This is something of an understatement, since the new acquisition remained in service with the company for a further twenty-six years! Curiously, although *Salt* had been renamed, *Steel* kept her original name throughout her lifetime. In 1932 the Blyth Tug Co.Ltd. dispensed with agents who had been used at home and abroad to further their business. With the above changes, the fleet had thus reduced to three:-

Earl of Beaconsfield (2)	(114/1889)	Steel	Paddle
Dunera	(91/1892)	Iron	Screw
Steel	(118/1889)	Steel	Paddle

all of them over forty years of age.

The company chairman, Matthew Coulson James, died on 5th May 1934 and was succeeded by Robert Bell. The vacancy on the board was filled by Robert Ingo James, the son of the late chairman. Sadly, Robert Bell's time as chairman was very short, since he, too, died just over a year later. Frank Batey, the managing director, then took on the additional position of chairman. In July 1936, Dr. John Thomas Batey, chairman of Lawson-Batey Tugs Ltd. and recently-retired managing director of R. & W. Hawthorn, Leslie & Co.Ltd., was appointed to the board.

The screw tug *Dunera II*, which the company had owned for the past eleven years, was sold, at the end of 1937, for £300 for breaking-up at Dunston on the River Tyne. Rather curiously, her name had been changed from *Dunera* only a few months before she was scrapped. It is assumed that this was done to release the name *Dunera* for the British India liner of that name which was completed by Barclay Curle & Co.Ltd., Glasgow, in August 1937. A third ex-North Eastern Railway Co. paddle tug *Greatham* (132/1893) was purchased for £675-12-6 (£675.625) in October 1937. She was very similar to *Earl of Beaconsfield* (2) and *Steel* but, surprisingly, her hull was constructed of iron even though she had been completed four years later than the other two. However, it will be found elsewhere in this volume that, in the same year that *Greatham* was completed, the NER built a tug, for operation at Middlesbrough, which had a wooden hull! Also, the same builders, Rennoldson & Sons of South Shields, were involved. Thus, by these changes, the company, in the period just before the outbreak of the Second World War, had reverted to an all-paddle fleet of three virtually standard tugs.

John Thompson, a director of the Blyth company, died on 19th September 1940. A year later, Cyril Wilfred Bell, who had previously been with the Lawson-Batey Co., was appointed company secretary. During 1941 and 1942 the tugs were beginning to show their age of around fifty years. *Earl of Beaconsfield* (2) needed boiler repairs which cost £2,122, and *Greatham* suffered broken wheel brackets and a fracture in her port condenser. Also, *Granville*, on charter from Ridley, had a damaged cylinder and needed engine bedplate repairs at about the same time. For a while, another Ridley tug, *King Edward VII* worked in her place. In 1941, a 'wiping' station had been set up at Blyth, whereby the magnetic influence of ships was modified so that there was less likelihood of their exploding magnetic mines in their vicinity. This operation was also referred to as degaussing. The large number of vessels calling for this treatment clearly meant more work for the tugs.

Yet another ex-NER tug was considered for use at Blyth when *Ceylon* was inspected at Hartlepool in 1943. Unfortunately, her length of 110 feet made her unsuitable. In May 1944 the hire rate for *Granville* was increased to £20 per week. The Blyth Tug Co.Ltd. office at this time was located at 1, Sea View, Blyth. R.I. James, one of the directors, died on 13th December 1944, and in the following year his place was taken by Major John Francis Batey, who, although serving in the Forces, was already a director of Lawson-Batey Tugs Ltd. In 1947, after representations from Blyth Harbour Commissioners, because of the poor tug service being given at Blyth, the company hired *Langton* from Lawson-Batey for a period of two years at a rate of £30 per week. It has been reported that Langton was the first screw tug to operate at Blyth, but if this is true, then *George Dinsdale* and *Dunera* (later *Dunera II*) must have been employed somewhere other than Blyth Harbour. In the event, the experiment of using *Langton* at Blyth was considered to be a success.

Frank Batey, who had been the company chairman since 1935, died on 14th March 1948 and in October 1948 Samuel Fisher retired from the board of directors. Samuel Fisher had worked for the Blyth Tug Co.Ltd. and its predecessor for sixty-one years after starting as a boy of fourteen straight from school. In 1949 Dr. J.T. Batey was appointed chairman of the company and two new directors, Stanley Spencer Woodbridge Dalgliesh and Frank Sewell Batey, were appointed directors as representatives of the Lawson-Batey company. It will be noted that, since the retirement of Samuel Fisher, there were now no members of the board with connections going back to the original Blyth Steam Tug Co. All the board members were now nominees of the majority shareholders, Lawson-Batey Tugs Ltd.

Company letterhead, 1939.

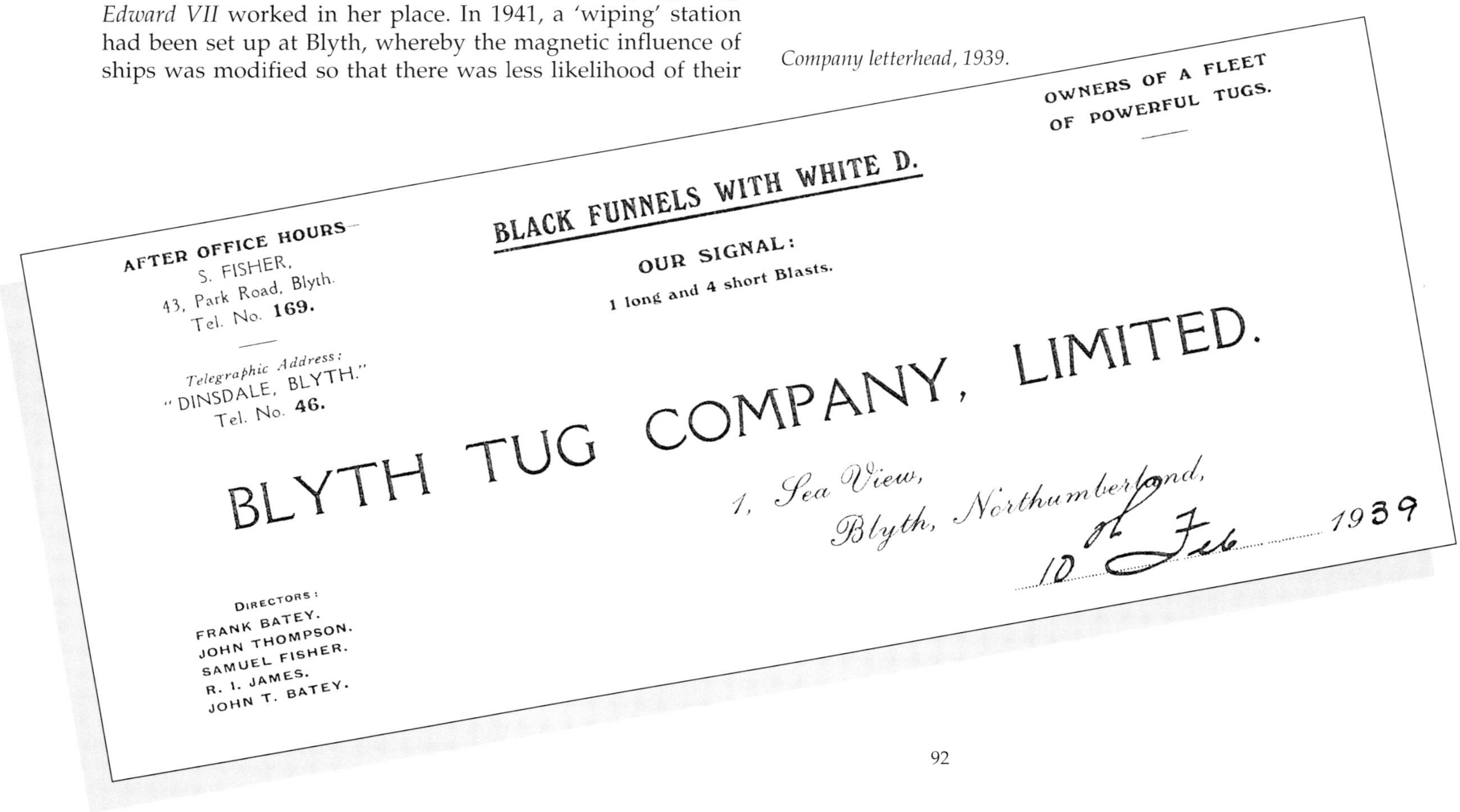
OWNERS OF A FLEET OF POWERFUL TUGS.

BLACK FUNNELS WITH WHITE D.

AFTER OFFICE HOURS
S. FISHER,
43, Park Road, Blyth.
Tel. No. 169.

OUR SIGNAL:
1 long and 4 short Blasts.

Telegraphic Address:
"DINSDALE, BLYTH."
Tel. No. 46.

BLYTH TUG COMPANY, LIMITED.

1, Sea View,
Blyth, Northumberland,
10th Feb 1939

DIRECTORS:
FRANK BATEY.
JOHN THOMPSON.
SAMUEL FISHER.
R. I. JAMES.
JOHN T. BATEY.

The 'Ridley' Steam Tug Co.Ltd. were not satisfied with the hire rate of £20 per week for *Granville*. As a result, the Blyth company made an offer to accept an increase to £30 per week provided the owners paid for all repairs and insurances instead of 50% as heretofore. Ridley would not accept this proposal and later in 1949 agreement was reached that the rate should be increased to £22-10-0 (£22.50), but subject to the original terms. The two-year time charter of *Langton* expired in May 1949 and Lawson-Batey were keen to have her back on the Tyne. However, they were prepared to sell the screw tug *Seasider* (1) (154/1919) to the Blyth company for £10,450. The transfer took place on 30th September 1949.

Earl of Beaconsfield (2) had problems again in September 1949, when it was found that new engine bed-plates were required. Whilst under repair at Blyth Drydocks and Shipbuilding Co.Ltd., not surprisingly further problems were uncovered - literally! Her beams and keelsons were in a very poor state and had to be renewed. As a result the *'Beaky'*, as she was fondly known, was out of service until December 1950. In 1951, when it became apparent that *Granville* could not be kept in service economically for very much longer it was agreed that *Comet* would be transferred from Lawson-Batey.

On 23rd July 1951, the chairman, Dr. J.T. Batey, died at the age of eighty- eight and soon thereafter he was succeeded by Frank Sewell Batey. In turn, George Harold Readhead Towers, who had, at the same time, joined the Lawson-Batey board, was appointed a director of Blyth Tug Co.Ltd. Samuel Fisher, who had retired in 1948, died on 12th August 1951 aged seventy-eight. The company reported a successful trading position for 1951.

In April 1952 *Granville's* charter ended and she went to the breakers. As had been decided previously, the iron paddle tug *Comet* (132/1876) was purchased from Lawson-Batey for £1,650. Consideration was given to purchasing *Moorcock* (151/1919), a similar screw tug to *Seasider* (1), for £10,000 from Liverpool owners, but this did not take place due to the condition of her boiler. A further tug which was surveyed with a view to purchase was *Swansea* (147/1918), a sister of *Seasider* (1), and owned by the British Transport Commission at Swansea. The asking price was £6,000, but, again, she was found to be in very poor condition, so nothing came of the proposal. More serious, and expensive, repairs to the fleet were required in 1952. First *Seasider* (1) needed a new propeller, then *Earl of Beaconsfield* (2) broke her port paddle shaft and, as a result, the surrounding frames as well. The latter tug was out of action for four months. On *Steel* the port wheel balance weights fell out. Those from *Earl of Beaconsfield* (2) (then under repair) were transferred to *Steel!*

Complaints were still being received from Blyth Harbour Commissioners and the Blyth Pilotage Authority regarding the towage service. No doubt this is why the company had investigated the possibility of buying *Moorcock* and *Swansea*. The outcome was that the company resolved to have a tug on duty each night, from 1st November 1952. There was also an approach to Tees Towing Co.Ltd. at Middlesbrough regarding the possibility of chartering a tug, but nothing came of this. Thus, at the end of 1952, the Blyth Tug Co.Ltd. fleet consisted of the following tugs:-

Earl of Beaconsfield (2)	(114/1889)	Steel	Paddle
Steel	(118/1889)	Steel	Paddle
Greatham	(132/1893)	Iron	Paddle
Seasider (1)	(154/1919)	Steel	Screw
Comet	(132/1876)	Iron	Paddle

with an average age of almost fifty-nine years! No wonder there were complaints.

In March 1953 *Comet* was sold for £950 for breaking-up on the River Tyne. The following month *Langton* (157/1892) was transferred from Lawson-Batey at a valuation of £5,000. Later in the year *Seasider* (1) worked on the Tyne for a short period.

At the 30th Annual General Meeting on 1st April 1953 it was reported that the company had made a loss on the year, due, primarily, to the six weeks' strike by coal trimmers, and increased running costs. Once again, Blyth Harbour Commission complained that more powerful tugs were required to handle the increasing size of ships using the port. John Francis Batey was appointed managing director from 17th April 1953.

In April 1954 William France, Fenwick & Co.Ltd. added their voice to the Harbour Commissioners when they, too, complained about the inadequate power of the tugs for handling their colliers. The company responded by saying that they were doing all they could to purchase more powerful tugs, but none could be found. When *Seasider* (1) was purchased it was known that her plates were corrugated. In addition her stem had been damaged due to a collision with the collier *Hudson Strait* (3,105/1946). Repairs at Brigham & Cowan's, South Shields, cost £2,850. In August 1954, whilst her crew were on holiday, *Langton* was taken to the Tyne to be operated by Lawson-Batey at a hire charge of £100 per week. Because Blyth Harbour was very slack she was to stay on the Tyne, where there was more work for her.

At the 32nd Annual General Meeting in April 1955 the directors reported a trading loss due to the heavy cost of repairs. The coal trade was depressed and had not recovered from the last strike. The hire rate for *Langton*, working under the control of Lawson-Batey on the Tyne, was increased to £120 per week. A contract was secured for towing two barges per day loaded with colliery waste from the Bates Colliery jetty to sea. The rate negotiated was £14-10-0 (£14.50) per barge per trip for the first year. The sixty-six-year-old paddle tug *Steel* was near to the end of her working life, and it was decided that she would be kept working until a major breakdown occurred. One would hope that this did not happen whilst towing a ship in confined waters.

In 1956 the condition of the old paddle tugs was reported as: *Earl of Beaconsfield* (2) - good hull and machinery but poor boilers; *Steel* - good boilers but poor hull; and *Greatham* - fairly good. Blyth Drydocks & Shipbuilding Co.Ltd. were the latest organisation to complain about the poor power of the tugs. The contract arranged in 1955 for the towage of barges did not, in fact, commence until 14th June 1956 at an increased rate of £17. By 26th July one hundred barges had been handled, indicating that this was a worthwhile regular utilisation of the tugs. However, by May 1957, the work ceased because it was found that the barges were very poor seaboats. The National Coal Board thereafter used a steam-powered hopper for this work. Nevertheless, this contract had brought in earnings of £11,626 over the eleven months of its existence.

It will be remembered that the funnel colours of the original Blyth Steam Tug Co. were black with a white 'D'. There is photographic evidence that these colours were still in use about 1956. Somewhere around this time the funnel colours were changed to black with a white Maltese cross, thus indicating the controlling interest of the Lawson-Batey company whose own funnel mark was, of course, a blue

AFTER OFFICE HOURS—
G. JACKSON
20 Wellington Street, Blyth
Tel. No. 169

Telegraphic Address :
"DINSDALE, BLYTH"
Tel. No. **46**

REGISTERED OFFICE—
QUAYSIDE, BLYTH

NEWCASTLE OFFICE—
LAWSON-BATEY TUGS Ltd.
Tel. No. : Newcastle **20558**
QUAYSIDE
NEWCASTLE UPON TYNE 1

Blyth Tug Company Ltd.

Directors : J. F. BATEY F. S. BATEY S. S. W. DALGLIESH G. H. R. TOWERS

QUAYSIDE, BLYTH, NORTHUMBERLAND

★

BLACK FUNNELS

OUR SIGNAL : I long and 4 short Blasts

Company advertisement of 1957.
Note the reference to funnel colour and whistle signal.

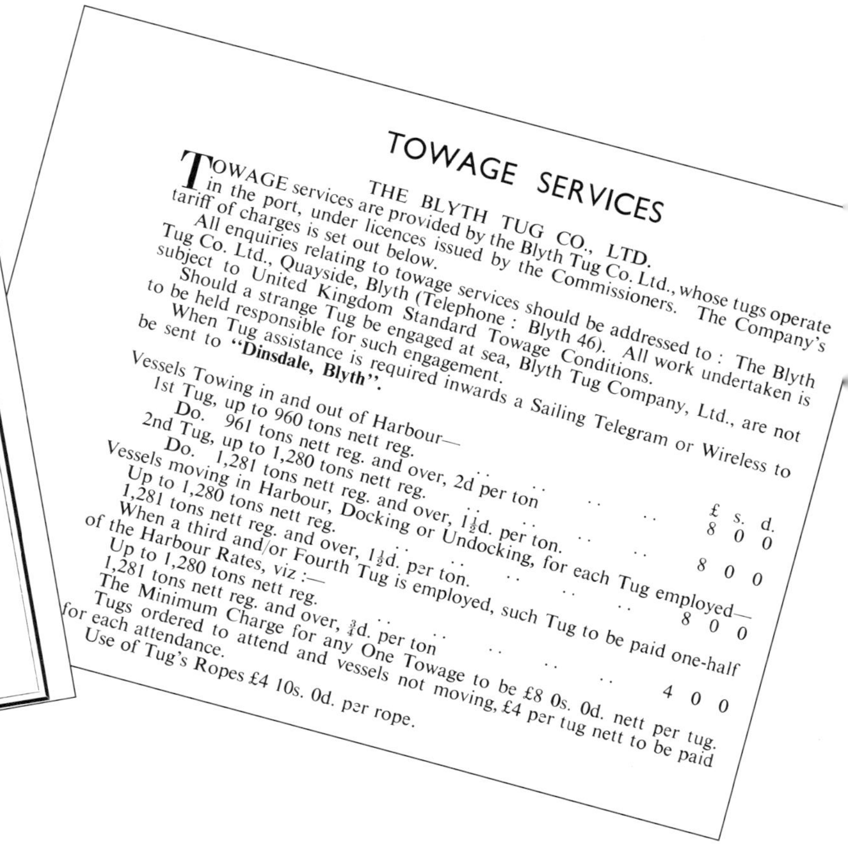

TOWAGE SERVICES

THE BLYTH TUG CO., LTD.

TOWAGE services are provided by the Blyth Tug Co. Ltd., whose tugs operate in the port, under licences issued by the Commissioners. The Company's tariff of charges is set out below.

All enquiries relating to towage services should be addressed to : The Blyth Tug Co. Ltd., Quayside, Blyth (Telephone : Blyth 46). All work undertaken is subject to United Kingdom Standard Towage Conditions.

Should a strange Tug be engaged at sea, Blyth Tug Company, Ltd., are not to be held responsible for such engagement.

When Tug assistance is required inwards a Sailing Telegram or Wireless to be sent to **"Dinsdale, Blyth"**.

	£	s.	d.
Vessels Towing in and out of Harbour—			
1st Tug, up to 960 tons nett reg.	8	0	0
Do. 961 tons nett reg. and over, 2d per ton			
2nd Tug, up to 1,280 tons nett reg.	8	0	0
Do. 1,281 tons nett reg. and over, 1½d. per ton.			
Vessels moving in Harbour, Docking or Undocking, for each Tug employed—			
Up to 1,280 tons nett reg.	8	0	0
1,281 tons nett reg. and over, 1½d. per ton.			
When a third and/or Fourth Tug is employed, such Tug to be paid one-half of the Harbour Rates, viz :—			
Up to 1,280 tons nett reg.	4	0	0
1,281 tons nett reg. and over, ¾d. per ton			

The Minimum Charge for any One Towage to be £8 0s. 0d. nett per tug.

Tugs ordered to attend and vessels not moving, £4 per tug nett to be paid for each attendance.

Use of Tug's Ropes £4 10s. 0d. per rope.

Towage tariff of 1957.

Ready for the launch of Roughsider *on 7 February 1958 are, left to right: back row, F.S. Batey and Mrs. 'Bet' Batey, who christened the tug; front row, Edgar Heal, Mrs. P. Stocker, Mrs. Joan Batey, Mrs. Frances Heal, unknown, Mr. J.F. Batey, Mr. N. Stocker (with handbag!) and Mr. C. Bell. (Burnitt Photographs, Goole).*

Maltese cross. There are also photographs showing some tugs with a plain black funnel - this was an interim situation between the 'D' and the Maltese cross.

The parent company, Lawson-Batey, had purchased their first diesel tug, *Quaysider,* in September 1955. Now, in February 1957, Blyth Tug Co.Ltd. were considering building a new 750 BHP diesel tug for themselves. Enquiries showed that most of the British tug-building shipyards were quoting deliveries of 24 - 30 months. However, Richard Dunston Ltd. of Thorne said that they would be able to make delivery in 13 - 15 months. An order was placed with them at a price of £62,100 excluding the cost of the German-made gearbox, with delivery promised for May 1958.

The year 1958 saw great changes in the fleet of Blyth Tug Co.Ltd. Dealing first with the departures: *Steel* and *Greatham* were sold in May to Hughes Bolckow Ltd. at Blyth for breaking-up for £600 and £650 respectively and *Earl of Beaconsfield* (2) went to the same destination in November for £600. The new tug from Richard Dunston, *Roughsider* (143/1958), was launched sideways at Thorne on 7th February after being christened by Mrs. Bet Batey, the wife of F.S. Batey the company chairman. *Roughsider* was completed, on time, on 29th May 1958 but, due to the falling-off of towage work at Blyth, she was set to work instead on the Tyne. When the level of business did not improve she was transferred to Lawson-Batey permanently in November 1958, at a book value of £71,000. To make up for the depletion of the Blyth fleet Lawson-Batey transferred the screw tugs *Homer* (157/1915) and *Northsider* (1) (149/1896) to the Blyth company at valuations of £8,000 and £1,500 respectively. Thus the year 1958 saw the end of Blyth's paddle tugs. One may think that the company soldiered on for too long with these ancient vessels in the face of increasing complaints and maintenance costs. It should be remembered, though, that in the years following the Second World War there was a serious shortage of materials, and probably money too, to allow the building of new tugs. The last tow by *Earl of Beaconsfield* (2) took place when she towed the Admiralty tanker *War Hindoo* (5,565/1919) to Hughes Bolckow for demolition.

At one period it had been argued that screw tugs would not be suitable for employment at Blyth, but the Blyth Tug Co.Ltd. owned the following steam screw-propelled fleet at the end of 1958:-

Seasider (1)	(154/1919)	
Langton	(157/1892)	Working on the Tyne
Homer	(157/1915)	
Northsider (1)	(149/1896)	

- still quite old at an average age of over fifty-two years.

Roughsider *entering the water sideways at Thorne on 7 February 1958. She is flying the rarely-photographed Blyth Tug Co. Ltd. houseflag.*
(Burnitt Photographs, Goole)

Headman *arriving at Blyth in 1962 before being renamed* Hillsider *(2).* *(John Clarkson, Longton).*

In 1959 *Seasider* (1) was involved in a serious accident when handling the Stephenson Clarke collier *James Rowan* (2,947/1955). The bridge of the tug was demolished by a mooring wire, and the skipper, Thomas Robson, had a lucky escape when he was pinned against the funnel by the bridge floor. After repairs the tug went back into service, but, in the following year, it was found necessary to renew the whole of her foredeck. The company was looking for a replacement for *Langton* since it was felt that it was not worth spending very much money on her. In October 1961 the United Towing Company's tug *Headman* was on the market at a price of £15,000, and it was agreed that she should be purchased if found to be in a satisfactory condition. After negotiations she was actually purchased for £11,250 in February 1962 and, on arrival on the north-east coast, she was renamed *Hillsider* (2) (177/1924). She was a heavily-built tug with a triple-expansion steam engine developing 800 IHP. When renamed she took the Lawson-Batey funnel colours of buff with a blue Maltese cross.

When berthed at Blyth the tugs used the Commissioners' No.11 jetty which was in a serious state of disrepair and required complete renewal. In 1962 the tug company offered to contribute £2,500 (later raised to £3,000) towards the work. The company were trying to sell *Langton* in November 1962 and she was laid-up at that time. There were no buyers for this seventy-year-old vessel and she eventually went to shipbreakers J.J. King & Co.Ltd. at Gateshead in May 1963. *Langton's* bell was given to the British Leprosy Relief Association for use at Kumasi Castle in Uganda. To make up for the loss of *Langton, Francis Batey* (151/1914) was transferred from Lawson-Batey at a valuation of £4,500 in November 1962.

A serious decline in earnings was reported in mid-1963 with concern that one crew might have to be paid off. This, in fact, did happen later in the year. *Northsider* (1) was sold for £1,150 to Hughes Bolckow for breaking-up after a sixty-seven-year life. *Francis Batey,* after only one year with the Blyth Tug Co.Ltd., was laid-up, leaving only three tugs in operation.

The Harbour Commissioners' tug/tender, *Chipchase* (106/1953), sank in the western channel of the harbour in July 1964. After being raised by lifting camels brought from Rosyth by Lawson-Batey tugs, *Chipchase* was towed to her berth by *Seasider* (1) and *Hillsider* (2). The year 1966 brought further problems, largely due to the strike of seamen which lasted for more than six weeks. Shipments of coal by the National Coal Board were greatly reduced resulting in two tugs being laid-up. Two were working, employing three crews of six men each. Six men and one boy had to be paid off. On 20th August 1966, Cyril W. Bell, the company secretary since 1941, died at the age of sixty-five. He was succeeded in that post by Douglas Frank Hamilton, an employee of Lawson-Batey Tugs Ltd. Partly due to the reduced level of business the towage rates were raised by 25% from 7th March 1966.

In March and April 1968 two more tugs were sent to the shipbreakers: *Homer* for £2,950, and *Francis Batey* for £2,576-3-6.(£2,576.175), to Hughes Bolckow at Blyth. Because of the poor trading conditions and results the Blyth Tug Co.Ltd. directors considered the following possibilities:-

a) That Blyth prospects were unlikely to improve.

b) That diesel tugs should be employed.

c) That the existing tugs should be sold to Blyth Harbour Commissioners.

It was felt that the cost of second-hand diesel tugs, at about £40,000 each, would be prohibitive. In the event, the

Commissioners were not interested in taking over the towage at Blyth, so that little scope was left for the company to make improvements. Consideration was given to transferring *Roughsider* from the Tyne, but concern was expressed that this might affect the charter arrangements with Tyne Tugs Ltd., especially since Jessel Securities Ltd. were now involved. In November 1968, *Tynesider* (1) (262/1942) was transferred from Lawson-Batey at a value of £5,000. This move left the Blyth company with three tugs, but not for long, since *Seasider* (1) was sold for £2,100, early in 1969, for breaking-up.

Throughout the period 1968 - 1970 the company traded at a loss. As an example, in 1970, the annual cost of wages at £17,000 and coal at £6,000 went a long way to swallowing up the towage revenue of £29,731. After allowing for other costs of, for example, maintenance, insurance and administration, there was no profit. At this time Lawson-Batey held more than 75% of the Blyth Tug Co.Ltd. share capital. After only sixteen months at Blyth *Tynesider* (1) was sold, for £4,150, to Hughes Bolckow for breaking-up, leaving *Hillsider* (2) as the only tug stationed at Blyth. When a second tug was required one would be brought round from the Tyne. Six more crewmen were made redundant, leaving only eight employed at Blyth. It was expected that, in July 1971, vessels would start to use the new Alcan (UK) Ltd. terminal for the import of alumina and the export of aluminium ingots. In view of this, the 825 BHP diesel tug *Maximus* (141/1956) was transferred from Ridley Tugs Ltd., for £37,500, after having been laid-up on the Tyne for some time. This transfer was possible because Ridley was a wholly-owned subsidiary of Lawson-Batey.

In 1972 the crew of the *Maximus* was laid-off for six weeks due to the miners' strike, and *Hillsider* (2), now forty-eight years old, was sold for scrap for £3,000. *Maximus*, therefore, now held the fort on her own.

Since 1970 Lawson-Batey had been trying to purchase the shares held by the minority shareholders in Blyth Tug Co.Ltd. Thus, by 1973, they held 81%, and, in the following year, 98.5% of the shares. In 1975 there was only one outstanding shareholder who held 100 shares, i.e. 1% of the total.

Consideration was given to returning *Maximus* to the Tyne, and working Blyth from there. The decision was postponed, as soon it was found that trading conditions had improved due to the Alcan activities.

On 13th November 1975, the managing director, John Francis Batey, was tragically killed whilst on business in London. Details of this awful occurrence are related in the Lawson-Batey chapter of this volume. J.F. Batey had been managing director since 1953 and a director since 1945. Early in 1976, he was succeeded as managing director by Thomas Anthony Greenwell. At the same time D.F. Hamilton, the company secretary, was, in addition, appointed to the board of directors. In December 1976 the last independent shareholder sold out, and thus Blyth Tug Co.Ltd. became a wholly-owned subsidiary of Lawson-Batey Tugs Ltd. During 1976 the company had handled 450 jobs - on average 1.23 jobs per day. This was not sufficient to keep even one tug fully occupied since, due to the compactness of the harbour, the time taken per job was very small compared with, say, that on the River Tyne.

In 1978 *Maximus* was sold to Lawson-Batey for £22,500 and then demise-chartered back to Blyth Tug Co.Ltd. D.F. Hamilton was appointed deputy managing director and Michael Francis Pyman joined the board in September 1980. At the end of 1982, T.A. Greenwell became vice-chairman, being succeeded as managing director by D.F. Hamilton, and S.S.W. Dalgliesh retired as a director. On 1st January 1983 H.R. Gilbert took over as company secretary. F.S. Batey, then eighty-four years of age, retired as chairman on 31st December 1983. He had held this position since 1951 and on his retirement was appointed Honorary Life President of the Lawson-Batey Group. T.A. Greenwell took over as group chairman.

In November 1983 *Westsider* (2) (151/1964) was transferred from Lawson-Batey, and *Maximus*, her charter completed, went to lay-up in South Dock, Sunderland. The level of trade at Blyth was now so low that when the miners' strike took place and crippled the port the tug company felt that there was no alternative to closing down their operations at Blyth and servicing that port from the Tyne. *Westsider* (2), without work for 35 weeks, was transferred back to Lawson-Batey in December 1984 and went to lay-up at Sunderland. The tug crew at Blyth was made redundant on 10th November and the office closed on 30th November 1984. The towage manager at Blyth, Gordon Jackson, who had been with the company since 1945, was transferred to a similar post with Tyne Tugs Ltd. at South Shields. The contract with Alcan for handling their shipping at Blyth was transferred to Tyne Tugs Ltd., and, with no operations based at Blyth, the company was officially declared dormant.

G.H.R. Towers retired from the board in 1985 and M.F. Pyman the following year. The secretary, H.R. Gilbert, retired in 1986 and was succeeded by J.Gordon Mucklow, who was also appointed a director. The chairman, T.A. Greenwell, died on 23rd June 1987 and the position was filled by J.G. Mucklow.

The Blyth Tug Co.Ltd. was struck off the Register of Companies on 8th June 1988 after trading under that name since 1923. Its predecessor, the Blyth Steam Tug Co., had commenced operations in 1887, so the two companies together had provided 101 years of towage service at the port of Blyth. Over these years the companies had owned a total of twenty-four tugs, half of them having paddle propulsion. The needs of the port are now serviced by tugs being sent ten miles from the Tyne as and when required.

APPENDIX A TO CHAPTER FIVE

Blyth Tug Co. Ltd. 1923-1988.

Board of Directors

Chairmen	M.C. James	1923-1934	Died
	R. Bell	1934-1935	Died
	F. Batey	1935-1948	Died
	Dr. J.T. Batey	1949-1951	Died
	F.S. Batey	1951-1983	Retired
	T.A. Greenwell	1983	Vice-Chairman
		1984-1987	Died
	J.G. Mucklow	1987-1988	
Managing Directors	J. Lawson } Joint	1923-1930	Disqualified
	F. Batey } Joint		Became sole M.D.
	F. Batey	1930-1948	Died
	J.F. Batey	1953-1975	Died
	T.A. Greenwell	1976-1982	Became Vice-Chairman
	D.F. Hamilton	1980-1982	Deputy M.D.
	D.F. Hamilton	1983-1988	
Directors	W. Thompson	1923-1930	Resigned
	R. Bell	1923-1934	Became Chairman
	J. Thompson	1930-1940	Died
	S. Fisher	1930-1948	Retired
	R.I. James	1934-1944	Died
	Dr. J.T. Batey	1936-1949	Became Chairman
	J.F. Batey	1945-1953	Became M.D.
	S.S.W. Dalgliesh	1949-1982	Retired
	F.S. Batey	1949-1951	Became Chairman
	G.H.R. Towers	1951-1985	Retired
	D.F. Hamilton	1976-1980	Became Deputy M.D.
	M.F. Pyman	1980-1986	Retired
	J.G. Mucklow	1987	Became Chairman
Secretaries	C.W. Bell	1941-1966	Died
	D.F. Hamilton	1966-1982	Became M.D.
	H.R. Gilbert	1983-1986	Retired
	J.G. Mucklow	1987-1988	

APPENDIX B TO CHAPTER FIVE

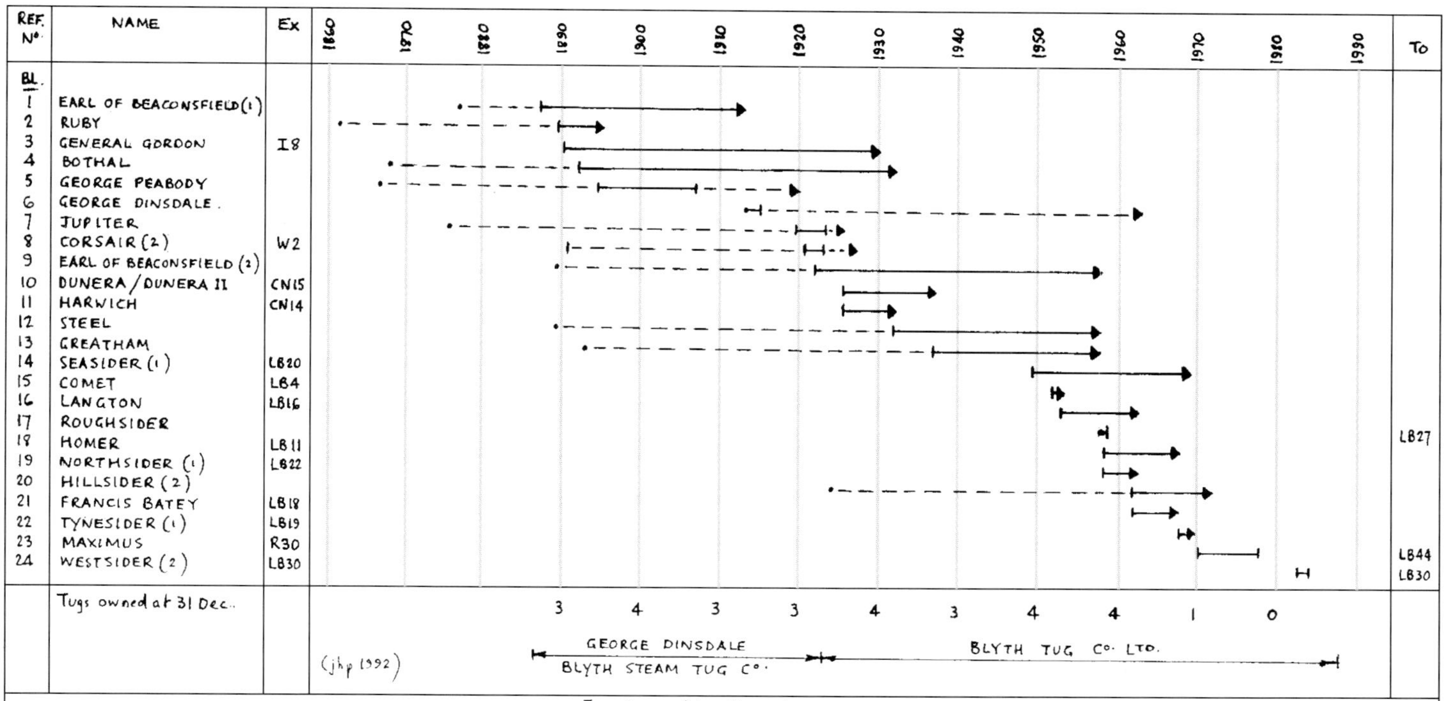

Tug Ownership Chronology.
THE BLYTH COMPANIES 1887-1988

APPENDIX C TO CHAPTER FIVE

The Blyth Companies Fleet List

Earl of Beaconsfield *(1), right, with* Corsair, *left. (Paul Hood)*

BL 1 **EARL OF BEACONSFIELD (1)** **(1887-1913)**
ON 72539 75 g 83.6 x 16.7 x 9.1 ft. Wood Paddle
B: J.P. Rennoldson, South Shields.
E: L. 1 Cy. 30" x 48". 30 NHP. By builder.

PR: 1877 Cardiff; 1887 North Shields; 1913 Blyth

02.1877: Completed for W.H. Martin, Cardiff.
06.1887: Purchased by Blyth Steam Tug Co. (G. Dinsdale et al)
1913: Sold to Hughes Bolckow & Co.Ltd., Blyth, for breaking up.

BL 2 **RUBY** **(1889-1895)**
ON 44288 98 g 90.2 x 17.9 x 9.8 ft. Wood Paddle
B: Not known. At South Shields.
E: L. 2 Cy. 46 NHP. J.P. Almond, North Shields.

PR: 1862 ?; 1863 Wexford; 1889 South Shields; 1890 North Shields.

1862: Completed for William Brown et al, Shields.
1863: Sold to Wexford Harbour Commissioners.
1889: Purchased by Blyth Steam Tug Co. (George Dinsdale et al)
10.1895: Vessel broken up.

BL 3 **GENERAL GORDON** **(1890-1930)**
See Irwing fleet list, I 8

BL 4 **BOTHAL** **(1892-1932)**
ON 60829 149 g 109.4 x 20.0 x 10.1 ft. Iron Paddle
B: R. & H. Green, Blackwall, London. (YN 366)
E: L. 2 Cy. 80 NHP

PR: 1868 London; 1875 Cork; 1886 Glasgow; 1892 North Shields; 1913 Blyth.

1868: Completed, as FERGUS MACIVOR, for Caledonian Steam Towing Co. London.
1873: Re-engined. L. 2 Cy., 30" x 54". 70 NHP. by J. Stewart, London.
1874: Sold to Henry Collings, London.
1875: Sold to Queenstown Screw Tug & Water Tank Co.Ltd.
1880: Sold to Queenstown New Tug Co.Ltd.
1886: Sold to Clyde Shipping Co., renamed FLYING SQUADRON.
02.1892: Purchased by Blyth Steam Tug Co., renamed BOTHAL.
07.1923: Owners restyled Blyth Tug Co.Ltd.
11.1932: Sold, for £75, to Clayton & Davie Ltd., Dunston, for breaking up.

BL 5 **GEORGE PEABODY** **(1894-1907)**
ON 56905 115 g 100.4 x 18.3 x 9.9 ft. Iron Paddle
B: Readhead, Softley & Co., South Shields (YN 23)
E: L. 2 Cy. 27½" x 48". 55 NHP. By builders

PR: 1868 London; 1887 Llanelli; 1895 North Shields; 1897 Blyth; 1907 Middlesbrough

12.1867: Completed for W.J. Homewood et al, Gravesend.
1871: Sold to J. Martin, Gravesend.
1879: Sold to T.A.F. Banks, Gravesend.
07.1885: Sold to W. Bowen, Jnr., Llanelli.
09.1894: Purchased, for £2,400, by Blyth Steam Tug Co.
04.1907: Sold to W. Slater and J. Robson, Middlesbrough. Later W. Slater only.
01.1916: Sold to The Robinson Tug Co.Ltd., Middlesbrough.
07.1920: Sold to Tees Towing Co. Ltd., Middlesbrough.
10.1920: Sold for breaking up on River Tyne.

BL 6 **GEORGE DINSDALE** **(1913-1915)**
ON 133224 105 g 90.0 x 18.1 x 8.4 ft. Steel Screw
B: Scott & Sons, Bowling. (YN 236)
E: T. 3 Cy. 12", 18" & 30" x 21". 400 IHP. Fishers Ltd., Paisley.

PR: 1913 Blyth; 1915 Grangemouth; 1924 Port Adelaide.

03.1913: Completed, for £3,670, for Blyth Steam Tug Co.
10.1914: Hired by Royal Navy
07.1915: Sold to Grangemouth & Forth Towing Co.Ltd.
12.1919: Returned to owners by Royal Navy
1924: Sold to Ritch & Smith Ltd., Port Adelaide, Australia.
1963: Broken up at Port Adelaide.

BL 7 **JUPITER** **(1919-1923)**
ON 68184 102 g 96.1 x 17.9 x 9.2 ft. Iron Paddle
B: J.T. Eltringham, South Shields. (YN 51)
E: L. 1 Cy. 33½" x 50". 40 NHP. Hepple & Co., South Shields.

PR: 1876 Grangemouth; 1914 Blyth

07.1876; Completed for J. & G. Mackey & G. Pederson, Grangemouth
10.1891: Sold to Grangemouth Towing Co.
1895: Owners restyled Grangemouth & Forth Towing Co.Ltd.
06.1914: Sold to O.W. Jewels, Blyth
11.1919: Purchased by Blyth Steam Tug Co.
07.1923: Owners restyled Blyth Tug Co.Ltd.
10.1923: Sold, for £275, to Robson & Cross, Lemington.
02.1926: Broken up at Sunderland.

BL 8 **CORSAIR (2)** **(1921-1923)**
See White fleet list, W 2

Earl of Beaconsfield *(2) off Battleship Wharf, Blyth. (Author's collection).*

BL 9 **EARL OF BEACONSFIELD (2)** **(1922-1958)**
ON 97374 114 g 95.0 x 19.2 x 9.9 ft. Steel Paddle
B: J.T. Eltringham, South Shields. (YN 150)
E: L. 2 Cy. 28" x 50". 350 IHP. 10 K. J.P. Rennoldson & Sons, South Shields.

PR: 1889 West Hartlepool; 1922 Blyth

1889: Completed, as SALT, for The North Eastern Railway Co.
05.1922: Purchased by Blyth Steam Tug Co., renamed EARL OF BEACONSFIELD
07.1923: Owners restyled Blyth Tug Co.Ltd.
11.1958: Sold, for £600, to British Iron & Steel Corporation. Allocated to Hughes Bolckow Ltd., Blyth for breaking up.

BL 10 **DUNERA / DUNERA II** **(1926-1937)**
See Coulson fleet list, CN 15

BL 11 **HARWICH** **(1926-1932)**
See Batey fleet list, B 15

Steel. *(John Clarkson, Longton).*

BL 12 **STEEL** **(1932-1958)**
ON 97383 118 g 95.3 x 19.1 x 9.8 ft. Steel Paddle
B: J.T. Eltringham, South Shields. (YN 151)
E: L. 2 Cy. 28" x 50". 350 IHP. 10.25 K. J.P. Rennoldson & Sons, South Shields.

PR: 1890 West Hartlepool; 1932 Blyth

1889: Completed, for £5,025, for The North Eastern Railway Co.
01.1923: Transferred to London & North Eastern Railway Co.
09.1932: Purchased, for £286, by Blyth Tug Co.Ltd.
05.1958: Sold, for £600, to British Iron & Steel Corporation. Allocated to Hughes Bolckow Ltd., Blyth for breaking up.

Greatham. *(A. Duncan)*

BL 13 **GREATHAM** **(1937-1958)**
ON 102703 132 g 95.0 x 19.1 x 9.8 ft. Iron Paddle
B: J.P. Rennoldson & Sons, South Shields. (YN 149)
E: L. 2 Cy. 28" x 50". 350 IHP. 10.25 K. By builders.

PR: 1893 West Hartlepool; 1937 Blyth

1893: Completed, for £5,450, for The North Eastern Railway Co.
01.1923: Transferred to London & North Eastern Railway Co.
10.1937: Purchased, for £675, by Blyth Tug Co.Ltd.
05.1958: Sold, for £650, to British Iron & Steel Corporation. Allocated to Hughes Bolckow Ltd., Blyth for breaking up.

Seasider *(1). (A. Duncan)*

BL 14 **SEASIDER (1)** **(1949-1969)**
See Lawson-Batey fleet list, LB20

BL 15 **COMET** **(1952-1953)**
See Lawson fleet list, L 22

Langton. *(John Clarkson, Longton).*

BL 16 **LANGTON** **(1953-1963)**
See Batey fleet list, B 17

BL 17 **ROUGHSIDER** **(1958 only)**
ON 186876 143 g 92'6" x 23'0" x 99'-5" Steel Screw
B: Richard Dunston Ltd., Thorne. (YN T.978)
E: Oil. 4SA. 6 Cy. 320 x 450 mm. 750 BHP. 11 BP. Klöckner-Humboldt-Deutz A.G, Cologne.

PR: 1958 Newcastle

05.1958: Completed for Blyth Tug Co.Ltd.
06.1958: Commenced service on River Tyne.
11.1958: Transferred, for £71,000, to Lawson-Batey Tugs Ltd.
01.1959: Chartered to Tyne Tugs Ltd.
06.1980: Chartered to Wear Tugs Ltd.
02.1986: Sold, for £4,700, to J.A. White & Co.Ltd., Inverkeithing, for breaking up.

Homer *and* Francis Batey. *(World Ship Photo Library).*

BL 18 **HOMER** **(1958-1968)**
See Lawson fleet list, L 29

Northsider *(1). (Maritime Photo Library).*

BL 19 **NORTHSIDER (1)** **(1958-1963)**
See Crosthwaite fleet list, CE 7, as JOSEPH CROSTHWAITE

Hillsider *(2). (Author's collection).*

BL 20 **HILLSIDER (2)** **(1962-1972)**
ON 147150 177 g 96'6" x 24'7" x 12'5" Steel Screw
B: Cochrane & Sons Ltd., Selby. (YN 800)
E: T. 3 Cy. 14½", 24" & 40" x 27". 800 IHP. Earle's Shipbuilding & Engineering Co.Ltd., Hull.

PR: 1924 Hull; 1962 Newcastle.

06.1924: Completed, as HEADMAN, for United Towing Co.Ltd., Hull.
02.1962: Purchased, for £11,250, by Blyth Tug Co.Ltd., renamed HILLSIDER.
10.1972: Sold, for £3,000, to Hughes Bolckow Ltd., Blyth for breaking up.

BL 21 **FRANCIS BATEY** **(1962-1968)**
See Batey fleet list, B 19

BL 22 **TYNESIDER (1)** **(1968-1970)**
See Lawson-Batey fleet list, LB 19

Maximus. *(Paul Hood).*

BL 23 **MAXIMUS** **(1971-1978)**
See Ridley fleet list, R 30

BL 24 **WESTSIDER (2)** **(1983-1984)**
See Lawson-Batey fleet list, LB 30.

• CHAPTER SIX •

The Crosthwaite Companies

Joseph Crosthwaite. c.1854 - 1924

The time when Joseph Crosthwaite Jr. commenced business is usually quoted as about 1889. This is probably the date on which he formally set up a company or partnership. However, an earlier Joseph Crosthwaite Sr., of North Shields, purchased a wooden paddle tug named *Honor* (38/1853) in 1854, but kept her for only about one year. There was then a period of fifteen years, during which no tugs owned by Joseph Crosthwaite have been discovered. This is not to say that Joseph Crosthwaite Sr. did not own, or part own, tugs in this period. He may well have held a minority share in a vessel which would not be apparent in, say, the Mercantile Navy Lists. Short of searching through all the possible Custom House, Christie's or Turnbull's Registers, all of which are not readily accessible, there is little a researcher can do, that will not take a disproportionate amount of time, to fill in such gaps in the history.

The wooden paddle tug *Selina* (1) (66/1870) was built new for Joseph Crosthwaite Sr. and another, in 1870, and remained in their ownership until 1876 when she was sold to South Shields owners. In the year following the disposal of *Selina* (1) another slightly larger tug was built new for Joseph Crosthwaite Sr. She was also named *Selina* (2) (78/1877), and thus there were to be two *Selinas* on the River Tyne for at least the next forty-three years. The iron-hulled paddle tug *Albert Edward* (76/1853) was purchased in 1885 by Joseph Crosthwaite Sr. and associates. This tug had been built abroad, possibly in Germany, as *Courier,* and had been renamed in 1884 whilst being re-engined at North Shields. Her new name was no doubt chosen to commemorate the opening, also in 1884, of the Albert Edward Dock on the north bank of the Tyne at North Shields. Notwithstanding having had a new engine fitted, *Albert Edward* only remained in Crosthwaite ownership for five years.

In 1894, Joseph Crosthwaite Jr., a son of Joseph Crosthwaite Sr., purchased the twin-funnelled iron paddle tug *Warrior* (112/1871) from Thames owners. She had been built at South Shields by Readhead Softley & Co., and had a double side-lever engine of 70 nominal horsepower. She was a sister of *Ulysses* (118/1874) and *Vigilant* (115/1874), which were also brought to the Tyne from the Thames by Joseph Lawson and George Brown respectively. *Champion* (3) (140/1877), a tug of similar style to *Warrior,* but of greater power, was purchased in 1900 from C.W. Duncan of Middlesbrough. So, just after the turn of the century, the Crosthwaite fleet at North Shields was:-

Name	Gross Tonnage/Year	Hull construction	Means of Propulsion
Selina (2)	(78/1877)	Wood	Paddle
Warrior	(112/1871)	Iron	Paddle
Champion (3)	(140/1877)	Iron	Paddle

with an average age of twenty-five years.

On 1st August 1906 *Selina* (2) had the misfortune to be struck by lightning and slightly damaged. The boy on *Champion* (3) moored alongside was knocked down by the force of the strike - fortunately more frightened than injured.

There were no further changes in the fleet until 1911, when the compound-engined steel screw tug *Huskisson* was bought from The Alexandra Towing Co.Ltd. and renamed *Joseph Crosthwaite* (149/1896). At the same time *Selina* (2) was disposed of to other Tyne owners who ran her for a further nineteen years. One year after the outbreak of the First World War *Joseph Crosthwaite* was hired to the Royal Navy and did not return to her home river until 1920. During her period of requisition she travelled far afield to the Dardanelles, where she assisted in the evacuation from 'Clyde' Beach, Gallipoli. Subsequently a plaque was carried on the bridge to commemorate the occasion. Later in the war, *Joseph Crosthwaite* sailed even further from home when she went, via the Suez Canal, to Basrah at the head of the Persian Gulf.

Joseph Crosthwaite Sr. died at North Shields on 7th April 1918. The crews of *Warrior* and *Champion* (3) were present at his funeral.

Joseph Crosthwaite (Tugs) Ltd. 1924 - 1950

On 17th March 1924 the company of Joseph Crosthwaite (Tugs) Ltd. was incorporated with the object of acquiring the assets of 'Joseph Crosthwaite'. The first directors were Joseph Crosthwaite Jr., Nicholas Brown and Septimus Crosthwaite. The tugs transferred to the new company were:-

Warrior	(112/1871)	Iron	Paddle
Champion (3)	(140/1877)	Iron	Paddle
Joseph Crosthwaite	(149/1896)	Steel	Screw

with an average age of forty-two years. The funnel colour used by the new company was plain black and, it is believed, had been so throughout the company's history.

By 1928 the directors were Joseph Crosthwaite Jr., Captain Frederick Brown West, and Lewis Crosthwaite, the latter-named being the third generation of Crosthwaites involved in the management of their company. It is worth pointing out that this Crosthwaite family, contrary to popular belief, is not connected with Crosthwaite of Middlesbrough who managed Tees Towing Co.Ltd. About 1929 an offer was received from The 'Ridley' Steam Tug Co.Ltd. to purchase the Crosthwaite company, as a going concern, for £16,000, but this approach was turned down. Similarly, in 1930, Lawson-Batey Tugs Ltd. offered £12,000 for the company, and again the offer was declined. The following year Lawson-Batey made an offer to purchase 16,500 £1 shares from the executors of the late Joseph Crosthwaite Jr. who had died on 29th September 1930 and the late Nicholas Brown for £9,900, this being 97% of the issued share capital. The executors, Captain F.B. West and Lewis Crosthwaite, did not accept the offer. In May 1933 the directors offered the company, for £8,500, to Ridley, who responded by making a counter-offer of £6,375, in true market-stall fashion. Two months later France, Fenwick Tyne & Wear Co.Ltd., Lawson-Batey Tugs Ltd., and Redhead & Dry's Tugs Ltd. made a new equally-shared offer for the Crosthwaite company of £8,500. This joint offer was accepted by the directors on 31st July 1933.

Although Joseph Crosthwaite (Tugs) Ltd. was now owned by the three joint-buyers it was not liquidated, and continued to trade under its existing name. Its principal contracts were with British shipowners. During the period of negotiations over the future of the Crosthwaite company there were changes to the tug fleet. In 1932 *Warrior,* which had been with the fleet for thirty-eight years, was sold for £85 to a Hartlepool shipbuilder, who soon scrapped her. In the same year the company purchased the twenty-year-old screw tug *Colonel Thys* from the Belgian Congo. She was a powerful tug with a triple-expansion steam engine of 700 IHP and had been built in Holland. On arrival on the Tyne she was renamed *Lewis Crosthwaite* (234/1912). Therefore, at the time of the take-over of the company, the fleet consisted of:-

Champion (3)	(140/1877	Iron	Paddle
Joseph Crosthwaite	(149/1896)	Steel	Screw
Lewis Crosthwaite	(234/1912)	Steel	Screw

After the change of ownership there were, quite naturally, changes to the Board of Directors. France, Fenwick were represented on the Crosthwaite board by John Edward Dawson and Frank Dawson, who became chairman and secretary respectively. Robert Ingo James and Frank Batey were the Lawson-Batey representatives, and James Patton and Alexander Patton looked after the interests of Redhead & Dry's. The Crosthwaite company had, as far as is known, always operated from North Shields, with addresses over the years at Prudhoe Street, Grey Street, Alma Place and, latterly, at 8 New Quay.

The company's last paddle tug, *Champion* (3), was sold for breaking-up at Sunderland in June 1935, leaving the two screw tugs in operation on the River Tyne. In 1944 France, Fenwick Tyne & Wear Co.Ltd. took-over Redhead & Dry's Tugs Ltd., which meant that the ownership of Joseph Crosthwaite (Tugs) Ltd. was now with France Fenwick and Lawson-Batey only. A transfer of shares took place so that these two companies then had an equal holding in the Crosthwaite company.

At this time, the company's office was at the Mill Dam, South Shields, and it would appear that the North Shields base had been closed down. During the Second World War further changes to the board took place. A.M.H. Wardlaw of France Fenwick had replaced J.E. Dawson, and R.I. James had died in 1944. Major John Francis Batey was appointed to the board by Lawson-Batey in September 1945.

In April 1949 France Fenwick and Lawson-Batey agreed to wind up Joseph Crosthwaite (Tugs) Ltd., and to divide the business between themselves. As for the two tugs, *Joseph Crosthwaite* went to Lawson-Batey at a valuation of £4,150, and *Lewis Crosthwaite* went to France Fenwick at £10,500 in December 1949. They were renamed *Northsider* (1) and *Earsdon* respectively. The final board meeting took place on 10th November 1950, after which the company ceased to exist. In the minutes of that meeting reference was made to the fact that the company had been managed by three generations of Crosthwaites since about 1889. A total of eight tugs had been owned since the middle of the nineteenth century - a small number compared with many other Tyne-based owners.

APPENDIX A TO CHAPTER SIX

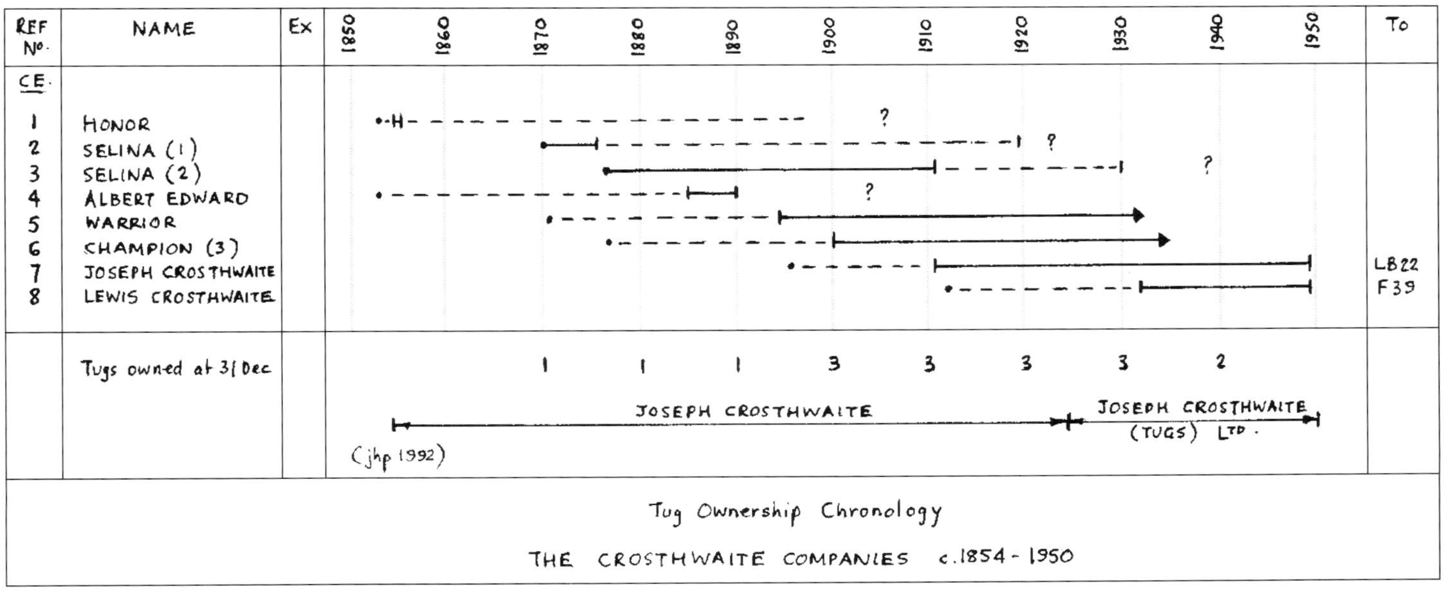

APPENDIX B TO CHAPTER SIX

The Crosthwaite Companies Fleet List

CE 1 **HONOR** **(1854-1855)**
ON 2072 38g. 69.4 x 13.9 x 7.1 ft. Wood Paddle
B: At North Shields.
E: L. 1 Cy. 18 NHP.

PR: 1853 Shields; 1855 Newcastle; 1858 Newcastle; 1882 Grimsby

05.1853:	Completed for J. Tweedy, Shields.
1854:	Purchased by Joseph Crosthwaite, North Shields.
1855:	Sold to W. Brown & G. Brewis, Newcastle.
1858:	Sold to T. Bright et al, Newcastle.
1879:	Sold to J. Thompson, Newcastle.
1882:	Sold to Hugh Ward, Cleethorpes.
1888:	Sold to E.G. Haigh, Weelsby.
1889:	Sold to J. Munson, Cleethorpes.
1892:	Sold to W. Stephenson, Cleethorpes.
1894:	Sold to F. Hazelgrove, Grimsby.
1897:	Not in Mercantile Navy List.

CE 2 **SELINA (1)** **(1870-1876)**
ON 60568 66 g. 79.7 x 16.5 x 9.0 ft. Wood Paddle
B: Heckels Bros., North Shields.
E: L. 1 Cy. 28½" x 47". 26 NHP. By Builder.

PR: 1870 North Shields.

1870:	Completed for Joseph Crosthwaite and another, North Shields.
1876:	Sold to J.P. Bickell, South Shields.
1884:	Sold to M. Pearson et al, South Shields.
1892:	Sold to T.W. Finch, South Shields.
c.1896:	Sold to R. Legg, South Shields.
1919:	Register closed.

Selina (2) with Corsair (2) behind. (Local History Library, South Shields Central Library)

CE 3 **SELINA (2)** **(1877-1911)**
ON 76649 78 g. 85.6 x 17.7 x 9.3 ft. Wood Paddle
B: Wouldhave & Johnston, North Shields.
E: L. 1 Cy. 31" x 48". 35 NHP. Baird & Barnsley, North Shields.

PR: 1877 North Shields

11.1877:	Completed for Joseph Crosthwaite, North Shields.
07.1911:	Sold to Ralph Robson, St. Anthonys.
09.1930:	Registry closed.

CE 4 **ALBERT EDWARD** **(1885-1890)**
ON 79214 76 g. 81.7 x 16.7 x 8.6 ft. Iron Paddle
B:
E: Oscillating. 1 Cy. 27½" x 34". 45 NHP. In Stettin.

PR: 1853 ?; 1878 North Shields.

1853:	Completed, as COURIER, for foreign owners.
12.1878:	Sold to T. Lambert & J. Chase, South Shields.
09.1881:	Sold to J. Hodgeman, Rochester, T. Hopper, Newcastle and H.W. Brightman, North Shields.
1883:	Sold to W. Findley, North Shields.
10.1884:	Renamed ALBERT EDWARD.
1885:	Re-engined: L. 1 Cy. 26¼" x 38", 30 NHP by R.B. Findley & Son, North Shields.
1885:	Purchased by Joseph Crosthwaite et al, North Shields.
03.1890:	Registry closed.

retary was W.M. Graydon but he resigned after one month and was succeeded by Francis Christian Batey, at a salary of £4-10-0 (£4.50) per week. F.C. Batey was a great-grandson of John Batey, the founder of the Batey line of companies. Joseph Ridley was appointed chairman and his son, Robert Ridley, managing director of the company, the latter with a salary of £5 per week. F.C. Batey was appointed to the board in August 1920. The two tugs transferred to the company were:-

Name	Gross Tonnage/Year	Hull construction	Means of propulsion
Oscar	(77/1876)	Iron	Paddle
Haakon	(110/1874)	Iron	Paddle

- both having double side-lever engines.

Hardback *towing the Bergen Steamship Co.'s cruise vessel* Meteor *(3,717/1904). It is interesting to note that both vessels have the same funnel colours.* *(George Scott).*

Of the £20,000 capital in £10 shares Joseph Ridley held 900, E.W. Gjemre 100, and Robert Ridley 100 shares.

It is known that the Ridley company had contracts with Scandinavian mail-boat companies to handle their ships when they were in the Tyne, and that the tugs wore the funnel colours of these regular visitors. There is photographic evidence of Ridley tugs in the 1920s with the funnel colours of the Svenska Lloyd company of Gothenburg i.e. a yellow star on a blue circle set on a white band with the remainder of the funnel black. It is possible that Joseph Ridley held the Svenska Lloyd contract even earlier and this may have been the reason for the choice of the name *Haakon* for one of his tugs in 1913. One of the directors, E.W. Gjemre, was also of Scandinavian origin. From 1922 the company's office was in Dean Street, South Shields.

In 1923 the Ridley company increased the size of their fleet dramatically with the purchase of no less than four second-hand tugs. In January, the steel paddle tug *Isaac Wilson* (123/1889) came from the Tees Conservancy Commissioners and cost £1,250. She is known to have had a board over her paddle box with the name 'Ridley Steam Tug Co.Ltd.' on it. This advertising of a company name on a Tyne tug was uncommon and may have been unique. Next, in March, came the paddle tug *Hardback* (125/1901) from The Hodbarrow Mining Co.Ltd., Millom, for £800, and she was followed by the steel screw tug *Premier* (83/1894) from the Grimsby Steam Tug Co. Finally, in October, the iron paddle tug *William Fallows* (122/1888) was bought from Tees Towing Co. Ltd. for £1,800. She was, apart from the material of her hull, a sister to *Isaac Wilson*. Thus, at the end of 1923, the Ridley company owned six tugs which was to be the size of their fleet for the next eight years. During 1923 Knud Zimmer was appointed a director of the company representing the Bergen Steamship Co. (BDS). It would thus appear that the Ridley company had a contract to handle the BDS passenger and mail boats in the Tyne from at least this time. The BDS funnel colours of black with three narrow white bands then began to appear on the Ridley ship-handling tugs and continued to do so right through to the formation of the charter company, Tyne Tugs Ltd., in 1959. By this time a total of £16,150 capital had been subscribed to the company by its shareholders.

In February 1924 the paddle tug *Granville* (160/1876) was purchased for £775 from Seaham Harbour Dock Co. She was reconditioned by the Middle Docks & Engineering Co.Ltd. at South Shields where, amongst other work done, a new boiler from Riley Bros., costing £800, was fitted. Conversely, in June, *Hardback*, which had only been in the Ridley fleet for fifteen months, was sold to Seaham Harbour Dock Co. for £3,750. This price represented a profit of £2,950

Premier *towing* Thule *(2,037/1892) of Svenska Lloyd in the mid 1920s. Both vessels carry the same funnel colours.* *(P.N. Thomas).*

on the vessel. Just why this exchange of tugs between the two companies took place has not been established, though possibly *Granville* was too large for Seaham Harbour. Whatever the reason both tugs had very long lives ahead of them before going to the breakers.

Just over one year after her purchase attempts were being made to sell *Granville* but nothing came of it. At the end of 1925 the Ridley fleet consisted of the following six tugs valued as indicated:-

				Value
Oscar	(77/1876)	Iron	Paddle	£2,100
Haakon	(110/1874)	Iron	Paddle	£3,000
Isaac Wilson	(123/1889)	Steel	Paddle	£2,500
Premier	(83/1894)	Steel	Screw	
William Fallows	(122/1888)	Iron	Paddle	£2,750
Granville	(160/1876)	Iron	Paddle	£3,000

- with an average age of forty-two years.

During 1925 another member of the family, Joseph William Ridley, joined the company as office boy at a salary of 10s.0d. (50p.) per week.

During the coal strike of 1926 Robert Ridley worked on *William Fallows*. When the strike was over, the company decided to send two of their tugs to work at Blyth. Photographic evidence shows that *William Fallows* was one of these two. Operations at Blyth did not continue for very long since in 1930 the Ridley contracts at that port were handed over to Blyth Tug Co.Ltd. Simultaneously *Granville* was chartered to the Blyth company for £10 per week so it seems likely that this tug was the second of the Ridley tugs stationed at Blyth in 1926.

For the year ending 30th June 1927 the company reported a trading profit of £312 on the Tyne, and a loss of £289 at Blyth. To say the least, a combined profit of £23 for a year's work cannot have been very encouraging. Knud Zimmer, the BDS representative on the Board of Directors, resigned in July 1927. Because of the contracts with the Norwegian company there was still a desire to have a BDS man on the board. Thomas S. Falck was, therefore, elected to the Ridley board in May 1928. E.W. Gjemre died in 1928 and was succeeded as a director by Thomas O. Gjemre in August of that year. Curiously, T.O. Gjemre is not thereafter mentioned again in the board meeting minutes.

In April 1928 the steel screw tug *Victor* was purchased, via a broker, from the Glasgow company of Steel & Bennie Ltd., for £850 plus *Premier* which was taken in part exchange at a valuation of £1,150. On arrival in the Tyne the new acquisition was renamed *Marty* (175/1906). *Marty* was a much larger screw tug than *Premier*, being over twice the gross tonnage, one third longer and having a compound steam engine of 675 IHP. To whom or to what the name *Marty* referred has not been discovered. In an attempt to gain more business on the Tyne, F.C. Batey went to Bergen in 1929 to try to acquire more contracts with Norwegian shipping companies and to consolidate the Ridley company's already strong Scandinavian connections. Towards the end of 1929 Robert Ridley, the managing director, died at the early age of 42 years, and was succeeded in that post by F.C. Batey. Edward Lawson was appointed a director.

The iron paddle tug *Oscar* was sold for £75 in 1932 to Grahamsley Ltd. for breaking up. There was also a payment of £80 received from her insurers which indicates that the tug had probably met with some misfortune. The drop in value from £2,100 in 1925 to £155 at that time is quite dramatic.

The North Shields tug company of Joseph Crosthwaite (Tugs) Ltd., in 1933, offered to sell their complete business to the Ridley company at an asking price of £8,500. This was equivalent to 10/- (50p.) per share. The Crosthwaite tug fleet consisted of *Champion* (3), *Joseph Crosthwaite*, and *Lewis Crosthwaite* and their business lay largely with British shipowners. Some years previously the Ridley company had made an offer of £16,000 for Crosthwaite but the negotiations fell through. Also Lawson-Batey Tugs Ltd. had made an offer for the same company in 1931; again this had not then progressed any further. In response to the Crosthwaite company's approach Ridley, in May 1933, made a counter offer of £6,375. This also proved to be unacceptable. The outcome of all these offers and counter-offers did not affect The 'Ridley' Steam Tug Co.Ltd. at all, since Joseph Crosthwaite (Tugs), Ltd. was sold jointly to Lawson-Batey Tugs Ltd., France, Fenwick Tyne & Wear Co.Ltd. and Redhead & Dry's Tugs Ltd. in July 1933. In the midst of all these negotiations France Fenwick had also made a tentative offer to purchase a controlling interest in the Ridley company, but this was not pursued further.

The steel screw tug *Trover* (156/1920) was purchased in 1933, for £2,800, from 'The Liverpool & Glasgow Association for the Protection of Commercial Interests as respects wrecked and damaged property'. She had originally been German-owned and had a triple-expansion steam engine developing 500 IHP. She had been purchased specifically to work on the disposal of the wreck of the Greek steamer *Archon* (5,000/-?—), five miles north-east of Margate. Her first task in this respect was to tow the salvage vessel *Disperser* (139/1871), owned by the North East Salvage & Shipbreaking Co. of West Hartlepool to the site of the wreck. Unfortunately the vessel foundered in a gale off the Yorkshire coast in February 1934, the crew of eight being rescued by *Trover* and the Runswick Bay lifeboat. Later some consternation was caused when *Trover*, loaded with explosives ran aground in Ramsgate Harbour. In October 1933 *Haakon* was sold for £200 for breaking-up at Bo'ness and in December *William Fallows* went to the quarry owners Rolmac Ltd. of Lemington-on-Tyne. Thus the fleet of four tugs was now:-

				Value
Isaac Wilson	(123/1889)	Steel	Paddle	£1,000
Granville	(160/1876)	Iron	Paddle	£2,592
Marty	(175/1906)	Steel	Screw	£3,000
Trover	(156/1920)	Steel	Screw	£2,800

- with an average age of about thirty-five years.

Her job off Margate completed, *Trover* was sold to Russia for £5,675 through the agency of Arcos Ltd.. Her new owners renamed her *Tuman*. Therefore, as well as carrying out a useful job, the company made a handsome profit on her after only thirteen months. *Marty*, too, was involved in the dispersal of *Archon*. In March 1935 the French-owned twin-screw tug *Bacalan* was purchased for £2,350. She was a powerful tug of some 620 IHP powered by twin triple-expansion steam engines. Strangely, she does not appear in Lloyds Register whilst French-owned. Her name clearly suggests a close relationship with her builders, Dyle & Bacalan at Bordeaux. Perhaps her duties had been in their dockyard only. On her arrival on the north-east coast Ridley renamed her *Wonder* (2) (169/1921), a reference back to the first tug known to have been owned by John Ridley.

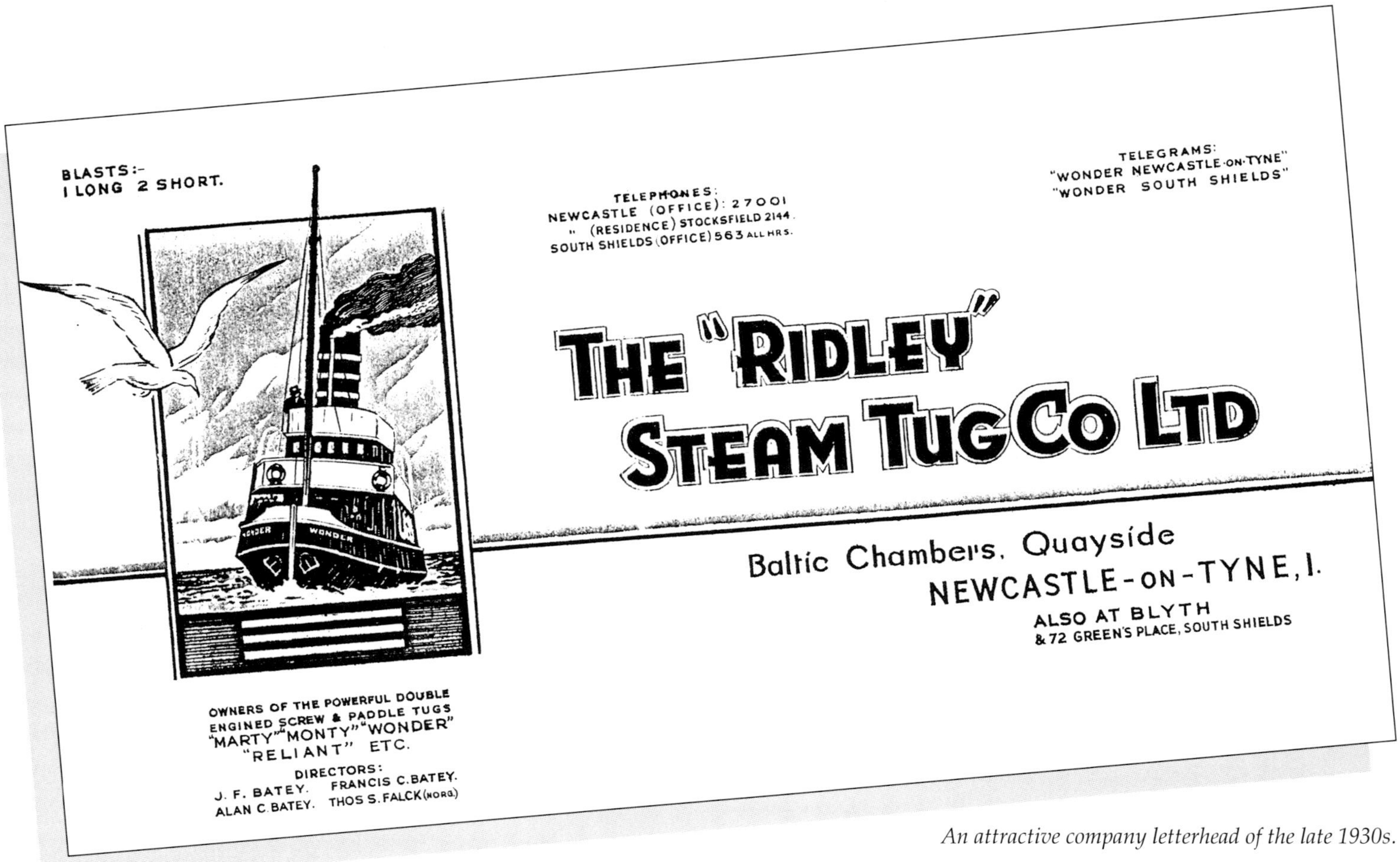

An attractive company letterhead of the late 1930s.

In July 1935 Ridley purchased the dumb lighters *Staithes* (237/1896) from Rolmac Ltd. for £1,000 and *Loftus* (250/1882) from Jarrow Dredging & Salvage Co.Ltd. for £825. Later in the year another barge, *Ella* (219/1905), also from Rolmac, was acquired for £275. All of these lighters had been in the past employed by Palmers Shipbuilding & Iron Co.Ltd. to bring iron ore from the Cleveland mines to their blast furnaces at Hebburn. It is strange that, having sold *William Fallows* to Rolmac Ltd., the Ridley company then bought lighters with the intention of carrying whinstone from Seahouses and Budle Bay on the Northumberland coast to Howdon and Jarrow on behalf of Rolmac. It is possible that Ridley still operated *William Fallows* even though they did not own her. In addition, a contract was secured to carry up to 4,000 tons of stone from Seahouses to Scarborough over a period of six months. The lighter *Loftus* was used for this purpose and was modified by the addition of winches and boilers at a cost of £450. To assist with the towing of the lighters the steel paddle tug *King Edward VII* (138/1901) was purchased from the Irwing Steam Tug Co.Ltd. of Sunderland for £425.

In October 1937 the tug owner Axel H. Andersen, who had been in business since the turn of the century, offered to sell his one tug and the goodwill of his business to The 'Ridley' Steam Tug Co.Ltd. for £1,500. There was an important condition attached to the offer and that was that Andersen was to take permanent employment with Ridley at a salary of £416 per annum. Negotiations followed, which resulted in the value of the concern being reduced to £1,400 and the conditions modified to include a salary of £364 per annum plus £1 per week for expenses. If the reader's mathematics is up to scratch then this will also be seen as amounting to the same £416 as was previously requested! However, there was an addition. If, in the event, the revenue from the Andersen contract was more than £520 per annum then Andersen would get 2.5% of the excess. By the end of the month the take-over was agreed and the Andersen steel screw tug *Overgarth* (97/1907) was transferred to the Ridley company on 1st November 1937. A.H. Andersen was born in Denmark in 1878 and his family came to Tyneside in the following year. It is said that he formed a tug company in 1899. If so it has not been discovered which tugs he owned at that time. He did, however, purchase the wooden paddle tug *Tyne Dock No.4* from the North Eastern Railway Co. in 1911 and renamed her *Enable* (98/1889). *Overgarth* was purchased in 1936 when *Enable* was broken up.

The lighter *Ella* was lost in Hartlepool Bay on 17th February 1938 whilst carrying two 50-ton boilers from Hartlepool to Middlesbrough and in tow of *King Edward VII*. She heeled over and sank in a heavy swell. Fortunately her crew of three were saved. An insurance payment of £1,089 was made against the loss.

The chairman of the company, Joseph Ridley, died on 11th April 1938 at the age of 82 years and was succeeded by Edward Lawson. In May 1938 Joseph William Ridley was elected to the Board of Directors. In June 1938 the small screw tug/workboat *Mildred* (26/1893) was purchased from William Cross of Lemington-on-Tyne. At the board meeting in August 1939 it was agreed that two of the directors were to be paid performance bonuses. The managing director, F.C. Batey, was to get 10% of the profits and J.W. Ridley 5% when the dividend, less tax, was 5% or more. The chairman was appointed to represent the company on the board of the Tyne Mutual Marine Insurance Association.

The Ridley company were keen to buy a coasting vessel, and initially they had had ideas of acquiring *Camroux I* (324/1934) for £9,000, or having a new vessel of 450 tons gross built in Holland for about £10,300. None of these possibilities came to fruition and, in the event, the motor coaster *Dawlish* (248/1937) was purchased for £8,750 in September 1939 just after the outbreak of the Second World War. Just why the tug company wanted a small coaster is not known.

In September 1939, the fifty-eight-year- old small tug/workboat *Scalpa* (32/1881) was purchased for £100 from Mrs. Alice Campbell of Newcastle. Earlier, in August, the two remaining lighters *Staithes* and *Loftus* were disposed of to the Liverpool Lighterage Co.Ltd. At the end of 1939, The 'Ridley' Steam Tug Co.Ltd.'s fleet of eight tugs was the largest in number that the company was ever to have:-

Isaac Wilson	(123/1889)	Steel	Paddle
Granville	(160/1876)	Iron	Paddle
Marty	(175/1906)	Steel	Screw
Wonder (2)	(169/1921)	Steel	Twin Screw
King Edward VII	(138/1901)	Steel	Paddle
Overgarth	(97/1907)	Steel	Screw
Mildred	(26/1893)	Steel	Screw
Scalpa	(32/1881)	Composite	Screw

- with an average age of forty-two years.

The size of the fleet was soon, however, to be reduced by the sale, in October 1940, of *Isaac Wilson*. She went to Grangemouth & Forth Towing Co.Ltd. for £2,000. Later, whilst working for that company she was damaged and sank, eventually going to the breakers in 1950. Whilst with Ridley *Isaac Wilson* was not only used for towage work. On 15th September 1938, skippered by Robert Clasper, she left Wallsend carrying a Norwegian wedding party. Eina Abrahamson and Theodor Henricksen were married at sea by Pastor Sverre Nilsen. They had not allowed for the considerable swell which was running at the time and the Pastor and most of the guests were sick! The contemporary reports state that the bride wore an ankle-length black dress of figured taffeta and that the bridegroom was an engineer on a Norwegian ship then berthed in the Tyne. After returning to port it was found that the wedding party had left behind several crates of beer. These were not wasted!

Sadly, one of the company's directors, Joseph William Ridley, died in 1940 on active service whilst serving with the British Forces. In 1942 France, Fenwick Tyne & Wear Co.Ltd. were suggesting ideas for the unification of towage activities on the River Tyne but at this time nothing came of the proposal. *Granville*, which had been on charter to Blyth Tug Co.Ltd. since 1930, suffered an engine breakdown in the spring of 1942. Temporarily, *King Edward VII* was sent to Blyth to replace her. *Scalpa*, after only three years' service with the Ridley company, must have suffered a misfortune for she was considered a total loss and was broken up at Dunston. The underwriters paid her owners the grand sum of £131-18-5. (£131.92)! She was sixty-one years old and had had seven previous owners since being built at Aberdeen as *Lizzie*, and, of course, only cost Ridley £100 when they acquired her. In March 1944 the coastal cargo vessel *Dawlish* was bare-boat chartered to the Ministry of War Transport and left the Tyne for 'an operational port'. *Granville* continued on charter to Blyth Tug Co.Ltd., her hire rate being increased to £20 per week in 1944.

In May 1945 *Dawlish* was sold for £17,200, less £300 commission, to Duff, Herbert & Mitchell Ltd. of Liverpool. This small coaster had an interesting subsequent history, having ten further owners, being fitted with two new engines, being renamed twice, having six further ports of registry and surviving a sinking in Norway. What is more, as far as is known, she is still in service at the time of writing!

Alan Christian Batey, the son of the managing director F.C. Batey, was appointed to the Ridley Board of Directors in September 1945 following his demobilisation from the Royal Air Force. He had had a lucky escape in 1944 when the Halifax bomber in which he was flying as rear-gunner flew into a hillside at Epinac in France near the Swiss border. Fortunately, he very soon found his way back to England.

During September and October 1945 Lawson-Batey Tugs Ltd. were able to acquire a sufficient number of shares in The 'Ridley' Steam Tug Co.Ltd. to give them a controlling interest in the company. Consequent upon that action there were a number of changes to the Board of Directors. The chairman, Edward Lawson, retired and was succeeded by Frank Batey, and John Francis Batey, Frank's son, joined the board although he was to remain in the army until March 1946.

An offer of £4,750 from Peter Foster & Co.Ltd. of Hull for *Overgarth* was accepted and she left the Ridley fleet in March 1946. She was replaced by *Cervia*, a 550 IHP steam-powered screw tug bought from the well-known William Watkins Ltd. of London. She was one of the many small vessels which took part in the rescue of the British Army from Dunkirk in June 1940. At first it was decided that she should be renamed *Pax* but before that happened the name *Monty* (157/1925) was chosen instead. The price was £11,000, indicating the effect of inflation and of the younger and better quality tug now being acquired. *Monty* commenced work on the Tyne on 21st June 1946. In 1947 the company reported a

Marty *at full speed passing a First World War Flower-class escort (either* Foxglove, Lupin *or* Rosemary) *during the Second World War.* *(University of Newcastle upon Tyne).*

trading loss in the operation of five of their tugs, - the sixth, *Granville*, was still at Blyth earning her hire charge of £30 per week.

On 13th March 1948 the chairman, Frank Batey, died aged sixty-seven, and was succeeded shortly afterwards by John Francis Batey. In January 1949 the company purchased another small tug/workboat *Birchrock* from Harrisons (London) Ltd., who had, no doubt, used her for towing lighters on the River Thames. She cost £2,000 and was soon renamed *Mildred II* (66/1902). After £1,000 had been spent on tidying her up and as soon as she was operational on the Tyne, *Mildred* was sold for £250 to Matthew Kirton & Sons of Newburn, who retained her original name. At the end of 1949 it was decided that two of the company's tugs should be fitted with radio telephone equipment, and at that time the office was moved from Dean Street to 72 Greens Place, South Shields. At the beginning of 1950 the Ridley fleet consisted of the following tugs:-

Granville	(160/1876)	Iron	Paddle
Marty	(175/1906)	Steel	Screw
Wonder (2)	(169/1921)	Steel	Twin screw
King Edward VII	(138/1901)	Steel	Paddle
Monty	(157/1925)	Steel	Screw
Mildred II	(66/1902)	Steel	Screw

- with an average of about forty-four years - still very high and two years worse than that indicated ten years earlier.

An etching showing a Ridley tug (probably Mildred *or* Mildred II*) at Newcastle Quay c.1950.* *(James Alder)*

The steel paddle tug *Old Trafford* which had worked on the Manchester Ship Canal for the past forty-three years was purchased for £1,650 in September 1950. For her voyage from Manchester to the Tyne it was necessary to stack coal on her deck so that she would have sufficient for the journey. Her journey of 886 miles took no less than thirty-two days through a succession of severe storms in which the galley was wrecked, leaving only a tiny cabin stove on which to do the cooking. In addition, her sponsons and lifeboat were damaged. Leaving the calm waters of Manchester on 7th October she must have wondered what she had done to deserve the wrath of the sea gods for during her voyage she had to shelter for four days at Liverpool, three at Holyhead, six at Ramsgate and nine at Dover. On arrival she was fitted with a new funnel around the old one, perhaps to make her look more powerful, and was renamed *Reliant* (156/1907). The company had intended to purchase her sister *Acton Grange* (156/1907) from the same source for £750 but this did not, in fact, take place. In the following month *King Edward VII* suffered a major breakdown which resulted in the underwriters paying £500 in full settlement, after which she went to the shipbreakers at Dunston. In 1951 it was decided that no major repairs were to be made to *Granville*, and if she should break down, then Lawson-Batey's *Comet* (132/1876) would be sent to Blyth to replace her. Later in the same year there was a proposal to purchase *Flying Foam* (217/1917) from Clyde Shipping Co.Ltd., but she was sold elsewhere. Enquiries were also made about her sister tug, *Flying Spray* (217/1917), but nothing further came of that proposed purchase either.

The hire of *Granville* to Blyth Tug Co.Ltd. ceased on 4th April 1952 and she was sold for £1,600, via the British Iron & Steel Corporation, for breaking up at Dunston. She was then seventy-six years of age! In the same year, *Wonder* (2) was out of commission for a while for her boiler to be retubed. No doubt an expensive operation. All told, the company were reporting a successful year in 1952.

In August 1953 a bid of £31,000 was made for the tug *Empire Dolly* (257/1943), then lying at Suez, but this was not successful. Along with the Lawson-Batey company Ridley were considering the possibilities of purchasing a new diesel-powered tug, bearing in mind the success which France Fenwick had enjoyed with their diesel tugs since 1950. It was expected that a diesel tug would cost about £50,000 and that this could be financed by taking £23,000 from investments, by selling *Monty* for an estimated £15-20,000, and borrowing the difference. The managing director, F.C. Batey, and F.S. Batey, the chairman of Lawson-Batey, went to Hamburg for discussions with prospective builders. The outcome was that Ridley placed an order in September 1953 with Mützelfeldwerft GmbH of Cuxhaven for an 825 BHP (14.25 tons bollard pull) diesel tug for delivery in June 1954, at a price of £53,240. At the time no British yard could match this delivery period. Her keel was laid in February 1954, she was christened and launched by Mrs. Ella Batey, wife of the managing director, on 6th May 1954, and completed trials on 24th July. She was named *Impetus* (141/1954) and was the first diesel tug to operate on the River Tyne. She had a six-cylinder Klöckner-Humboldt-Deutz engine and was fitted with a Kort Rudder around her propeller. Her owners were so proud of her that, before going into service, she was put on show on the Thames, near London Bridge, and at Newcastle during August 1954.

In 1953, F.C. Batey was elected chairman of the British Tugowners Association, a post he was to hold for the traditional two years.

Meanwhile other changes to the Ridley fleet were taking place. An offer of £8,500 had been made and accepted for the thirty-two-year old Swedish-owned tug *Diana*. She was delivered at Oskarsham and departed from there for the

F.C. Batey, second from left, managing director of The Ridley Steam Tug Co. Ltd. c.1953. On his right is A.M.H. Wardlaw of France Fenwick and on his far left is J.F. Batey of Lawson-Batey. (Company archives).

Momentum *at work during her short stay in the Ridley fleet in 1954.*

Diana *on arrival in the Tyne from Sweden in 1954 before being renamed* Momentum. *(Les Turner).*

United Kingdom, via the Kiel Canal, on 30th April 1954. Unfortunately, she sprang a leak when half-way across the North Sea and had to be towed part of the way by the tanker *Sepia* (6,214/1936), eventually arriving in the Tyne on 4th May. The total cost of the vessel, after repairs, came to £10,674. Although photographic evidence shows her running as *Diana* in the Ridley colours, she was very soon renamed *Momentum* (213/1922). The new acquisition replaced *Monty*, which was sold to J.H. Pigott & Son Ltd. of Grimsby for £17,500, who renamed her *Lady Elsie*. Problems soon arose over the manning level necessary on diesel tugs. The arguments for and against, and the outcome of the Industrial Disputes Tribunal, has already been related in the Lawson-Batey chapter.

After having been in the Ridley fleet for only nine months *Momentum* was sold for £21,500 to J.H. Lamey Ltd. of Liverpool, who renamed her *Marie Lamey*. No doubt the price obtained was the attraction which caused Ridley to dispose of her so quickly.

Ridley Tugs Ltd. 1955 - 1978

At an Extraordinary General Meeting of the company held on 21st March 1955 it was decided that the company title should be changed from The 'Ridley' Steam Tug Co.Ltd. to Ridley Tugs Ltd. This was formally put into effect on 30th March 1955. The directors of the company at this time were:-

J.F. Batey	Chairman
F.C. Batey	Managing director
T.S. Falck	Representing BDS
A.C. Batey	

and the tug fleet comprised:-

Marty	(175/1906)	Steel	Screw
Wonder (2)	(169/1921)	Steel	Twin screw
Mildred II	(66/1902)	Steel	Screw
Reliant	(156/1907)	Steel	Paddle
Impetus	(141/1954)	Steel	Screw

Due to the purchase of the new *Impetus* the average age of the fleet had at last begun to decrease significantly, to thirty-seven years.

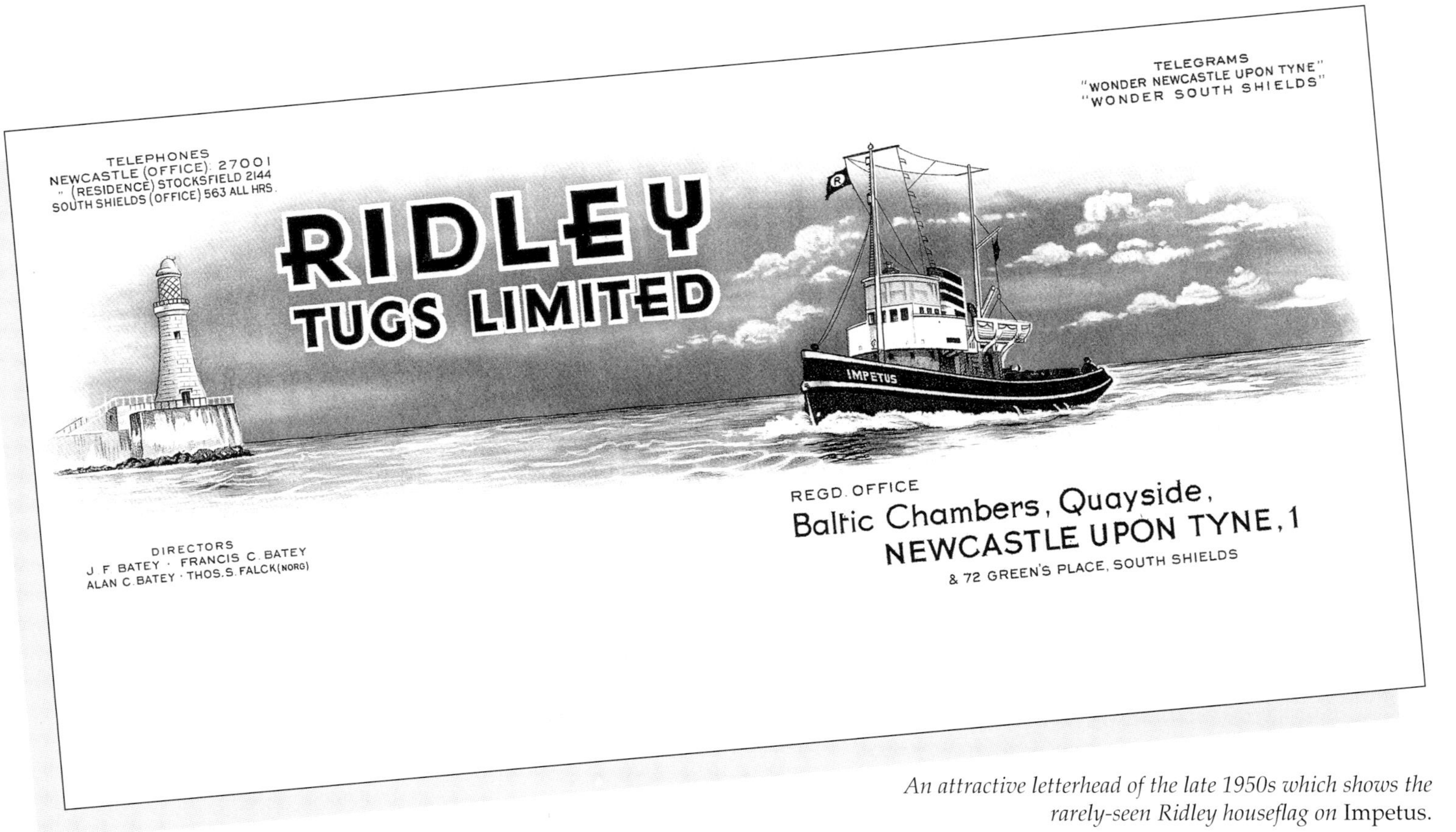

An attractive letterhead of the late 1950s which shows the rarely-seen Ridley houseflag on Impetus.

A second tug, a sister to *Impetus,* was ordered from Cuxhaven in the spring of 1955 for delivery about one year later. She was expected to cost about £1,600 more than the price of *Impetus.* Her launch took place on 17th October 1955 when she was christened *Maximus* (141/1956) by Mrs. Marjorie Courtney, secretary and assistant to F.C. Batey, and the tug arrived in the Tyne on 6th April 1956. Though substantially the same as *Impetus* she was not, however, fitted with a Kort Rudder, but did have a bow thruster. She was the first British tug to be fitted with this latter facility, but it was little used and later removed. Before choosing her name, the company's employees were asked for suggestions and over two hundred were received. Those put to the Ministry of Transport, in order of preference, were *Maximus, Pegasus, Taurus* and *Petasus.* The first skipper of *Maximus* was Albert Thompson, who had transferred from *Impetus.* He had been in tugs for thirty-nine years, thirty-five with Ridley, and had been a skipper since 1931.

As had been the case with Lawson-Batey and France Fenwick crew members, examples were made of those who had been found stealing coal from their respective tugs. In the Ridley company the skipper of *Reliant* was fined £2 for stealing 1/-(5p.)worth of coal! Due to the falling off in hopper towage *Mildred II* was laid up from 25th September 1955 and, in the following year, she was put up for sale.

Rather surprisingly, by July 1956, the company were trying to sell *Maximus* and/or *Impetus.* Figures of around £90,000 were reported as being offered, no doubt due to the long delivery dates being quoted by British shipbuilders. In the event, neither was sold at that time. The last paddle tug working on the Tyne, *Reliant,* was sold for £4,000 to

Maximum *on trials flying the German flag.* *(Company archives).*

Seaham Harbour Dock Co. in September 1956. After working for them for a further thirteen years she ended her days as an exhibit in the National Maritime Museum, Greenwich. At times, whilst in the company's ownership, she had worked as a relief tug at Methil.

Mildred II had been up for sale, at an asking price of £2,000, since October 1956, without attracting any offers. In June 1957 she was, therefore, sold via an agent to C.W.Dorkin Ltd., Gateshead for breaking up. The 820 IHP steam tug *Battleaxe* (163/1911) was purchased, for £8,550, in October 1957 from Steel & Bennie Ltd., Glasgow. Ridley wished to rename her *Romulus*, but apparently this name was not available. *Rumpus* was suggested as an alternative, but, in the event, she did not take any Ridley name, for she was transferred at a valuation of £8,000 to Lawson-Batey Tugs Ltd. in December 1958 and then renamed *Hillsider* (1). In November 1957 agreement was reached with Seaham Harbour Dock Co. that they would purchase *Wonder* (2) for £5,500 provided that new furnaces were fitted before delivery at a cost to the seller of £1,700. Formal acquisition by the Seaham company took place on 14th April 1958.

Lawson-Batey Tugs Ltd., who already held a majority shareholding in Ridley Tugs Ltd., wished to purchase the remaining shares so that the Ridley company would become a wholly-owned subsidiary of Lawson-Batey. Discussions and negotiations took place throughout 1958, resulting in the acquisition of the company by Lawson-Batey in October of that year. Further details of this transaction are given in the Lawson-Batey chapter. At this time Axel H. Andersen, whose tug business had been absorbed into the Ridley concern in 1937, retired at the age of eighty years.

Concurrently, negotiations had been progressing since early in 1958 regarding the amalgamation of the Ridley Tugs Ltd., Lawson-Batey Tugs Ltd. and France, Fenwick Tyne & Wear Co.Ltd. towage services on the River Tyne. Details of these negotiations are given fully in the Lawson-Batey and France Fenwick chapters. The outcome was the formation of Tyne Tugs Ltd., who would charter twenty-four tugs in all; two from Ridley, ten from Lawson-Batey and twelve from France Fenwick. Each of the three companies would own a proportionate shareholding in the new charter company. The two Ridley tugs which would enter into the scheme were:-

Impetus	(141/1954)	Both diesel-powered	825 BHP
Maximus	(141/1956)		13 BP

It had been intended that the third Ridley tug, *Marty*, should be broken up, but before this could take place one of the France Fenwick tugs, *Earsdon*, which was listed for chartering to Tyne Tugs Ltd., sank. It thus came about that *Marty* was chartered to France Fenwick so that she could replace *Earsdon*. However, *Marty* was not saved from the breakers for long, since, in May 1959, she was sold for £1,000 to J.J. King & Co. Ltd., Gateshead.

The changes in company organisation which had taken place in 1958 resulted in A.C. Batey and T.S. Falck resigning, and Frank Sewell Batey, chairman of Lawson Batey, joining the Board of Directors. Consideration was given to the voluntary liquidation of Ridley Tugs Ltd., but this was thought not to be appropriate at the time.

For the continuing history of Ridley Tugs Ltd. we will not consider further the operational activities of their tugs since these are covered in Chapter Twenty dealing with the Charter Companies. Just to complicate matters, and as will be seen later, Ridley Tugs Ltd. themselves later became one of these charter companies. With the transfer of the two Ridley tugs to the operational control of Tyne Tugs Ltd., the familiar black and white Ridley funnel colours disappeared, to be replaced by the red and black of the charter company. Lawson-Batey and France Fenwick lost their individual colours too, much to the regret of many people.

When the Ridley tugs were transferred on charter to Tyne Tugs Ltd. so also were the employees. Boswell Whitaker, who had managed the tugs using latterly his home address 72 Greens Place, South Shields as the towage office, was one of those transferred. He had joined Lawson-Batey Tugs Ltd. in March 1927 when fourteen years of age as an office boy for a wage of 5/- (25p.) per week. Two years later, in order to earn more money, he moved to The 'Ridley' Steam Tug Co. Ltd. for 10/- (50p.) per week plus expenses, and by February 1938 he had advanced to the position of riverside manager at South Shields. Having joined the Tyne Electrical Engineers branch of the Territorial Army in 1936 he was called into the army in August 1939 just before the outbreak of the Second World War. He served with 307 Company, 37th Battalion, Royal Engineers until demobilised in November 1945 when he was reinstated in his pre-war job two weeks after returning home and continued so until the end of 1958.

Bos Whitaker has related how Joseph Ridley and he frequently used to fight with each other verbally, and how he was given the sack 'every week' for some imagined misdemeanour. On one occasion he was dismissed 'twice in two hours'! When at work as riverside manager for Ridley he recalls how the local newspaper reporters would from time to time ring up asking what the situation on the River Tyne was and whether there was any news worth publishing. His stock answer was 'Hinny, there's no sinkings, no collisions, nowt on fire, nobody drowned, in fact nowt to laugh about at all'.

Bos went on to work for Tyne Tugs Ltd. for over eighteen years before retiring with fifty years service at the age of sixty five. Since then he has continued with his research into local history and has had three books published in 1979/80; 'Skuetender Lifeboat', 'South Shields Volunteer Life Brigade' and 'Tynemouth Volunteer Life Brigade'.

In February 1960, F.C. Batey, who had served the company as secretary since 1920 and managing director since 1929, retired. The chairman, J.F. Batey, took on, in addition, the post of managing director, and Arthur Scott became company secretary. The company's registered office was at Mercantile Chambers, Quayside, Newcastle. At the end of 1967 the secretary retired and Douglas Frank Hamilton succeeded him in that position from 1st January 1968 at a salary of £50 per annum. *Impetus* was sold in May 1969, for £38,000, to Lubeck owners where she was renamed *Joachim*. Her skipper at the time of her disposal was James C. Addison. *Maximus*, which had been laid-up on the Tyne since November 1968, was transferred to Blyth Tug Co.Ltd. at a valuation of £37,500. From this point in time the Ridley company did not own any tugs. Therefore the directors discussed the future of the company and itemised the following possible courses of action:-

a) Buy a tug from Lawson-Batey Tugs Ltd.
b) Take over Blyth Tug Co.Ltd.
c) Wind up Ridley Tugs Ltd.

- the outcome being that they decided on no change.

A model of Reliant, *in Seaham Harbour Dock Co. colours, seen at the Gateshead Garden Festival in 1990.* *(Les Turner).*

As described in the Lawson-Batey chapter, the chairman, J.F. Batey, was tragically killed in London on 13th November 1975. Early in 1976 F.S. Batey was appointed chairman of the board and Thomas Anthony Greenwell was elected to the board as managing director. At the Annual General Meeting on 14th April 1976 the secretary, D.F. Hamilton, was elected a director.

Wear Tugs Ltd. 1978 - 1989

When, in 1977, Lawson-Batey Tugs Ltd. took over France, Fenwick Tyne & Wear Co.Ltd. five of the latter company's tugs were on charter to Tyne Tugs Ltd. and remained so, and the other five operated on the Wear at Sunderland. One of these latter five was soon sold and Lawson-Batey wished to charter the others to an operating company similar to Tyne Tugs Ltd. This was achieved by Ridley Tugs Ltd. changing its name on 24th January 1978 to Wear Tugs Ltd. and becoming an operating company for the four ex-France Fenwick tugs working at Sunderland. Thus the tugs demise-chartered to Wear Tugs Ltd. by Lawson-Batey were:-

Cornhill	(176/1943)
Prestwick	(119/1955)
Marsden (4)	(122/1956)
Dunelm (2)	(132/1964)

These tugs retained their France Fenwick funnel colours but wore the Lawson-Batey house-flag.

For the story of Wear Tugs Ltd., through to its demise in 1989, the reader is referred to the second section of the chapter on the Charter Companies.

APPENDIX A TO CHAPTER SEVEN

The 'Ridley' Steam Tug Co. Ltd. 1920-1955
Ridley Tugs Ltd. 1955-1978
Wear Tugs Ltd. 1978-1989

Board of Directors

Chairmen	J. Ridley	1920-1938	Died
	E. Lawson	1938-1945	Retired
	F. Batey	1945-1948	Died
	J.F. Batey	1948-1975	Died
	F.S. Batey	1976-1983	Retired
	T.A. Greenwell	1983	Vice-Chairman
	T.A. Greenwell	1984-1987	Died
	J.G. Mucklow	1987-1989	
Managing Directors	R. Ridley	1920-1930	Died
	F.C. Batey	1929-1960	Retired
	J.F. Batey	1960-1975	Died
	T.A. Greenwell	1976-1982	Became Vice-Chairman
	D.F. Hamilton	1980-1982	Deputy M.D.
	D.F. Hamilton	1983-1989	
Directors	E.W. Gjemre	1920-1928	Died
	F.C. Batey	1920-1929	Became M.D.
	K. Zimmer	1923-1927	Resigned
	T.S. Falck	1928-1959	Resigned
	T.O. Gjemre	1928	
	E. Lawson	1930-1938	Became Chairman
	J.W. Ridley	1938-1940	Died on active service
	J.F. Batey	1945-1948	Became Chairman
	A.C. Batey	1945-1959	Resigned
	F.S. Batey	1959-1976	Became Chairman
	D.F. Hamilton	1976-1980	Became Deputy M.D.
	S.S.W. Dalgliesh	1978-1982	Resigned
	G.H.R. Towers	1978-1986	Retired
	M.F. Pyman	1980-1986	Resigned
	J.G. Mucklow	1986-1987	Became Chairman
Secretaries	W.M. Graydon	1920	Resigned
	F.C. Batey	1920-1960	Retired
	A. Scott	1960-1967	Retired
	D.F. Hamilton	1968-1982	Became M.D.
	H.R. Gilbert	1983-1986	Retired
	J.G. Mucklow	1986-1989	

APPENDIX B TO CHAPTER SEVEN

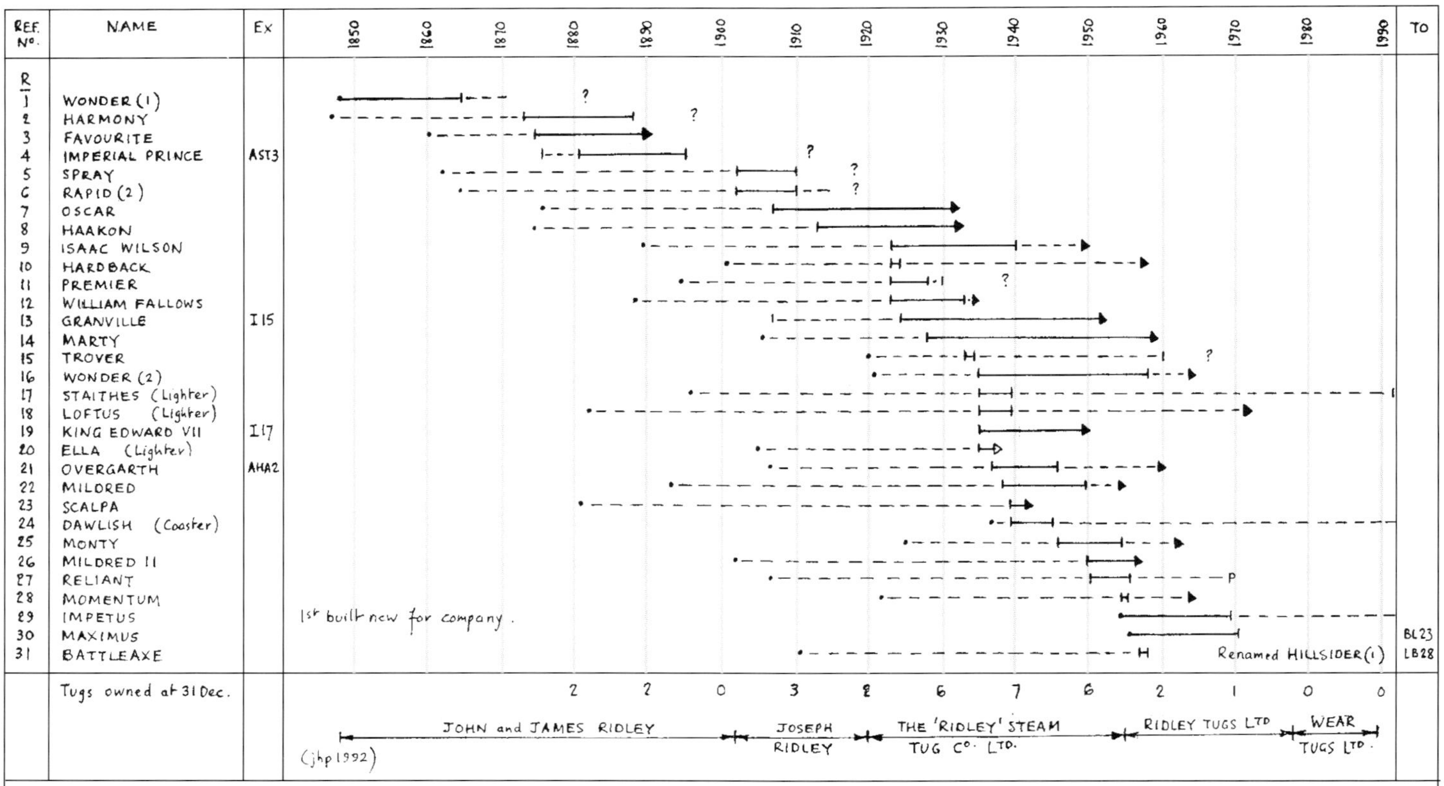

Tug Ownership Chronology.

THE RIDLEY COMPANIES c.1848 – 1989

APPENDIX C TO CHAPTER SEVEN

The Ridley Companies Fleet List

R 1 **WONDER (1)** **(1848-1864)**
ON ? 58 g. 75.0 x 15.8 x 8.3 ft. Wood Paddle
B: At Peter's Quay, Northumberland.
E: 130 NHP.

PR: 1848 Newcastle.

1848:	Completed for John Ridley.
1864:	Sold to J. Matthews, London. Renamed PRINCE OF HESSE.
1871:	Not in Mercantile Navy List.

R 2 **HARMONY** **(1873-c.1887)**
ON 10165 47 g. 66.1 x 13.2 x 8.0 ft. Wood Paddle
B: At Low Lights, North Shields.
E: 20 NHP.

PR: 1847 ?; 1854 Shields.

1847:	Completed for W. Hall, Newcastle.
1854:	Sold to Jarrow Chemical Co.
1860:	Sold to J. & W. Deal and R. Sanderson, North Shields.
1870:	Sold to H. Strachan, Howdon.
1873:	Purchased by James Ridley et al, Bill Quay.
1888:	Not in Mercantile Navy List.

R 3 **FAVOURITE** **(c.1874-1891)**
ON 29867 42 g. 70.5 x 14.7 x 7.6 ft. Wood Paddle
B: At South Shields.
E: 25 NHP.

PR: 1860 Sunderland; 1864 Shields; 1888 North Shields.

1860:	Completed for E. Brown, Sunderland.
1864:	Sold to J. Dodds, North Shields.
c.1874:	Purchased by James Ridley, Bill Quay.
1888:	Transferred to John Ridley, Bill Quay.
07.1891:	Sank following collision. Believed broken up.

R 4 **IMPERIAL PRINCE** **(1881-c.1895)**
See Anchor Steam Tug fleet list, AST 3

R 5 **SPRAY** **(c.1902-c.1910)**
ON 43639 47 g. 71.5 x 15.0 x 8.1 ft. Wood Paddle
B: J.D. Burdis, North Shields.
E: 30 NHP. J.P. Rennoldson, South Shields.

PR: 1862 South Shields.

1862:	Completed for J.B. Middleton, South Shields.
1869:	Sold to William Thompson, Newcastle.
c.1902:	Purchased by John Ridley, Bill Quay.
c.1910:	Not in register.

R 6 **RAPID (2)** **(c.1902-c.1910)**
ON 45633 74 g. 82.0 x 17.1 x 9.3 ft. Wood Paddle
B: At North Shields.
E: 31 NHP.

PR: 1864 Shields.

1864:	Completed for J. Heads and W. Heads, North Shields.
1882:	Sold to T. Boag, Shields.
c.1902:	Purchased by Joseph Ridley.
c.1910:	Sold to R. Jull, Percy Main.
1914:	Not in Mercantile Navy List.

R 7 **OSCAR** **(1907-1932)**
ON 73635 77 g. 82.0 x 16.6 x 8.9 ft. Iron Paddle
B: G. Butchard, Milton, nr. Gravesend.
E: L. 2 Cy. 22" x 45". 45 NHP. By Builder.

PR: 1876 London; 1893 Dundalk; 1923 Newcastle.

1876:	Completed for J. Martin, London.
c.1879:	Sold to G.H. Jewell, Kings Lynn.
1881:	Sold to Dundalk Harbour Commissioners.
11.1907:	Purchased by Joseph Ridley.
04.1920:	Transferred to The 'Ridley' Steam Tug Co.Ltd.
03.1932:	Sold, for £75, to Grahamsley Ltd., for breaking up.

Oscar *in Richardson & Brock's wooden floating dock which lies alongside the remains of the Stanhope & Tyne Railway coal drop at South Shields. The tug is wearing the funnel colours of the Svenska Lloyd company.* *(Local History Library, South Shields Central Library).*

Haakon *as* Goole No.1 *when owned by the Goole & Hull Steam Towing Co.*

R 8 **HAAKON** **(1913-1933)**

ON 67810 110 g. 96.6 x 18.4 x 9.4 ft. Iron Paddle
B: J. Softley & Co., South Shields.
E: L. 2 Cy. 25⅜" x 46". 40 NHP. T. Hepple & Co., Low Walker.

PR: 1874 Goole.

1874:	Completed, as GOOLE No. 1, for Goole & Hull Steam Towing Co.
04.1913:	Purchased by Joseph Ridley, renamed HAAKON.
04.1920:	Transferred to The 'Ridley' Steam Tug Co.Ltd.
10.1933:	Sold, for £200, to Thomson & McGregor & Co., Bo'ness, for breaking up.

Isaac Wilson *at Newcastle Quay. (Author's collection).*

R 9 **ISAAC WILSON** **(1923-1940)**

ON 95923 123 g. 101.0 x 18.9 x 9.5 ft. Steel Paddle
B: Hepple & Co., North Shields. (YN 432)
E: L. 2 Cy. 26" x 48". 45 NHP. By Builder.

PR: 1889 Stockton.

1889:	Completed for Tees Conservancy Commissioners.
01.1923:	Purchased, for £1,250, by The 'Ridley' Steam Tug Co.Ltd.
10.1940:	Sold, for £2,000, to Grangemouth & Forth Towing Co.Ltd.
12.1950:	Sold to G.W. Brunton, Grangemouth, for breaking up.

Hardback *when owned by Seaham Harbour Dock Co. (George Scott).*

R 10 **HARDBACK** **(1923-1924)**

ON 113132 125 g. 90.5 x 19.1 x 9.4 ft. Steel Paddle
B: J.P. Rennoldson & Sons, South Shields. (YN 208)
E: L. 2 Cy. 25" x 45". 50 NHP. By Builder.

PR: 1901 Whitehaven.

1901:	Completed for The Hodbarrow Mining Co.Ltd., Millom.
03.1923:	Purchased, for £800, by The 'Ridley' Steam Tug Co.Ltd.
06.1924:	Sold, for £3,750, to Seaham Harbour Dock Co.
07.1958:	Sold, for £625, to British Iron & Steel Corporation. Allocated to C.W. Dorkin Ltd., Gateshead, for breaking up.

R 11 **PREMIER** **(1923-1928)**

ON 102869 83 g. 79.0 x 17.6 x 9.7 ft. Steel Screw
B: A.W. Robertson & Co., Canning Town.
E: C. 2 Cy. 15½" & 32" x 20". 350 IHP. By Builder.

PR: 1894 London; 1918 Grimsby.

01.1894:	Completed for Parkinson & Edwards, Gravesend.
1917:	Sold to Grimsby Steam Tug Co.
06.1918:	Hired by Royal Navy.
01.1919:	Returned to owners.
1923:	Purchased by The 'Ridley' Steam Tug Co.Ltd.
04.1928:	Sold, at value of £1,150, to Peter McLeod, Alloa, in part exchange for MARTY.
1930:	Not in Mercantile Navy List.

R 12 **WILLIAM FALLOWS** **(1923-1933)**

ON 95921 122 g. 101.0 x 19.0 x 9.6 ft. Iron Paddle
B: Hepple & Co., North Shields. (YN 426).
E: L. 2 Cy. 25" x 48". 45 NHP. By Builder.

PR: 1889 Stockton; 1923 Middlesbrough; 1928 Newcastle.

1888:	Completed for Tees Conservancy Commissioners.
03.1923:	Sold to Tees Towing Co. Ltd., (Robinson & Crosthwaite, Managers).
10.1923:	Purchased, for £1,800, by The 'Ridley' Steam Tug Co.Ltd.
12.1933:	Sold to Rolmac Ltd., Lemington-on-Tyne.
1935:	Sold to Clayton & Davie Ltd., Dunston, for breaking up.

R 13 **GRANVILLE** **(1924-1952)**

See Irwing fleet list, I 15.

Marty *with shorter funnel. (John Clarkson, Longton).*

R 14 **MARTY** **(1928-1959)**

ON 121336 175 g. 106.0 x 22.1 x 10.4 ft. Steel Screw
B: A. Stephen & Sons Ltd., Linthouse. (YN 418).
E: C. 2 Cy. 20" & 40" x 27". 675 IHP. By Builder.

PR: 1906 Glasgow; 1928 Newcastle.

1906:	Completed, as VICTOR, for Steel & Bennie Ltd., Glasgow.
1926:	Sold to Peter McLeod, Alloa.
04.1928:	Purchased, for £850 plus PREMIER, by The 'Ridley' Steam Tug Co.Ltd., renamed MARTY.
03.1955:	Owners restyled Ridley Tugs Ltd.
01.1959:	Chartered to France, Fenwick Tyne & Wear Co.Ltd., who in turn chartered her to Tyne Tugs Ltd. to replace the sunken EARSDON.
05.1959:	Sold, for £1,000, to British Iron & Steel Corporation. Allocated to J.J. King & Co.Ltd., Gateshead, for breaking up.

Trover *with* Marty *behind.* *(George Scott).*

R 15 **TROVER** **(1933-1934)**
ON 145958 156 g. 99.8 x 21.4 x 11.7 ft. Steel Screw
B: F. Schichau, Elbing.
E: T. 3 Cy. 350, 580 & 900 x 550 mm. 500 IHP. 11 K. By Builder.

PR: 1920 Flensburg; 1922 Liverpool; 1933 Newcastle; 1934 Leningrad.

1920:	Completed, as KONIGSTEIN, for H.C. Horn, Hamburg.
1922:	Sold to 'The Liverpool & Glasgow Association for the protection of Commercial Interests as respects wrecked and damaged property'; renamed TROVER.
09.1933:	Purchased, for £2,800, by The 'Ridley' Steam Tug Co.Ltd.
10.1934:	Sold, for £5,675, to Arcos Ltd. who then sold her to Russia; renamed TUMAN.
1960:	Not in Lloyds Register.

Wonder *(2).* *(John Clarkson, Longton).*

R 16 **WONDER (2)** **(1935-1958)**
ON 161577 169 g. 100.3 x 23.0 x 10.1 ft. Steel Twin screw
B: Société Anonyme de Traveaux Dyle & Bacalan, Bordeaux.
E: 2- T. 3 Cy. 10½″, 17″ & 28″ x 15″. 620 IHP. 11 K. Les Hériteurs de G. Cochot, Paris.

PR: 1921 ?; 1935 Newcastle.

1921:	Completed as BACALAN. Owners not known and vessel not listed in Lloyds Register. Possibly owned by builders as dockyard tug.
02.1935:	Purchased, for £2,350, by The 'Ridley' Steam Tug Co.Ltd. Renamed WONDER.
03.1955:	Owners restyled Ridley Tugs Ltd.
04.1958:	Sold, for £5,500, to Seaham Harbour Dock Co.
11.1964:	Sold to Clayton & Davie Ltd., Dunston, for breaking up.

R 17 **STAITHES (Lighter)** **(1935-1939)**
ON 105867 237 g. 125.0 x 21.2 x 8.9 ft. Steel Unpowered
B: Sir Raylton Dixon & Co., Middlesbrough. (YN 433).
E: None.

PR: 1896 London; 1920 Newcastle.

1896:	Completed, as SAPPHIRE, for William Cory & Son Ltd.
c.1899:	Sold to S. Pearson & Son Ltd., London.
1906:	Sold to Goole & Hull Steam Towing Co.Ltd., Goole.
04.1920:	Sold to Palmers Shipbuilding & Iron Co.Ltd., renamed STAITHES.
11.1920:	Sold to Hebburn Transport Co.Ltd.
03.1932:	Sold to George McLaren, Craster, et al.
01.1934:	Sold to Rolmac Ltd., Lemington.
07.1935:	Purchased, for £1,000, by The 'Ridley' Steam Tug Co.Ltd.
08.1939:	Sold to Liverpool Lighterage Co.Ltd., Liverpool.
04.1971:	Sold to Bulk Cargo Handling Services Ltd., Liverpool.
1992:	Not in Lloyds Register.

R 18 **LOFTUS (Lighter)** **(1935-1939)**
ON 86084 250 g. 130.0 x 26.6 x 8.9 ft. Iron Unpowered
B: Wallsend Slipway & Engineering Co.Ltd., Wallsend.
E: None originally and finally.

PR: 1882 Newcastle; 1929 Newcastle; 1934 Newcastle.

08.1882:	Completed for Palmers Shipbuilding & Iron Co.Ltd. as a dumb lighter.
1929:	Fitted with engine: Oil. 2 SA. 1 Cy. 335 x 330 mm. 50 IHP. 4.6 K. Made by Kromhaut Engineering Works, Amsterdam.
02.1934:	Sold to G.W. Cook, Jarrow.
04.1934:	Engine removed.
05.1934:	Sold to Jarrow Dredging & Salvage Co.Ltd.
07.1935:	Purchased, for £825, by The 'Ridley' Steam Tug Co.Ltd.
08.1939:	Sold to Liverpool Lighterage Co.Ltd., Liverpool.
04.1971:	Sold to Bulk Cargo Handling Service, Liverpool.
1972:	Broken up.

R 19 **KING EDWARD VII** **(1935-1950)**
See Irwing fleet list, I 17.

R 20 **ELLA (Lighter)** **(1935-1938)**
ON 122706 219 g. 105.3 x 22.1 x 9.95 ft. Steel Unpowered
B: Joseph Scarr & Son, Beverley.
E: None.

PR: 1905 Grimsby; 1910 Kings Lynn; 1919 Grimsby; 1920 Newcastle.

1905:	Completed.
By 1913:	Owners R. & W. Paul, Ipswich.
05.1920:	Sold to Palmers Shipbuilding & Iron Co. Ltd.
11.1920:	Sold to Hebburn Transport Co.Ltd.
05.1932:	Sold to Rolmac Ltd., Lemington.
11.1935:	Purchased, for £275, by The 'Ridley' Steam Tug Co.Ltd.
02.1938:	Foundered in Hartlepool Bay.

R 21 **OVERGARTH** **(1937-1946)**
See Andersen fleet list, AHA 2 (Appendix D to Chapter 7)

R 22 **MILDRED** **(1938-1949)**
ON 98878 26 g. 61.0 x 11.5 x 6.3 ft. Steel Screw
B: Pool, Skinner & Williams, Falmouth.
E: C. 2 Cy. 10″ & 18″ x 14″. 98 IHP. 10 K. R. Arthur & Son, Newport.

PR: 1893 Newport, Mon.; 1899 Falmouth; 1902 Plymouth; 1920 Sunderland.

1893:	Completed for W.D. Arthur, Newport,, Mon.
1899:	Sold to F.W. Hall, Falmouth.
1902:	Sold to W.T. Cundy, Plymouth.
08.1914:	Requisitioned by Royal Navy.
08.1919:	Returned to owners.
06.1920:	Sold to Osbourne, Graham & Co.Ltd., Hylton.
03.1932:	Sold to R.H. Gibson and K. Gibson, Corbridge.
12.1932:	Sold to William Cross, Lemington.
06.1938:	Purchased by The 'Ridley' Steam Tug Co.Ltd.
03.1949:	Sold, for £250, to Matthew Kirton & Sons, Newburn.
07.1955:	Registry closed. Vessel broken up by J.J. King & Co. Ltd., Gateshead.

R 23 **SCALPA** **(1939-1942)**
ON 84353 32 g. 65.0 x 13.1 x 6.1 ft. Composite Screw
B: At Aberdeen.
E: 25 NHP.

PR: 1881 Aberdeen.

1881:	Built, as LIZZIE, for J.G. Mackie, Kircudbright.
1882:	Sold to J. Paley, Preston.
c.1885:	Sold to R. Kyrie, London.
c.1895:	Sold to W. Binnie, Malton, renamed SCALPA.
Pre 1913:	Sold to S. West, North Shields.
By 1924:	Sold to E. Cook, Felling.
By 1937:	Sold to Alice Campbell, Newcastle.
09.1939:	Purchased, for £100, by The 'Ridley' Steam Tug Co.Ltd.
11.1942:	Sold to Clayton & Davie Ltd., Dunston, for breaking up.

R 24 **DAWLISH (Coaster)** **(1939-1945)**

ON 165591 248 g. 130.7 x 23.6 x 8.5 ft. Steel Screw
B: A. Vuijk & Zonen, Capelle a/d Ysel. (YN. 640).
E: Oil. 4 SA. 4 Cy. 11¼" x 16 9/16". 225 BHP. Motoren-Werke Mannheim. A.G., Mannheim.

PR: 1937 London; 1953 Bergen; 1958 Alesund; 1963 Sandnes; 1965 Bergen; 1970 Trondheim; 1982 Panama.

10.1937:	Completed for H. Harrison (Shipping) Ltd., London.
09.1939:	Purchased, for £8,750, by The 'Ridley' Steam Tug Co.Ltd.
03.1944:	Bare-boat chartered to Ministry of War Transport.
05.1945:	Sold, for £17,200, to Duff, Herbert & Mitchell Ltd., Liverpool.
12.1947:	Fitted with new 4 SA. 5 Cy. 8¾" x 11½" engine made by R.A. Lister Ltd., Dursley.
1949:	Owners became Sir C.M.R.V. Duff (O.T. Williams, Managers), Port Dinorwic.
1953:	Owners became Dinorwic Slate Quarries Ltd. (O.T. Williams, Managers), Port Dinorwic.
1953:	Sold to H. & O. Haugland, Bergen, renamed FJELLMAN.
1958:	Sold to Red. Goald, Alesund.
08.1963:	Sold to O.C. Ostraadt A/S., Sandnes, and fitted with new 4 SA, 6 Cy. 5⅜" x 6½" engine, made by Caterpillar Tractor Co., Peoria.
1965:	Sold to Alf Nygard P/R., Bergen.
08.1969:	Sank off Grenaa, later refloated and repaired.
1970:	Sold to A/S Red. Tambur, Trondheim.
1971:	Sold to Olav and Haakon Johansen P/R., Bjugn, later Haakon Johansen only.
1975:	Sold to Grandes Red. A/S, Mosvik.
1982:	Sold to Louis-Jean Paschal, Panama, renamed NATALYNA.
1992:	Still in service.

Monty. *(John Clarkson, Longton).*

R 25 **MONTY** **(1946-1954)**

ON 148586 157 g. 90.0 x 22.0 x 10.6 ft. Steel Screw
B: J.P. Rennoldson & Sons Ltd., South Shields. (YN 326)
E: C. 2 Cy. 18" & 40" x 26". 550 IHP. 11 K. By Builders.

PR: 1925 London; 1946 Newcastle.

04.1925:	Completed, as TAMESA, for United Steam Tug Co., Gravesend.
06.1937:	Sold to William Watkins Ltd., renamed CERVIA.
1941:	Requisitioned by Royal Navy.
1945:	Returned to owners.
05.1946:	Purchased, for £11,000, by The 'Ridley' Steam Tug Co.Ltd., renamed MONTY.
04.1954:	Sold, for £17,500, to J.H. Pigott & Son Ltd., renamed LADY ELSIE.
09.1962:	Renamed LADY HAZEL.
09.1963:	Sold to Arie Rijsdijk Boss en Zoonen, Hendrik Ido Ambacht for breaking up. Arrived in tow of tug WORKMAN.

Mildred II *after alterations. (Les Turner).*

R 26 **MILDRED II** **(1949-1957)**

ON 115898 66 g. 71.0 x 17.1 x 9.4 ft. Steel Screw
B: Lobnitz & Co.Ltd., Renfrew.
E: C. 2 Cy. 14" & 30" x 22". 250 IHP. 9 K. By Builders.

PR: 1902 London; 1949 Newcastle.

1902:	Completed, as JUNO, for The Union Lighterage Co.Ltd., London.
By 1933:	Sold to Arther Heather, Margate.
By 1948:	Sold to Harrisons (London) Ltd., renamed BIRCHROCK.
01.1949:	Purchased, for £2,000, by The 'Ridley' Steam Tug Co.Ltd., renamed MILDRED II.
03.1955:	Owners restyled Ridley Tugs Ltd.
09.1955:	Laid up. Subsequently offered for sale, for £2,000, October 1956.
06.1957:	Sold to agents H.E. Moss & Co. and thence to C.W. Dorkin Ltd., Gateshead, for breaking up.

Reliant. *(Les Turner).*

R 27 **RELIANT** **(1950-1956)**

ON 124285 156 g. 100.0 x 20.2 x 9.9 ft. Steel Paddle
B: J.T. Eltringham & Co., South Shields. (YN 263).
E: L. 2 Cy. 30" x 54". 400 IHP. 11 K. Hepple & Co.Ltd., South Shields.

PR: 1907 Manchester; 1950 Newcastle.

10.1907:	Completed, for £6,400, as OLD TRAFFORD, for Manchester Ship Canal Co.
09.1950:	Purchased, for £1,650, by The 'Ridley' Steam Tug Co. Ltd.
11.1950:	Renamed RELIANT.
03.1955:	Owners restyled Ridley Tugs Ltd.
09.1956:	Sold, for £4,000, to Seaham Harbour Dock Co.
06.1969:	Sold to National Maritime Museum. Towed to Thames by DHULIA. Arrived 12 June 1969.
08.1969:	Moved to Cory's Barge Works, Charlton, where sectionalised for transfer to and re-assembly in Neptune Hall of NMM for permanent display as RELIANT.

Impetus *showing the rarely photographed Ridley house flag.* *(Company archives/ Turners).*

Momentum. *(Les Turner).*

R 28 **MOMENTUM** **(1954-1955)**
ON 169263 213 g. 110.0 x 23.0 x 10.2 ft. Steel Screw
B: Lübecker Maschinenbau Gesellschaft, Lubeck.
E: T. 3 Cy. 372, 590 & 890 x 540 mm. 57 NHP. By Builder.

PR: 1922 Hamburg; 1927 Stockholm; 1954 Newcastle; 1955 Liverpool.

1922:	Completed, as DIANA, for Bugsier Rederei Bergungs A.G., Hamburg.
1927:	Sold to Trafik A.B. Grängesburg-Oxelösund, Stockholm.
03.1954:	Purchased, for £8,500, by The 'Ridley' Steam Tug Co.Ltd.; renamed MOMENTUM.
01.1955:	Sold, for £21,500, to J.H. Lamey Ltd., Liverpool; renamed MARIE LAMEY.
05.1964:	Sold, for £1,360, to T.W. Ward Ltd., Preston, for breaking up.

R 29 **IMPETUS** **(1954-1969)**
ON 169261 141 g. 97′2″ x 22′5″ x 10′5¾″ Steel Screw Kort rudder
B: Mützelfeldtwerft Gmbh., Cuxhaven. (YN 136).
E:Oil. 4 SA. 6 Cy. 320 x 450 mm. 825 BHP. 13 BP. 12 K. Klöckner-Humboldt-Deutz, Cologne.

PR: 1954 Newcastle; 1969 Lubeck; 1984 Svendborg.

07.1954:	Completed, for £53,240, for The 'Ridley' Steam Tug Co.Ltd.
03.1955:	Owners restyled Ridley Tugs Ltd.
01.1959:	Chartered to Tyne Tugs Ltd.
05.1969:	Sold, for £38,000, to J. Johannsen & Sohn, Lubeck, renamed JOACHIM.
04.1971:	Re-engined. Details as above, but 5 Cy.
1984:	Sold to Niels Hendriksen, Svendborg, renamed STORESUND.
1992:	Still in service.

Maximus. *(John Clarkson, Longton).*

Battleaxe. *(A. Duncan).*

R 30 **MAXIMUS** **(1956-1971)**
ON 186853 141 g. 97'2" x 23'5" x 10'5¼". Steel Screw Kort rudder Bow thruster.
B: Mützelfeldtwerft Gmbh., Cuxhaven. (YN 144)
E: Oil. 4 SA. 6 Cy. 320 x 450 mm. 825 BHP. 13 BP. 12 K. Klöckner-Humboldt-Deutz., Cologne.

PR: 1956 Newcastle.

03.1956:	Completed, for £55,348, for Ridley Tugs Ltd.
01.1959:	Chartered to Tyne Tugs Ltd.
11.1968:	Laid up on River Tyne until March 1971.
05.1971;	Transferred, for £37,500, to Blyth Tug Co.Ltd.
03.1978:	Transferred, for £22,500, to Lawson-Batey Tugs Ltd. Chartered back to Blyth Tug Co.Ltd.
12.1983:	Laid up at Sunderland.
01.1984:	Sold, for £25,000, to 'M' Tugs Ltd., (Maureen Shirley McCann, Manager), Lowestoft, to be renamed A.P. McCANN but retained name MAXIMUS.
03.1985:	Sold to Satim Towage Ltd., Ipswich.
08.1987:	Sold to Paul Magnus Group Ltd.
1992:	Still in service.

R 31 **BATTLEAXE** **(1957-1958)**
ON 129580 163 g. 112'0" x 23'0" x 13'6". Steel Screw
B: A. Stephen & Sons Ltd., Linthouse. (YN 446).
E: C. 2 Cy. 21" & 42" x 27". 820 IHP. By Builder.

PR: 1911 Glasgow; 1957 Newcastle.

1911:	Completed, as CAMPAIGNER, for Steel & Bennie Ltd.
1957:	Renamed BATTLEAXE.
10.1957:	Purchased, for £8,550, by Ridley Tugs Ltd.
12.1958:	Transferred, for £8,000, to Lawson-Batey Tugs Ltd., renamed HILLSIDER.
01.1959:	Chartered to Tyne Tugs Ltd.
10.1960:	Sold, for £1,375, to British Iron & Steel Corporation. Allocated to Hughes Bolckow Ltd. for breaking up.

APPENDIX D TO CHAPTER SEVEN

The Andersen Company Fleet List

AHA 1 **ENABLE** **(1911-1936)**
ON 98277 98 g 76.3 x 19.0 x 9.6 ft. Wood Paddle
B: J.P. Rennoldson & Sons, South Shields.
E: L. 2 Cy. 27" x 45". 50 NHP. By Builder.

PR: 1890 South Shields.

1889:	Completed, for £4,300, as TYNE DOCK No. 4, for North Eastern Railway Co.
03.1911:	Purchased by Axel H. Andersen, North Shields, renamed ENABLE.
1936:	Broken up.

AHA 2 **OVERGARTH** **(1936-1937)**
ON 124069 97 g. 81.6 x 17.6 x 10.0 ft. Steel Screw
B: Lytham Shipbuilding & Engineering Co.Ltd., Lytham. (YN 369)
E: C. 2 Cy. 15" & 32" x 21". 320 IHP. By Builders.

PR: 1907 Liverpool; 1946 Hull.

05.1907:	Completed for Rea Transport Co. Ltd., London.
1924:	Owners restyled Rea Towing Co.Ltd., Liverpool.
10.1936:	Purchased by Axel H. Andersen, North Shields.
11.1937:	Transferred, for £1,400, to The 'Ridley' Steam Tug Co.Ltd.
03.1946:	Sold, for £4,750, to Peter Foster & Co.Ltd., Hull.
04.1961:	Sold, for £900, to British Iron & Steel Corporation. Allocated to C.W.Dorkin & Co.Ltd. Gateshead, for breaking up. Towed to Tyne by HEADMAN.

• CHAPTER EIGHT •

France, Fenwick Tyne & Wear Co. Ltd.

The Evolution of the Company and its First Twenty-one Years 1918-1939

In the latter part of the 19th century two collier-owning companies, Taylor & Sanderson and Fenwick, Stobart & Co.Ltd., were well established in Sunderland. To a certain extent they conducted a joint operation in that they worked together in the chartering and agency of their combined collier fleets and in managing a number of tugs owned by Taylor & Sanderson. Soon the Taylor & Sanderson interest declined and the Fenwick, Stobart company, who already owned William France & Co.Ltd., combined with H.C. Pelly & Co. to form William France, Fenwick & Co.Ltd. in July 1901. A year earlier, the towage part of the business became known as The Sunderland Towage Co.Ltd. with, originally, Fenwick, Stobart, and later William France, Fenwick holding a number of the shares. In 1918 William France, Fenwick bought out a number of the Sunderland Towage shareholders whilst the remainder transferred their shares into a holding in a new company entitled France, Fenwick Tyne & Wear Co.Ltd. which was set up as a subsidiary of William France, Fenwick. As well as operating the fleet of tugs, France, Fenwick Tyne & Wear were also fitters for the Wearmouth Coal Co. (The term 'fitters' meaning that they marketed and sold the production of the colliery as had their forebears since before 1843). In addition to their establishment at Sunderland they had also had for some years a branch office in Newcastle. This now became the head office of France, Fenwick Tyne & Wear, with that at Sunderland being referred to as the branch office. For the remainder of this chapter the reader should note that, for convenience, the France, Fenwick Tyne & Wear Co.Ltd. will be referred to simply as France Fenwick. When William France, Fenwick & Co.Ltd. is mentioned they will be described as William France Fenwick.

The new France Fenwick company was formally registered on 1st March 1918, by which time William France Fenwick had a controlling interest. Lt.Col. Hugh Morton Stobart was appointed chairman of the company. Shortly before he had been awarded the DSO, was soon to be made a CBE and was also chairman of Wearmouth Coal Co. Initially, the directors were John Edward Dawson (managing), Kenneth Raymond Pelly, MC., Frederick William Taylor and Edwin Edwards. The secretary was H.F. Carter. The eight tugs comprising the Sunderland Towage Co. fleet, which were transferred to France Fenwick on 17th May 1918, were in order of acquisition by their former owner:-

Name	Gross Tonnage/Year	Hull construction	Means of propulsion
Roker	(119/1905)	Steel	Paddle
Hendon (1)	(130/1906)	Steel	Paddle
Souter	(123/1910)	Steel	Screw
United	(87/1873)	Wood	Paddle
Sunderland	(172/1914)	Steel	Screw
Harwich	(125/1877)	Wood	Paddle
Marsden (2)	(195/1917)	Steel	Screw
Transit	(169/1884)	Iron	Paddle

J.E. Dawson, managing director of France Fenwick 1924-1938.

At an average age of about nineteen years the fleet was relatively modern by the standards of the time. Three of these tugs were still on hire to the Royal Navy, two of them not returning to their owners' control until 1919 and one not until 1920. Also *Transit* was hired by the Royal Navy in August 1918 for nine months during which time she was renamed *Transfer*. Initially the tugs retained the funnel and flag colours of The Sunderland Towage Co.; viz: a buff funnel with black top below which was the 'armlet' of pale blue and white vertical stripes, whilst the flag had the colours of the 'armlet' only. As will be seen later in this book the design originated from the days in the previous century when the Taylor & Sanderson company, who had used similar colours for their collier fleet, had an interest.

In 1919, *Sunderland, Transit* and *Marsden* (2) were returned from their period of war-time hire to the Royal Navy. *Transit* was, however, soon sold to Steel & Bennie Ltd. by whom she was re-named *Brigadier,* and *United* was sold for further service on the River Tyne. Although already thirty-five years old *Brigadier* was to remain in service on the Clyde for a further seventeen years. At about the same time two paddle tugs were acquired - *Wexford* (141/1888), which had until recently been on hire to the Royal Navy, from Wexford Harbour Commissioners, and *Iron* (118/1890), from the North Eastern Railway Co. at Hartlepool, for £5,025. They both retained their original names, though *Iron* did become *Seaburn* about twelve years later. *Hendon* (1) was released from Royal Navy hire in March 1920, making a fleet of eight tugs - five paddle and three screw-propelled.

In July 1920, France Fenwick took over the Anchor Steam Tug Co.Ltd. of North Shields, thus acquiring a presence on the Tyne in keeping with their company title chosen two years earlier. The tugs absorbed in this amalgamation were:-

Rainbow (2)	(125/1864)	Iron	Paddle
Lord Warden	(118/1868)	Iron	Paddle
Rescue (3)	(142/1869)	Iron	Paddle
Triton	(75/1883)	Iron	Screw

- with an average age of 49 years!

The funnel colours of the Anchor Steam Tug Co.were white with a black top and an anchor superimposed on the upper part of the white area. After the amalgamation France Fenwick cleverly incorporated the Anchor Steam Tug colours into their existing funnel design to form the familiar (until 1986) white funnel with black top below which was the blue and white striped 'armlet' with superimposed anchor. From now onwards France Fenwick were often referred to as 'Anchor Tugs' or 'The Anchor Line of Tugs' and the company did, in fact, use these terms in their advertising.

The recently-acquired *Lord Warden* was soon disposed of to C.W. Robinson on the River Tees, but she did not survive very long, being sent to the breakers in 1924. To replace her, France Fenwick bought the steel screw tug *Criccieth* (102/1905) from Welsh owners in 1920, so that by the end of the year they owned an enhanced fleet of seven paddle and five screw tugs.

In 1921 the screw tug *Marsden* (2) was sold to the British Tanker Co.Ltd. who re-named her *Mumbles,* presumably for operation in the Swansea area. Ten years later she was wrecked at Oxwich Point in Swansea Bay. It is surprising that she, being the youngest in the fleet, should be the one to be sold. No doubt the price was good and that would be all that mattered. In addition, another relatively modern paddle tug, *Hendon* (1), was sold to the Board of Trade. No further history of this tug has been discovered. Even though she was of 130 gross tonnage she does not appear either in Lloyds Register or the Mercantile Navy List. It may have been surprising that *Marsden* (2) and *Hendon* (1) should have been sold, but not when the forty-four-year-old wooden paddle tug *Harwich* was sold to The Coulson Tug Co.Ltd. for further service on the Tyne. This company was one of those subsequently to be absorbed by France Fenwick.

In 1922 the iron screw tug *Triton* had been sold to Cardiff owners. Then in March 1923 the paddle tug *J.C. Stevenson* was purchased from the Tyne Improvement Commissioners and renamed *Dunelm* (1) (165/1883). She was to give France Fenwick a further thirty-three years' service. In October 1923 a new screw tug was delivered for the company from J.P. Rennoldson & Sons Ltd., having received the name *Marsden* (3) (241/1923). Like her two predecessors of that name she was to spend only a short time in service on the north-east coast of England. *Marsden* (3) was followed in March 1924 by a sister vessel, *Hendon* (2) (241/1924), from the same builders, and in this case she stayed with the company for forty-two years. She was christened by Mrs. Frank Dawson. At this time, the screw tug *Sunderland* was sold to the Mersey Docks & Harbour Board, by whom she was re-named *Assistant.*

In 1924 Lt.Col. Frank Dawson, who had joined the company in 1919, became, with his father J.E. Dawson, joint managing director. The latter had recently been appointed chairman of the River Wear Commissioners. A major towage job involving the movement of the new Floating Dock, then the largest in the world, from the Tyne to Southampton took place between 17th and 21st April 1924. This tow was supervised by Robert Redhead & Son Ltd., but the tugs of several companies were involved, two of these being France Fenwick's *Marsden* (3) and *Hendon* (2).

Hendon *(2) with the Southampton Floating Dock in 1924. This illustration provides a good example of the bow decoration commonly applied to tugs at this time.* Hendon *(2) was about one month old at the time. (Harrington collection).*

The Coulson Tug Co.Ltd. was taken over by France Fenwick in 1925 and this resulted in the addition to the fleet of the four tugs *Hector, Corsair* (3), *Flying Scud* and *Panther.* Thus at the end of 1925 the France Fenwick fleet consisted of the following fourteen tugs:-

Roker	(119/1905)	Steel	Paddle
Souter	(123/1910)	Steel	Screw
Wexford	(141/1888)	Iron	Paddle
Iron	(118/1890)	Steel	Paddle
Rainbow (2)	(125/1864)	Iron	Paddle
Rescue (3)	(142/1869)	Iron	Paddle
Criccieth	(102/1905)	Steel	Screw
Dunelm (1)	(165/1883)	Iron	Paddle
Marsden (3)	(241/1923)	Steel	Screw
Hendon (2)	(241/1924)	Steel	Screw
Hector	(101/1882)	Wood	Paddle
Corsair (3)	(153/1892)	Iron	Paddle
Flying Scud	(74/1877)	Iron	Paddle
Panther	(150/1884)	Iron	Screw

- with an average age of about thirty-two years.

However, numbers were soon to be reduced when, in 1926, *Hector* and *Panther* were broken-up, and *Marsden* (3) was sold to Buenos Aires owners with whom she is still listed in Lloyds Register. Enquiries to check whether she is in fact still working unfortunately did not produce a reply. To offset this reduction in numbers the Wijsmuller tug *Cyclop* was purchased from the Netherlands and re-named *Cullercoats* (181/1898). She had been built on the Tyne by Rennoldson who also supplied her 600 IHP compound engine, and proved to be a sound purchase as she was to serve the company for over forty years. Her skipper Hector M. Stewart stayed with her for ten years and both he and the tug will be mentioned again in this story. Just over two years after arriving on the Tyne, *Cullercoats* was involved in a serious accident in Tyne Dock. She was in collision with a sculler boat which was crossing the dock with six men aboard. All six were thrown into the water. Five were rescued but unfortunately John I. Key, aged 20, was drowned. It was alleged that the tug was leaving the dock against the signal light, but it was revealed at the inquest that this was a common practice with all tugs. Also, it transpired that there was a 2-knot speed limit on vessels moving within the dock, yet *Cullercoats* had been travelling much faster than this, and that tug skippers had never been supplied with a copy of the dock regulations. A verdict of accidental death was recorded, with the rider that the regulations be brought to the notice of all craft using the dock.

In 1929 the France Fenwick head office was at Mercantile Chambers, Quayside, Newcastle and the branch office at 21 Bridge Street, Sunderland. In that year, *Rainbow* (2) was broken up at Sunderland after a life of sixty-five years, and was replaced in February 1929 by the newly-built screw tug *Wearmouth* (182/1929) which began work at Sunderland. After about eight years on the Wear she was transferred to the Tyne until later going back to Sunderland for the last two years of her 40-year life.

In 1930, France Fenwick took over yet another local tug company. This time the South Shields-based Brown Tug Co.Ltd. was acquired together with their three iron paddle tugs. Two of these tugs, *Contest* (87/1866) and *Vigilant* (115/1874), were immediately broken-up and did not become part of the France Fenwick fleet. The third tug, *President* (130/1876), did join their fleet and was to be a familiar sight on the Tyne for a further twenty-nine years. Later in the same year the company purchased the Admiralty screw tug *C.III* (148/1899) which had started life as *Toxteth* and owned by The

Wearmouth *on trials in 1929. (Harrington collection).*

Cullercoats *and* Wearmouth *towing* Greenwich *(3,554/1904) into the Tyne c.1930. (Harrington collection).*

Cullercoats *towing* Anglo-Chilean *(6,987/1916) in the late 1920s. (Harrington collection).*

An impressive view of Cleadon *leaving the Tyne at full speed shortly after her acquisition by France Fenwick.* *(Harrington collection).*

Alexandra Towing Co.Ltd. at Liverpool. In 1901, she was sold to the Admiralty where from c.1920 she was also known as *C.III*. When she arrived on the north-east coast photographs taken whilst she was re-fitting show her bearing *C.III* on the bows and *Toxteth* on the stern! However, the problem was soon solved when France Fenwick renamed her *Cleadon* and set her to work at Sunderland, where she was to operate for no less than thirty-eight years. *Rescue* (3) was broken-up in 1932 and in June of the same year *Iron* was renamed *Seaburn* after being in the service of France Fenwick since 1919. Just why she was not renamed when purchased from the North Eastern Railway Co. remains one of those little mysteries of no real consequence which often puzzle researchers. In 1933, or shortly afterwards, *Flying Scud* disappeared from the fleet, but details of her fate have not been discovered.

On 17th July 1933, looking for further acquisitions, France Fenwick made an offer to The 'Ridley' Steam Tug Co.Ltd. to purchase a controlling interest in their business. This approach was declined by Ridley. Only seven days later France Fenwick approached Lawson-Batey Tugs Ltd. and Redhead & Dry's Tugs Ltd. regarding a joint scheme between the three companies to purchase Joseph Crosthwaite (Tugs) Ltd., who operated on the Tyne with a base at North Shields. The Crosthwaite capital was £17,000 and France Fenwick already had an option to purchase at 10/- (50p.) per share. On 31st July the deal was settled at that price, with each company taking a one-third share in the Crosthwaite company, which continued to operate under its own name, with three tugs. Two France Fenwick directors, J.E. Dawson and F. Dawson, became chairman and secretary of Joseph Crosthwaite.

In June 1934 a sister tug to *Cleadon* was purchased from The Alexandra Towing Co.Ltd., Liverpool. She was *Nelson* and had remained with Alexandra for the thirty-five years since being built by John Jones & Sons, Liverpool. She was renamed *Fulwell* (149/1899) and was to give a further twenty-seven years service for France Fenwick before being broken-up in 1961.

Aerial view of the River Wear at Sunderland in 1930. The tier of tugs at centre right belong to the France Fenwick, Cook and Irwing companies. The down-stream side of the river is to the top of the picture and north is to the left. *(Harrington collection).*

Cleadon *towing the three-masted barque* Killoran *(1,817/1900) out of the River Wear in 1932.* Killoran *was owned by the well-known Gustaf Erikson of Mariehamn. Her hull appears to be in need of a coat of paint.* *(Harrington collection).*

Wearmouth *and* Corsair *(3) undocking the Eagle Oil tanker* San Gaspar *(12.910/1921) in the 1930s. (Harrington collection).*

Wearmouth *with a foy boat in tow in the early 1930s. (Harrington collection).*

Another take-over by the company took place in July 1934. This time it was the Irwing Steam Tug Co.Ltd. of Sunderland which was absorbed. They had, at that time, three paddle tugs, none of which appear to have joined the France Fenwick fleet - or if they did so it must have been for a very short period of time. *Conqueror* (1) (144/1861) and *Anthony Irwing* (119/1887), both ex-North Eastern Railway Co. tugs from West Hartlepool, were broken-up in 1934-35 and *King Edward VII* (138/1901) was sold to The 'Ridley' Steam Tug Co.Ltd. for further service. Thus, without acquiring further tugs, the company had no doubt gained the extra business at Sunderland which had previously been carried out by the Irwing Company.

By mid-1934 the company had the following fleet of thirteen tugs which was to remain unchanged for the next ten years:-

*	*Roker*	(119/1905)	Steel	Paddle
	Souter	(123/1910)	Steel	Screw
*	*Wexford*	(141/1888)	Iron	Paddle
*	*Seaburn*	(118/1890)	Steel	Paddle
	Criccieth	(102/1905)	Steel	Screw
	Dunelm (1)	(165/1883)	Iron	Paddle
	Hendon (2)	(241/1924)	Steel	Screw
	Corsair (3)	(153/1892)	Iron	Paddle
	Cullercoats	(181/1898)	Steel	Screw
*	*Wearmouth*	(182/1929)	Steel	Screw
*	*President*	(130/1876)	Iron	Paddle
*	*Cleadon*	(148/1899)	Steel	Screw
*	*Fulwell*	(149/1899)	Steel	Screw

* these tugs worked at Sunderland, the remainder on the Tyne.

Since 1925, the proportion of paddle tugs had reduced from 64% to 46% of the fleet. Over the same period the average age of the tugs had increased from thirty-two to thirty-four years and of course would go on increasing for a long time yet, due to the unchanged formation of the fleet.

To alleviate the severe unemployment being suffered in Jarrow and elsewhere the demolition of the Cunard-White Star liner *Olympic* (46,439/1911), the sister of the ill-fated *Titanic*, was allocated to the River Tyne. She arrived in the river on 13th October 1935 and France Fenwick's tugs *Hendon* (2) and *Wearmouth* were involved in her towage to Jarrow along with tugs of Lawson-Batey and Redhead & Dry's. The *Olympic* was a large vessel, 852 x 92 feet, and made an impressive, though sad, sight on that sombre October day. Since her arrival, in 1929, *Wearmouth* had usually operated at Sunderland, but soon after the assistance to *Olympic* she was transferred to the Tyne.

Hendon *(2), now with taller masts for radio aerials, towing the Norwegian tanker* Thorshavn *(6,869/1930) in the 1930s. (Harrington collection).*

Cleadon *towing the Royal Mail liner* Lochkatrine *(9,419/1922) in the 1930s. (Harrington collection).*

Wearmouth *and* George V *(2) assisting the White Star liner* Olympic *when she arrived in the Tyne on 13th October 1935 for dismantling at Jarrow. (Harrington collection).*

President *assisting* Fernmoor *(4,972/1936) following her launch at Sunderland. (Harrington collection).*

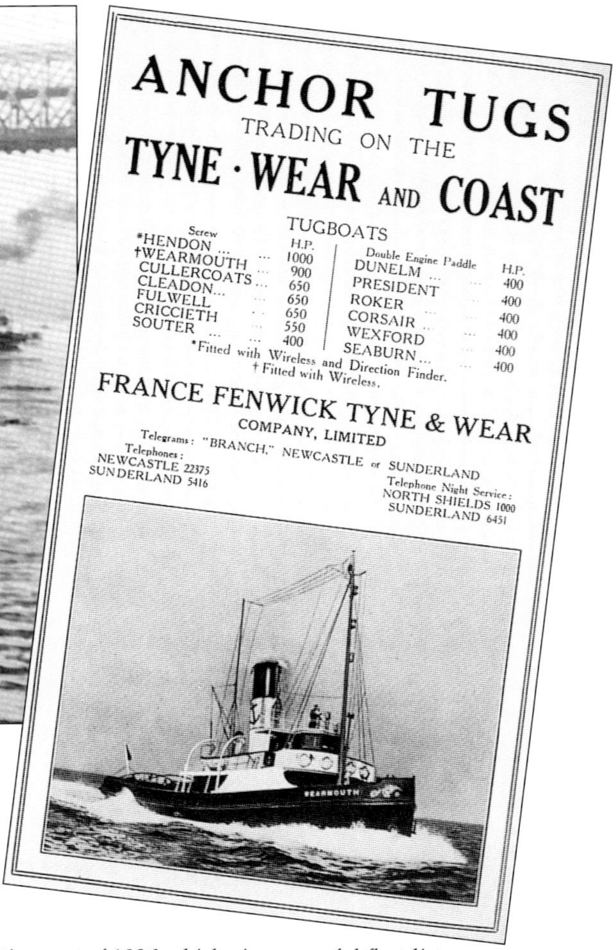

A company advertisement of 1936 which gives a useful fleet list.

Dunelm *about to make fast to the newly-launched* Sutherland *(5,083/1938) at South Shields.*
(Local History Library, South Shields Central Library).

In 1936, Hector Stewart, the skipper of *Cullercoats* since her purchase, was transferred to the command of *Hendon* (2), where he remained for no less than seventeen years. By 1937 the Board of Directors of France Fenwick included:-

Hugh Morton Stobart	(Chairman)
John Edward Dawson	} Joint managing directors
Frank Dawson	
Edwin Edwards	
Frederick William Taylor	
Kenneth Raymond Pelly	
John Leslie	

The marine superintendent was G.W. Blakey and the principal shareholders in the company were:-

William France, Fenwick & Co.Ltd.	£45,437	J.E. Dawson	£14,207
F.W. Taylor	£ 4,168	Mrs. E. Dawson	£ 3,579
F. Dawson	£ 3,291	R.L. Stobart	£ 2,223
H.M. Stobart	£ 2,223	E. Edwards	£ 2,015

The total paid-up capital being £100,000 in £1 shares. In the following year F. Dawson became sole managing director.

In September 1937 the hulk of the liner *Olympic* was towed away from Jarrow to Inverkeithing where she could be beached for the completion of her demolition. She was replaced in December of the following year by the Cunard-White Star liner *Berengaria* (52,101/1913), 883 x 98 feet, which had been damaged by fire. Again tugs of three companies shepherded her up-river to Jarrow, including France Fenwick's *Hendon* (2) and *Wearmouth.* In July 1939 a contract was obtained from Trinity House to lay buoys between the Tyne and Holy Island. For this job *Dunelm* (1) was fitted with a derrick on her foremast.

The Cunard liner Berengaria *arriving in the Tyne on 18th December 1938 for demolition. She is being assisted by the tugs* Wearmouth *and* Homer.
(Local History Library, South Shields Central Library).

Shortly before the Second World War began in September 1939 Edwin Edwards, a director of the company, died and Andrew Mylne Hogg Wardlaw, who had joined the company as office boy in 1923 was appointed secretary. Hector M. Stewart, skipper of *Hendon* (2) was made fleet commodore. Large liners were still being built on the Tyne and his tug was one of those which towed the newly-built Shaw Saville & Albion *Dominion Monarch* (26,462/1939) from the Swan Hunter yard at Wallsend to sea.

Through the Second World War. 1939 - 1945

As described in the Lawson-Batey chapter, the tugs had to contend with great difficulties and dangers in operating under war-time conditions. An example of the problems experienced at night is the occasion when a tug collided with the closed Newcastle Swing Bridge, believing it to be open to river traffic. The result was the loss of the mast and damage to the funnel, but fortunately there were no casualties.

The war was only three months old when the first of many major salvage operations took place off the north-east coast of England. On 6th December 1939 *Eskdene* (3,829/1934), carrying a cargo of timber, was reported damaged by mine or torpedo but still afloat about 70 miles north-east of the Tyne entrance. France Fenwick's *Hendon* (2) and Redhead & Dry's *George V* (2) left the Tyne in mid-afternoon to search for her. Both tugs carried a crew of eight and were escorted by the sloop HMS *Stork*. After steaming an indirect route to avoid the minefields, *Eskdene* was sighted at 7am. the following day. She was a sorry sight, her stern was under water and she had a list to starboard. It appeared that it was only the deck cargo which was keeping her afloat and that she had been abandoned. Three men from the *Hendon* (2), using the tug's boat, managed to board the *Eskdene* and they were able to rig tow lines to the two tugs. Towing began at 9.30am. with their escort being a Royal Air Force plane, since HMS *Stork* had left them shortly before. The job, difficult enough in the first place, was made worse when the wind rose to gale force, causing the tow to roll heavily, and reducing the speed to about 1.5 knots. As darkness fell their escort left and, because of the weather, *Eskdene* could not be seen from the tugs. Just before midnight, the captain of *Hendon* (2), Hector Stewart, was able to get a radio bearing which, to his horror, indicated that they were being driven ashore. Fortunately, they were able to change course to avoid this happening. By 7am. on 8th December the gale subsided somewhat and better progress was made towards the Tyne, only to be hampered again by rain and fog. Just as the vessels were nearing the Tyne an explosion occurred on the *Eskdene* and she began to settle even further in the water, giving rise to concern for the four men who had been put on board to handle the hawsers as they approached the river entrance. The only way to keep the ship afloat was to tow at maximum speed which enabled them to pass between the piers at 11.30am. and to beach the vessel on the Herd Sand, South Shields by 3pm. *Eskdene's* cargo was subsequently unloaded into barges and the ship was refloated and repaired. Thus, forty-eight hours of hard and dangerous work had saved a valuable ship and her cargo.

In February 1940, the Danish vessel *Prins Knud* (1,340/1922) was driven ashore on Goswick Sands near Holy Island on the Northumberland coast. She was believed to be a total loss but the Danish government thought otherwise and requested France Fenwick to try to salvage her. *Hendon* (2) was sent to make the attempt and after three visits to the area and forty-two days' work *Prins Knud* was refloated. The tugs *Wearmouth* and *Corsair* (3) were also involved in this success. After her salvage the Danish vessel was placed under the management of France Fenwick, so the effort was well worthwhile. The subsequent salvage award was £6,675. A few days after *Hendon* (2) was freed from this job, she was hired by the Royal Navy, for £168-15-0 (£168.75) per month, and sent to work at Scapa Flow for just over two years, thus effectively reducing the France Fenwick fleet on the Tyne and Wear to twelve tugs.

In October 1940 the destroyers HMS *Fame* and HMS *Ashanti* ran aground at high speed on to the beach at Whitburn, about two miles north of the River Wear entrance. Incredibly they ended up literally laid alongside each other with parts of their hulls touching. Five of the Wear-based tugs of the company were involved in refloating and towing the warships to the repair yards.

In December 1941 the management of one of the standard war-time-built tugs, *Empire Plane* (129/1941), was entrusted to France Fenwick for a period of about two years. *Hendon* (2) returned from her sojourn at Scapa Flow on completion of her hire to the Royal Navy in July 1942 and was soon back into her usual role of assisting ships in trouble near her home base on the Tyne. On 13th November 1942

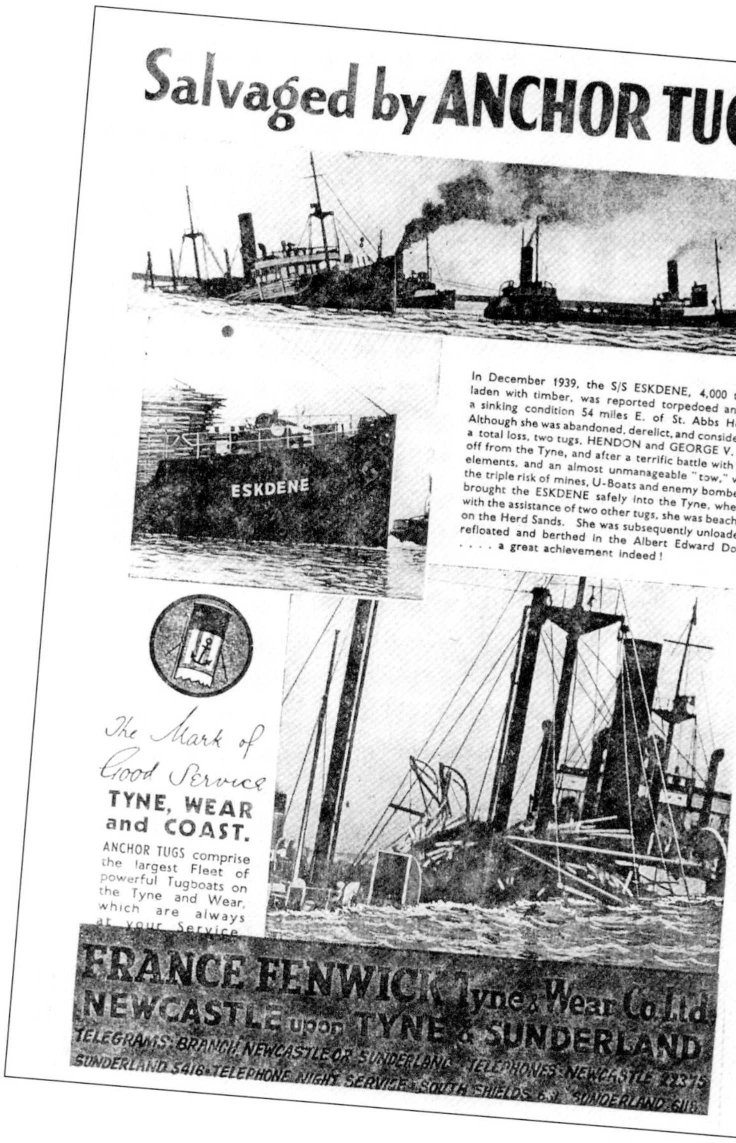

A company advertisement of 1946 describing the salvage of Eskdene *in the early days of the Second World War.*

Hendon (2) and *Wearmouth* were called to the tanker *Esturia* (6,968/1914) at 7am., on hearing the sounding of the tanker's siren. They found her seriously on fire towards the stern with a possibility that her ammunition would explode. As well as directing their hoses on to the fire along with several other tugs in attendance, *Hendon* (2), assisted by *Cullercoats*, found it necessary to move the Dutch tanker *Keilhaven* (2,968/1918) to a place of safety away from the heat of the fire. After 11am., when the fire was almost out, the tugs' crews helped in moving the hot and scorched boxes of ammunition to safety.

Although none of the company's tugs was lost due to enemy action during the Second World War there were two occasions at Sunderland in 1943 when tugs were damaged due to enemy bombing. These raids resulted in one fire-watcher being killed and one injured.

Early in the morning of 8th July 1943 *Hendon* (2) and *Wearmouth* were instructed to proceed to the assistance of *Framlington Court* (4,888/1924) which had been in collision and was holed in the engine room. The casualty had a cargo of iron ore and was already under tow by Redhead & Dry's *George V* (2) and the Dutch tug *Amsterdam* (368/1937). The tugs had slipped the tow and an attempt had been made to beach the cargo vessel, but without success. Her rudder was jammed and she was very low in the water, making her almost impossible to tow with only two tugs. However, by brute force, *Hendon* (2), *Wearmouth* and *George V* (2) were able to get her underway again and bring her into the Tyne, where they beached her on the Herd Sand at South Shields. The *Framlington Court* was later towed up river to be unloaded. It appears that the Herd Sand was a favourite place for beaching damaged ships temporarily until they were made fit for a later journey up river. Of course, the opposite side of the river at Tynemouth was no place for beaching a vessel for there lay the notorious Black Middens rocks.

The management of *Empire Plane* for the Ministry of War Transport (MOWT), ceased but was very soon replaced, in August 1943, by the allocation of the larger *Empire Sara* (276/1943) newly-built by Cochrane & Sons Ltd. at Selby.

France Fenwick had, since 1942, been discussing with other Tyne tug companies the possibility of unification of the towage businesses on the Tyne. There were at that time, in addition to themselves, the Lawson-Batey, Ridley, Redhead & Dry's and Joseph Crosthwaite companies. It will be remembered that, in 1933, the latter company had been bought jointly by Lawson-Batey, Redhead & Dry's and France Fenwick. It thus came about that France Fenwick purchased Redhead & Dry's Tugs Ltd. in May 1944 and this transaction, of course, included Redhead & Dry's one-third share in Joseph Crosthwaite. One half of this one-third holding was sold to Lawson-Batey so that they, and France Fenwick, became equal joint owners of the Crosthwaite company, which continued to trade under its own name. The Redhead & Dry's tugs which were transferred to France Fenwick ownership as a result of the amalgamation were:-

Conqueror (2)	(177/1884)	Iron	Paddle
George V (2)	(224/1915)	Steel	Screw
Washington	(122/1870)	Iron	Paddle
Malta	(113/1875)	Iron	Paddle
Great Emperor (2)	(171/1909)	Steel	Screw
Robert Redhead	(178/1900)	Steel	Screw

These tugs, together with the thirteen owned since mid-1934, made a company fleet of nineteen, nine paddlers and ten screw-propelled - much the same proportion as ten years earlier. The average age of the fleet was just over forty-six years!

As well as the transfer of the owned-tugs from Redhead & Dry's there was also the transfer of the management of *Empire Mustang* (242/1943). Alexander Patton, who had been towage manager of Redhead & Dry's, became manager of the France Fenwick tugs operating on the Tyne, and Robert Dryden managed the fleet of tugs at Sunderland. Further changes in 1944 were that the MOWT tugs *Empire Andrew* (138/1944) and *Empire Jane* (232/1944) were placed under the company's control, whilst the management of *Empire Sara* was terminated.

When the invasion of Normandy began in June 1944, and shortly afterwards, when a great many tugs were involved in the assembly of the multitude of components required for the building of the 'Mulberry' harbour at Arromanches, France Fenwick's managed tugs were among those present in this operation. From mid-1943 there is photographic evidence that at least *Hendon* (2), *George V* (2), three Lawson-Batey tugs *Plover*, *Langton* and *Francis Batey*, R.L. Cook's *Stag* (5) and Crosthwaite's *Lewis Crosthwaite*, carried a letter and number painted on their bows. These were, respectively, T.1485, T.741, T.393, T.391, T.1519, T.1233 and T.788. Only one of these six was

Hendon *(2) showing the mysterious letter and number which was applied to several of the Tyne tugs during the latter part of the Second World War.* *(Harrrington collection).*

hired by the Royal Navy so naval identification does not seem to have been the reason. Unless these were means of identification during the period of invasion of France when large numbers of tugs were in one concentrated area, so far no reason for the addition of these codes has been discovered. One of the known photographs of these tugs at this time has appeared in a publication with the codes deleted, presumably for security reasons - yet another case of a minor

Wearmouth *handling the new destroyer HMS* Trafalgar. *(Company archives).*

maritime mystery. During the latter part of the war *Stag* (5) (148/1897), belonging to R.L. Cook Towage Co.Ltd. of Sunderland, was working on the Tyne on charter to France Fenwick and *Corsair* (3) was away on charter to Tees Towing Co.Ltd.

In 1945 the Board of Directors consisted of:-

Hugh Morton Stobart,	Chairman
Frank Dawson,	Managing Director
Frederick William Taylor	
Kenneth Raymond Pelly	
John Leslie	
H.F. Charlton	

A.M.H. Wardlaw, company secretary since 1939, and G.W. Blakey, marine superintendent, continued in these posts. The former was, however, appointed to the board in the following year.

The management of the tugs *Empire Andrew* and *Empire Jane* was transferred elsewhere in mid-1945, but in September of that year one of the tugs taken as war reparations from Germany was placed under the company's management. She was *Expatriate* (134/1925), ex *Eversand* of Unterweser Reederei A.G. She did not stay with France Fenwick very long, being allocated to the United States in February 1946. At Sunderland the tugs of the Lambton, Hetton & Joicey Collieries Ltd. had for many years serviced the colliers which came to load coal at their staiths on the River Wear. In November 1945 their towage business, with three tugs, was taken over by the France Fenwick company. The tugs were:-

Houghton	(133/1904)	Iron	Paddle
Lumley	(135/1907)	Steel	Paddle
Eppleton Hall	(166/1914)	Steel	Paddle

These additions to the fleet brought the number of tugs owned by France Fenwick on the Tyne and Wear rivers up to twenty-two. Throughout the period of the Second World War they had suffered no losses due to enemy action and by amalgamations had increased the strength of the fleet by almost 70% and so owned by far the largest fleet of tugs on the north-east coast of England. Nevertheless, the fleet was to increase in numbers still further in the post-war years ahead.

Post-War Consolidation and Experience with the Company's first Diesel Tugs. 1946 - 1954

As will have been seen in the previous section, France Fenwick came through the Second World War in a very strong position, owning more tugs than all their competitors in the area put together. Surprisingly, as will be seen later, this did not mean that in the long-term they were to survive as a viable company.

When France Fenwick had absorbed Redhead & Dry's in 1944 they had by this acquisition become owners of the important contract to handle all Royal Navy and Royal Fleet Auxiliary vessels which required assistance on the River Tyne. Even after the war this resulted in a great deal of work for the company. Typical examples were the handling of the

Wearmouth *and* Robert Redhead *towing the new RFA* Olna *(12,695/1945) from her builders in May 1945.* *(Company archives).*

Hendon *(2) and* George V *(2) handling the Landing Ship (Gantry) HMS* Derwentdale *(16,750 disp./1941) on 14th February 1945.*
(Company archives).

R.L. Cook's Stag *(5) on charter to France Fenwick and* Hendon *(2) holding on to the aft end of the stern half of* Harpagus *(7,271/1942) as she is towed past Newcastle Quay downstream to Wallsend where a new bow section was fitted. The photograph is dated 22nd April 1945.* *(Company archives).*

Wearmouth, Hendon *(2) and* George V *(2) towing the aircraft carrier HMS* Hercules *on 18th July 1946.* Empire Mustang *is in attendnace to the left.* *(Company archives).*

incomplete Majestic Class light aircraft carriers HMS *Leviathan* and HMS *Hercules* when leaving their builders Swan Hunter and Vickers respectively.

The small ship-repair yard and slipway of R.B. Harrison & Son Ltd. at Bill Quay on the south bank of the Tyne was taken over in May 1946. The yard was in a very run-down state and had to be refurbished before being brought back into operation. It was, however, to be a valuable future asset for the repair of the company's and other tugs and various small craft which used the river. The yard manager was Sidney Armstrong who joined R.B. Harrison when the site was overgrown with grass and weeds. With a staff of seven men and three boys he restored the yard to use. The tug *Criccieth* was the first vessel to be slipped for repairs after the restoration of the yard.

Another large tow which took place in August 1946 involved towing the stern section of the cut-down hulk of the liner *Berengaria,* which, having been taken to Jarrow for breaking up in 1938, was transferred to Rosyth so that the scrapping could be completed. Five France Fenwick tugs were allotted to tow this unwieldy hulk when it (it doesn't deserve to be referred to as she!) left the Tyne. The management of *Empire Mustang,* which the company had carried out since taking over Redhead & Dry's in 1944, ceased in 1947 when she was sold by the Ministry of Transport to Tees Towing Co.Ltd. at Middlesbrough. By May 1947 all of the company's tugs had been fitted with radio telephone equipment with a range of 15 miles thus allowing much improved control of towage operations.

During 1947 Lt.Col. H.M. Stobart, CBE, DSO, resigned as company chairman but remained on the Board of

George V *(2) towing the incomplete aricraft maintenance ship HMS* Perseus *(ex* Edgar*) on 11th May 1945.* *(Company archives).*

Criccieth *ready to be hauled on to R.B. Harrison's slip on 27th August 1946.*
(Company archives).

A general view of R.B. Harrison's ship-repair yard at Bill Quay. (Company archives).

Company advertisement July 1946.

Wexford, Souter, Corsair *(3) and* Cleadon *docking* Lottinge *(2,468/1918) at Sunderland c.1946.*
(Company archives).

Great Emperor (2) towing Stanfirth (7,285/1944) in 1948. Note the two foyboats being towed by the tug. *(Company archives).*

Directors. His place was taken by K.R. Pelly, MC. In November 1947 the R.L. Cook Towage Co.Ltd. of Sunderland was acquired. Cook's two tugs, the screw *Stag* (5) (148/1897) and the paddler *Snowdon* (106/1885), joined the France Fenwick fleet. It will be remembered that, from 1944, Joseph Crosthwaite (Tugs) Ltd. had been jointly owned by France Fenwick and Lawson-Batey. In December 1949 it was decided that the Crosthwaite company should be liquidated and that one if its two tugs should be transferred to each of the owning companies. So, *Lewis Crosthwaite* (234/1912) joined the France Fenwick fleet and was renamed *Earsdon*. How the two companies decided who should get which tug has not been discovered. The Crosthwaite contracts were divided equally by mutual agreement.

Earlier in 1949, the company had purchased the screw tug *Queensgarth* from R. & J.H. Rea Ltd. She was one of the many war-time tugs built for the Ministry of War Transport, had originally been named *Empire Paul* and was soon to be renamed *Beamish* (242/1944). She was powered by a triple-expansion steam engine.

In October 1949 another partly-completed light aircraft carrier was due to leave the Tyne. This time it was HMS *Albion*; laid-up since her launch at Swan Hunter's shipyard in 1947, she was being taken to Rosyth by *Hendon* (2), *George V* (2) and *Beamish*. However, this turned out to be a difficult and hazardous contract, for, soon after leaving the Tyne, the tug crews found themselves experiencing gale conditions. Whilst off the Northumberland coast, with *Albion* yawing severely, the aircraft carrier came into collision with the collier *Maystone* (2,025/1945), which later sank with severe loss of life. *Albion* was holed towards her stern and appeared to be in danger of foundering. The tugs, therefore, attempted to

Company advertisement showing George V (2) towing HMS Belfast out of the Tyne in April 1945.

The cut-down stern section of the Cunard liner Berengaria *being towed away to Rosyth on 26th August 1946 for completion of her demolition.* Great Emperor *(2) and* Cullercoats *are nearest to the camera whilst the head tugs are* Empire Mustang, Hendon *(2) and* George V *(2). (Company archives).*

tow *Albion* into shallow water near St.Abbs Head as a precaution in case she did sink. In so doing *Hendon* (2) was put out of action when a wire rope became entangled around her propeller so that she was in danger of being driven ashore. *George V* (2) and *Beamish* found that, on their own, they could do no more than heave-to. By this time calls to Rosyth had alerted the Royal Navy and a large naval tug and a destroyer had appeared on the scene. The tug took hold of *Albion* and the destroyer took *Hendon* (2) in tow whilst her propeller was freed. The aircraft carrier reached Rosyth where she was repaired but it was not long before she was back on the Tyne where she was completed in 1954, ten years after being laid-down. Law court proceedings regarding this episode were not settled until July 1953.

A more successful venture involved *Souter* which towed the coaster *Crescence* (255/1936) from Seaham to Sunderland, for which a salvage award of £750 was received some time later.

In 1949, Col. Frank Dawson was honoured by being appointed chairman of the British Tugowners' Association for two years. The year 1950 was to prove a significant period for France Fenwick in many ways, both good and bad. Frank Dawson, the managing director, retired and he was succeeded by A.M.H. Wardlaw who had for a while held that post jointly with his predecessor. The new managing director had been appointed Secretary in 1939 and had become a director in 1946. He thus had already twenty-seven years' service with the company.

On 11th February 1950 the screw tug *Stag* (5) was assisting with the tow of the newly-built, but dead-ship, *City of Manchester* (7,585/1950) from the River Tyne to her builders at Sunderland after having turbines fitted at Wallsend. A strong westerly gale was blowing so that the tow was yawing from side to side and becoming very difficult to manage. When off Souter Point, almost half-way through the short sea passage, *Stag* (5) was pulled over by the tow and sank. Sadly, the mate William Brown, the engineman Alex Mollinson, the deckhand W. Noble and the boy R. Wigham were lost, only the skipper, Ernest Theodore Cook Baister, and the fireman, George Brown, being saved. Exactly one year later, a memorial service was held on board *Eppleton Hall* over the site of the tragedy.

Hendon *(2) passing the Bergen Line cruise vessel* Stella Polaris *(5,209/1927) on 7th July 1948.*
(Photo by Lawrence Dunn).

After the Second World War it was to be expected that a great deal of surplus equipment would come on the market and that some of it would either be new or little used. In 1950 a number of ex-United States Army diesel tugs were for sale, and, after a lot of thought, France Fenwick decided that there would be a lot of sense in introducing diesel propulsion to their fleet. This was a bold step, for it should be remembered that there were very few diesel ship-handling tugs in operation in the United Kingdom at that time. On 3rd October *ST.764* and *ST.761* were purchased, and, after modifications at R.B. Harrison's yard were renamed *Grangetown* (176/1943) and *Ryhope* (176/1943) respectively. Both were to give twenty years very useful service to the company. They each had eight-cylinder General Motors Corporation engines producing 700 BHP and were capable of 14 knots. Being of American design there were, of course, some features which did not suit British practice. One major change made to them was that the long deck housing was shortened aft so that the towing hook could be moved forward. When these tugs were purchased, five twelve-cylinder General Motors' engines of the same design as in *Grangetown* and *Ryhope* were also obtained. This purchase later proved invaluable when five newly-built tugs were to incorporate these engines, so that for a period of twelve years all the company's tugs had standardised engines with the same cylinder dimensions of 8.5" diameter x 10" stroke.

With two additional tugs available it was not surprising that *Snowdon* was sent to the breakers during 1950. Nevertheless, at the end of the year the company fleet was at its maximum strength of twenty-six tugs, thereafter to decline steadily in numbers, if not in effectiveness. The fleet consisted of the following tugs, with those working at Sunderland being marked with an asterisk:-

*	*Roker*	(119/1905)	Steel	Paddle
*	*Souter*	(123/1910)	Steel	Screw
*	*Wexford*	(141/1888)	Iron	Paddle
*	*Seaburn*	(118/1890)	Steel	Paddle
	Criccieth	(102/1905)	Steel	Screw
	Dunelm (1)	(165/1883)	Iron	Paddle
	Hendon (2)	(241/1924)	Steel	Screw
*	*Corsair* (3)	(153/1892)	Iron	Paddle
	Cullercoats	(181/1898)	Steel	Screw
	Wearmouth	(182/1929)	Steel	Screw
*	*President*	(130/1876)	Iron	Paddle
*	*Cleadon*	(148/1899)	Steel	Screw
*	*Fulwell*	(149/1899)	Steel	Screw
	Conqueror (2)	(177/1884)	Iron	Paddle
	George V (2)	(224/1915)	Steel	Screw
	Washington	(122/1870)	Iron	Paddle
	Malta	(113/1875)	Iron	Paddle
	Great Emperor (2)	(171/1909)	Steel	Screw
	Robert Redhead	(178/1900)	Steel	Screw
*	*Houghton*	(133/1904)	Iron	Paddle
*	*Lumley*	(135/1907)	Steel	Paddle
*	*Eppleton Hall*	(166/1914)	Steel	Paddle
	Earsdon	(234/1912)	Steel	Screw
	Beamish	(242/1944)	Steel	Screw
*	*Grangetown*	(176/1943)	Steel	Screw
*	*Ryhope*	(176/1943)	Steel	Screw

- with an average age of over forty-five years and 46% being paddle tugs - the same proportion as in 1934. The great age of *Washington* at eighty years should be noted; would any of the 'new-fangled' diesel-propelled tugs last anywhere near so long?

Conqueror *(2),* Corsair *(3) and* Lumley *c.1950. (Les Turner).*

Empire Paul *in 1946. She became France Fenwick's* Beamish *in April 1949.*

Hendon (2) (Maritime Photo Library)

Great Emperor *(2)* *(Les Turner)*

owning company with the long name is traditionally referred to as 'Red Funnel Steamers'!

In April 1952, two of the old paddle tugs, *Washington* and *Seaburn* were sold for breaking up on the Tyne at Gateshead and Dunston respectively. Not all salvage operations were dramatically successful, as was illustrated when *Hendon* (2), with Hector Stewart in command, towed the badly listing German *Hans Hoth* (374/1940) for forty-five miles only to have her sink within fifteen miles of the Tyne. Fortunately *Hendon* (2) was able to rescue the crew.

In 1953, the coronation of Her Majesty Queen Elizabeth II took place, and, during June of that year, as is traditional, there was a Review of the Fleet by Her Majesty. This event took place at Spithead, and although primarily for naval vessels, had usually included a number of representative mercantile ships. France Fenwick were honoured to be asked to provide a tug at the review to represent the British towage industry. The tug chosen was *Beamish,* which was now commanded by the fleet commodore Hector Stewart, following his 17-year sojourn in *Hendon* (2). *Beamish* was taken in hand and, it appears, she was painted from bow to stern and keel to truck by courtesy of British Paints Ltd. Few tugs before had looked so spick and span. Not only the tug but also the crew were tidied up! Her skipper wore full uniform and the crew were kitted out in spotless white overalls and all wore their medal

So successful were the ex-U.S.Army tugs that another identical one, *ST 767*, was purchased in April 1951, and was towed from Antwerp to the United Kingdom. After undergoing the same modifications as her sisters, she was renamed *Cornhill* (176/1943) and thereafter was to be seen around the north-east coast for the next twenty-nine years. *Malta* was sold to J.A. White in March 1951 for breaking up at North Queensferry after seventy-six years of active towing on the Thames and Tyne rivers. In August *Beamish* went away on charter for about three months to the Southampton, Isle of Wight and South of England Royal Mail Steam Packet Co.Ltd. at Southampton. Fortunately, this ferry- and tug-

Criccieth *and* Earsdon *(Company archives).*

Robert Redhead *with* City of Port Elizabeth *(13,363/1952) in tow.* *(Company archives).*

ribbons as appropriate. *Beamish* arrived at Spithead on 14th June 1953 in time for the review on the following day and was congratulated on her smart appearance by the Commander-in-Chief in HMS *Vanguard.* She left Spithead on 16th and was back in the Tyne the following day, no doubt all aboard being very pleased with themselves. Soon after her return *Beamish* is reported as having refloated *Westwood* (1,040/1935) at Walker and extinguished a fire in the Finnish vessel *Hildegaard* (2,468/1906) in Albert Edward Dock.

How rarely one sees a tug crew looking like this! Captain Hector Stewart and his crew are as smart as their tug prior to departure to the 1953 Coronation Review. (Company archives).

In 1953 a phased order was placed with P.K. Harris & Sons Ltd., Appledore, for no less than five new diesel-powered tugs. These were to utilise the five 12-cylinder General Motors' engines which had been purchased earlier. Consideration was also being given to converting *Beamish* to diesel propulsion but, in the event, this did not take place for another ten years! It was, in fact, *George V* (2) which was to be converted, her engine and boiler being in a very poor state. In February 1954 Hector Stewart transferred from *Beamish* to *George V* (2) so that he could take the latter to P.K. Harris & Sons Ltd., Appledore for the work to be carried out. Her previous skipper, Robert Hunter, and all the crew except the engineer, F. Morris, had declined to sail her that distance because of her condition. Perhaps they were the wise ones. Hector Stewart, always one for a challenge, was quite happy to make the voyage, but it turned out to be one which he never forgot. He picked a stand-in crew, none of whom gave much for their chances of reaching Appledore, and left the Tyne on 27th February. They limped their way around the east and south coasts of England, anchoring here and there to repair the many defects which arose. Lands End was rounded on 3rd March with a full gale blowing

Beamish *all spick and span and on her way to the Coronation Review at Spithead, June 1953.* *(Company archives).*

Hendon *(2) and* Great Emperor *(2), complete with smoke-screen, towing* City of York *on 26th October 1953.* *(Company archives).*

On the bridge of George V *(2) on her return to the Tyne are, from left to right: A.M.H. Wardlaw, the company's managing director, Mrs. Wardlaw, Mrs. Stewart and Hector Stewart, skipper of the tug. (Company archives).*

from astern and the tug's pumps unable to cope with the amount of water coming aboard. On top of this the fireman was exhausted by his attempts to keep up steam from a severely leaking boiler. After a long struggle they were able to round Hartland Point and drop anchor off Clovelly in the shelter of Barnstaple Bay. In the distance, to the north-east, huge seas could be seen breaking over the Barnstaple bar which marked the entrance to Appledore. The local pilots advised against entering in the prevailing conditions, but *George V* (2) was by now in such a poor mechanical state that it was imperative that they made the attempt on the next high tide. Against all the odds they made it to the Harris shipyard at 4am. on 4th March where they berthed her high and dry on a flat sandy bottom. There they left the ailing *George V* (2) and the shipbuilders began work, virtually having to strip out everything inside her. After a few days' rest Hector Stewart was back in command of *Beamish* by 8th March 1954. However, in August, he was again at Appledore to oversee the final work on the conversion and to take *George V* (2) on trials, on 27th and 28th August, in Barnstaple Bay. All tests being successful she returned on 31st August and was put on show at Sunderland, and later at Newcastle Quayside. Her new engine was a General Motors 12-cylinder diesel giving an output of 1086 BHP as compared with the 850 IHP of her original

George V *(2) arriving at Newcastle Quay in early September 1954 after her conversion. The bridges over the Tyne show up well in this photograph.* *(Company archives).*

triple-expansion steam engine. The France Fenwick marine superintendent responsible for the conversion was A.A. Boyd.

No sooner had *George V* (2) arrived in home waters than a dispute arose regarding the manning of diesel tugs. The management claimed that a lower manning level was desirable for diesel tugs, but, quite naturally, the crews did not agree since this was likely to put their jobs at risk. The outcome of this dispute, which affected all the local companies who owned or were about to own diesel tugs, has been told in the Lawson-Batey chapter.

At the end of 1954, with the first of the new tugs on order at Appledore ready for launching, the company asked their employees to choose names for them. Those initially suggested were *Prestwick, Ashbrooke, Alnwick, Tow Law* and *Marsden*. Later the fourth-named was changed to *Bamburgh*. Just what prompted the choice of *Tow Law*, the name of a small Durham town twenty miles inland, is not known, but presumably was because of the play on the word 'tow'. On 10th December 1954 *Prestwick* was christened by Mrs. Nan Wardlaw, the wife of France Fenwick's managing director. This tug was to be the first built new for the company since *Wearmouth* was completed in 1929. At the Sunderland office in 1954 the manager there, George Dryden, died and was succeeded by George Mons Martin.

George V *(2) at Appledore ready for her return to the Tyne.*
Why did her masts have to be curved? (Knights photographs, Barnstaple).

A detailed view of the aft end of the superstructure of George V *(2) before leaving Appledore.*
(Knights photographs, Barnstaple).

George V *(2) at Appledore nearing completion of her conversion to diesel power in August 1954.*
(Knights photographs, Barnstaple).

George V *(2) passing through the English Channel.* *(Skyfotos).*

Hendon *(2) and* George V *(2) towing a British Tanker Co.'s vessel through the English Channel.* *(Skyfotos).*

Inter-company Cooperation and the End of Steam Propulsion. 1955-1968

During 1955 and on into 1956 *George V* (2) was frequently contracted to tow dredging plant and barges for Westminster Dredging Co.Ltd. As well as work on the Tyne this involved tows to and from Inverness, Portsmouth, Southampton, Leith, Preston, Liverpool and Avonmouth, involving six transits of the Caledonian Canal within the space of a year. Other notable tows involving the same tug were the transfer of three tankers, *British Diligence* (8,408/1937), *British Destiny* (8,470/1937) and *British Resolution* (8,408/1937), from lay-up in Falmouth to the Tyne for dry-docking, and the towage of the collier *Cornwood* (2,787/1940) from London to Sunderland. *Beamish* had a lucky escape on 10th January 1955 at Wallsend, when she almost capsized whilst handling the tanker *San Silvestre* (10,953/1949), and, worse still, *Cleadon* sank off Sunderland Fish Quay for reasons unknown, after being tied up for the night. Only the top of her funnel and mast showed above water. She was later raised and continued to give useful service for another thirteen years.

Prestwick (119/1955) was completed in March 1955 and proceeded to London where, on the 23rd, she was proudly put on show, together with *George V* (2), for inspection by members of

Company advertisement on a match-book cover.

companies with towage interests in that area. She then sailed to Sunderland, arriving on 25th March, for further exhibition and finally, before getting down to serious work, headed a procession of diesel tugs up the Tyne to Newcastle with *Grangetown, Ryhope, Cornhill* and *George V* (2). Her skipper was Alec Hill.

As had happened within the Lawson-Batey and Ridley companies, examples were made of crew members who were involved in petty pilfering. Thus, in March 1955, one of the enginemen was fined £2 for stealing 1/- (5p.)-worth of coal from the tug *Great Emperor.* Amusing though such an episode appears to be nowadays, what followed in May was definitely not so. For the second time in this history a person was fatally injured when he fell into the mechanism of a a side-lever engine. This time, on the seventy-one-year old *Conqueror* (2), Edmund Wallace, aged twenty-three, fell, due to a broken section of platform over the engine.

In July, *Ashbrooke* (119/1955), and, in October 1955, *Alnwick* (119/1955), were completed at Appledore, and arrived in the Tyne on 27th July and 1st November respectively. *Bamburgh* was launched, without ceremony, at the same yard on 7th December. In November, the paddle tugs *Corsair* (3) and *Lumley* were sold, for a total of £3,200, to the British Iron & Steel Corporation who allocated them to J.J. King at Gateshead for breaking up. These recent changes left the company with a fleet of seven paddle and eighteen screw tugs.

Prestwick *during building at Appledore in the spring of 1955.* *(S.J. Turner).*

George V *(2) and* Prestwick *on display at London on 23rd March 1955.* *(Fox Photos Ltd.)*

The procession of diesel tugs up the river Tyne to Newcastle on 30th July 1956. Left to right, Marsden *(4),* Bamburgh, Alnwick, Grangetown, Ashbrooke, Prestwick *and* Ryhope. *(Local History Library, South Shields Central Library).*

was re-engined and re-boilered at R.B. Harrison's yard at Bill Quay, using the triple-expansion steam engine taken from the French trawler *Bearn* (294/1946). Whilst mentioning Harrison's yard, it is worth noting that they now employed 140 men and that the person responsible for the rejuvenation of the slipway and equipment, Sidney Armstrong, had retired in April 1956.

Late in the evening of 2nd March 1956, in a north-westerly gale, the new and incomplete Shell tanker *Volvula* (20,732/1956) measuring 660 feet long x 84 feet beam, broke loose from her moorings at the Swan Hunter shipyard at Wallsend, with only a single watchman on board. She began to drift, with the wind, down river causing absolute turmoil by damaging moored vessels, breaking others loose and by colliding with quays and wharfs and other riverside equipment. There was an emergency call-out of as many tug crews as possible, but naturally late at night this took some time to organise. Hector Stewart received his call-out and had his tug *George V* (2) ready to leave her South Shields mooring by 11pm. By this time *Volvula* had careered, stern first, some two-and-a-half miles down river and was between Readhead's and Middle Docks' shipyards, South Shields, striking all in her path. The tide had begun to ebb. If something was not done soon the tanker's speed would increase and there would be a real danger of the river being totally blocked. *George V* (2) was positioned on the starboard bow of the tanker, but of course there was no possibility of getting a line aboard with only the single watchman to handle it, nor could foyboatmen do anything on the water in the severe conditions. At 2am., with only *George V* (2) in close attendance, *Volvula* approached the

Changes in the fleet went on apace through 1956. In March *Bamburgh* (122/1956) was completed and arrived in the Tyne on the 10th of that month and was on view at Newcastle Quayside on the 14th. Her skipper was John Softley, who had started with the company as a fireman in 1931. *Marsden* (4) (122/1956) was launched without ceremony on 23rd May, carried out trials on 20th July and arrived in the Tyne on 30th July, when there was a grand procession of all of France Fenwick's diesel fleet up river to Newcastle. The parade included *Grangetown, Ryhope, Cornhill, George V* (2), *Prestwick, Ashbrooke, Alnwick, Bamburgh* and *Marsden* (4) - three ex-U.S. Army, one conversion and five newly built for the company. At this time all of the five recently-built tugs sported their names across the front of the bridge structure. As well as looking attractive this no doubt made identification easier when seen head-on. This feature, from photographic evidence, does not seem to have lasted very long and one wonders why. The design of both *Bamburgh* and *Marsden* (4) was modified from that of the previous three tugs so that the accommodation could cope with a crew of ten rather than six. This was to allow them to be used for sea towage work, although their engine power output of 1080 BHP was no different from their sisters. In May and June the paddle tugs *Conqueror* (2), *Dunelm* (1) and *Wexford*, seventy-two, seventy-three and sixty-eight years old respectively, were sold for breaking up. *Wexford* had been laid-up since September 1954. *Cullercoats*, rather surprisingly because of her fifty-eight year age,

For a while the France Fenwick diesel tugs carried their names across the bridge front. With the names also on lifebelts and the bell there should have been no difficulty in identifying this tug when she was in service. (Company archives).

The complete France Fenwick diesel-powered fleet arrived at Newcastle Quay following the procession of 30th July 1956. From bottom to top George V *(2)*, Marsden *(4)*, Bamburgh, Alnwick, Ashbrooke, Grangetown, Prestwick, Ryhope *and* Cornhill. *(Turners (Photography) Ltd.)*.

Empress of England *(25,585/1957) leaving the Tyne in 1957 in charge of four France Fenwick tugs. The Hawthorn-Leslie shipyard at Hebburn is in the background.* *(Turners (Photography) Ltd.).*

Mill Dam at South Shields where the tug, using full power, was able to push her aground and pin her against the quay. Once in that position the foyboatmen were able to secure her stern to the shore, whilst the tanker settled onto the mud as the tide fell. By now many other tugs were in attendance making sure that *Volvula* did not break free again. *George V* (2) kept a tight wire on her bow right through until 1pm. the following afternoon when she was refloated and returned to Swan Hunter's shipyard. Episodes of this nature underline the point that tugboatmen have always had to expect the unexpected and have to be willing to turn out at unsocial hours and to work for long periods when necessary.

Regarding domestic matters, the tug crews on the Wear were to work a five-day week of nine hours per day for a trial period of six months. Previously they had worked a fifty-hour week including Saturday morning. By the nature of the business there was, of course, plenty of overtime beyond these hours. In August 1956, A.M.H. Wardlaw, the France Fenwick managing director, resigned from his post and from the board. He was succeeded by William Richard Nicholson. William Rutherford, who had begun work with Redhead & Dry's in 1938, was appointed towage manager at South Shields.

The year 1957 presented many problems. In March, due to a strike of shipyard workers, part of the France Fenwick tug fleet had to be laid-up. On 7th September a fatal accident occurred when a tow rope parted on *Prestwick*. Sadly a deckhand, George Green, aged twenty-three, was killed and a deckboy, John Coates, aged eighteen, was injured. Once again it had been shown that, even with a very new tug, towing operations involved an element of danger which could never be removed entirely. On 22nd September 1957, after having been negotiating with the management for some time over wages and conditions, the tug crews operated an overtime ban which, a week later, evolved into a strike of all crews in the area. The men asked for a 10% rise but eventually settled for 7.5% after a dispute lasting twenty-nine days. Agreement was also reached that each man would have a guaranteed twenty-four hour period off each week. This was achieved by providing an extra man in each crew. Away from the rivers, contracts were still being obtained for sea-work: *George V* (2) and Lawson-Batey's *Quaysider* towed a coal-boring tower from off Methil to Granton, and the former was again chartered to Westminster Dredging, on and off, for the next three years.

Late in 1957, and continuing into 1958, the chairman, K.R. Pelly, had made overtures to the Lawson-Batey and Ridley companies regarding co-operation between the three of them for operations on the River Tyne. The details of how this amalgamation was engineered and put into operation under the title of Tyne Tugs Ltd. has been fully described in the Lawson-Batey chapter and need not, therefore be repeated here. The new arrangements were to be effective from 1st January 1959. Whilst these discussions were taking place,

being slipped on 15th December. Examination showed that she was beyond economical repair and so she was sold, for £1,500, for breaking up at Gateshead in March 1959.

On Saturday evening, 15th November 1958, Hector Stewart was at home preparing for an evening out with his wife, when the telephone rang, and he was requested to collect his crew and proceed with *George V* (2) to the assistance of the Swiss vessel *Nyon* (4,956/1952) which was reported aground some seventy miles north of the Tyne just north of St.Abb's Head. The crew of *Beamish* had likewise been alerted. *George V* (2), without her normal mate, who could not be located in time, left the Tyne at 10.30pm. but *Beamish* was delayed because the whereabouts of her skipper, L. Harbutt, were unknown; in the event James Stratford took command. The weather was poor with considerable fog, but, fortunately, little wind. *George V* (2) arrived at the casualty at 5am., and *Beamish* at 8.30am., on 16th November, to find *Nyon* held hard and fast by the bow on the rocks close under high cliffs. Her stern, however, was still afloat in relatively deep water. Two hours towing by *George V* (2) had no effect, which was not surprising since the tide was now falling, so she continued to hold the *Nyon's* stern steady to prevent further damage to the bow. Radio communications were not possible from the casualty because of her closeness to the cliffs so that *George V* (2) had to be used as a conduit to all the interested parties ashore. Eventually a decision was reached that the only way to save the greater part of the ship was to cut through her hull, just aft of the bow section which was pinned on the rocks. This could only be done down to water level, the remainder having to be achieved by the tugs towing and twisting the vessel apart.

Two Dutch tugs now appeared on the scene. They were *Hector* (200/1958) and *Simson* (206/1958), and naturally they wanted, if possible, to be involved in the salvage operations. Fortunately for France Fenwick, the *Nyon's* captain had signed Lloyds Open Form of Agreement which Hector Stewart had arranged shortly after his arrival on the scene. Thus the Dutch tugs, with their considerable salvage equipment, could only be employed as co-salvors, sub-contracted by France Fenwick. By 11am. on Tuesday 18th November, France Fenwick's managing director, W.R. Nicholson, and towage manager, A. Patton, were on the scene, having arrived by fishing boat from Eyemouth. *George V* (2) then

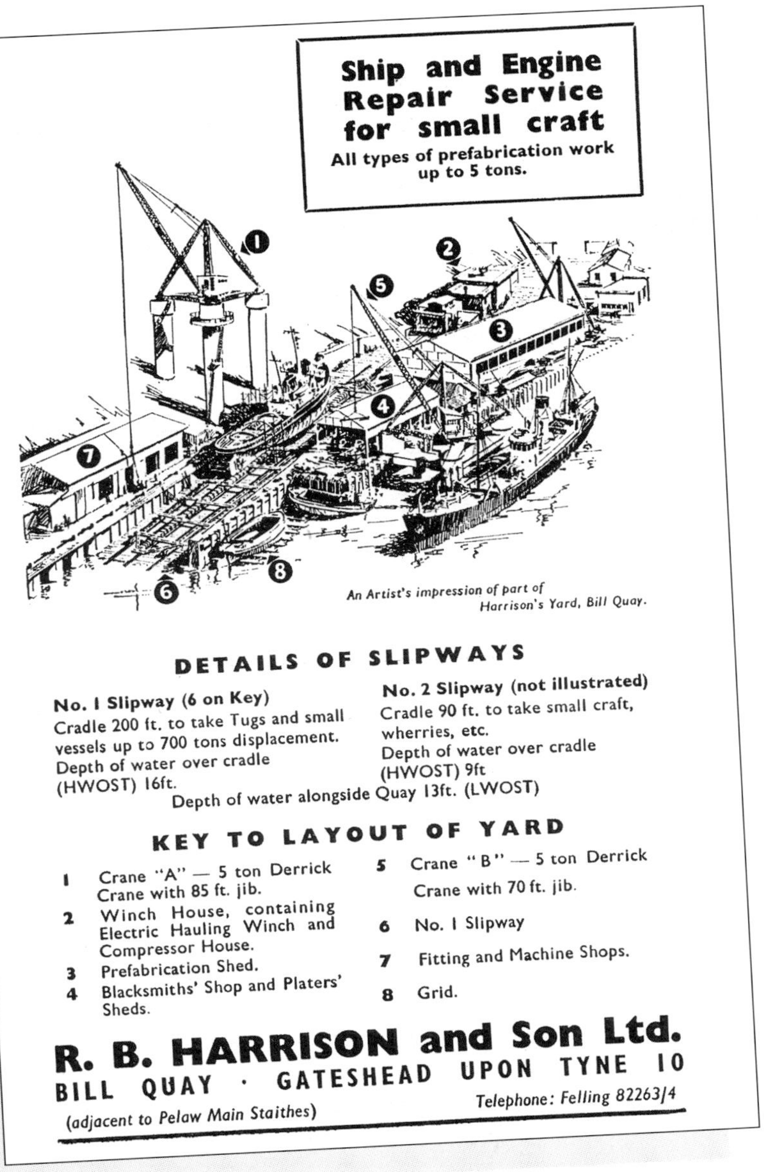

An advertisement of 1958 for R.B. Harrison & Son's ship repair yard at Bill Quay.

and details were being finalised for the smooth working of the new charter company, France Fenwick were, of course, still operating as an independent company both on the Tyne and on the Wear.

The fifty-three-year-old screw tug *Criccieth* was sold for £800 for breaking up at Gateshead in March 1958. On 11th June, Frank Dawson, MC and Bar, the former managing director of the company, died, aged sixty-three. With his father he had been involved with the company since its inception in 1918. He was also a former Wear Commissioner. At about this time, Andrew Desmond Pelly, DFC, joined the board as a director of the company. A further management change which occurred at the end of 1958 was the retirement of Alexander Patton as towage manager. He had transferred to France Fenwick from Redhead & Dry's when the latter company was taken over in 1944.

At 3am. on 23rd October 1958 *Earsdon* sank in the River Tyne after being in collision with the collier *James Rowan* (2,947/1955) at St. Anthony's Beacon. The skipper, Jack Bewick, his crew of six and a pilot rowed to safety and got ashore at Felling. The tug was raised by Risdon, Beazley & Co. of Southampton on 11th December and temporarily beached before

Nyon *ashore near St. Abb's Head on 16th November 1958 with* George V *(2) in the background. It is easy to see why the vessel had to be cut in two before the stern section could be towed away. (Les Turner).*

spent four days ferrying equipment from the Royal Navy at Rosyth to the site of the wreck. Meanwhile the cutting-off of the bow section was underway, and was down to the water level by Sunday 23rd November. Fortunately, the salvors were favoured with reasonable weather, otherwise work could not have gone ahead at all and the ship could have become a total loss. *George V* (2) and Hector now began to attempt to twist off the stern two-thirds of the ship. This went on for four days, success occurring at 6pm. on Thursday 27th November. *Beamish* and *Simson* had been standing-by and dealing with equipment all the time. The portion of *Nyon* which had been saved arrived in the Tyne at 9am. on Friday 28th, and was taken to Smiths Dock, North Shields for temporary repair. Later, she was taken to Holland for a new bow to be fitted. This episode began with Hector Stewart intending to go out for the evening twelve-and-a-half days earlier, and it is a fine example of how the wives and families of the tugboatmen had to put up with disruption of their day-to-day activities and social lives. Of course, there would be compensations, not only in the high wages received by the crew for their continuous work, but also in their share of the salvage awards.

After a short rest, *George V* (2) returned the hired equipment to Rosyth, and, in between, towed from and to Rosyth the four lifting camels required for the salvage of the tug *Earsdon*, which has been referred to earlier.

On 1st January 1959 the charter company, Tyne Tugs Ltd., became operative, when the tugs of Lawson-Batey, Ridley and those working on the Tyne for France Fenwick, joined together. As described in the Lawson-Batey chapter no change of ownership of tugs took place - they were on ten-year demise charters to the newly-formed operating company. The crews of the France Fenwick tugs involved, and some of the office staff, now became employees of Tyne Tugs Ltd. The arrangement was for Lawson-Batey to provide ten tugs, Ridley two, and France Fenwick twelve. As originally planned when the agreement was drawn up, the France Fenwick tugs were to be:-

Hendon (2)	(241/1924)	Steam
Cullercoats	(181/1898)	Steam
Wearmouth	(182/1929)	Steam
Fulwell	(149/1899)	Steam
George V (2)	(217/1915)	Diesel
Robert Redhead	(178/1900)	Steam
Earsdon	(234/1912)	Steam
Beamish	(242/1944)	Steam
Ashbrooke	(119/1955)	Diesel
Alnwick	(119/1955)	Diesel
Bamburgh	(122/1956)	Diesel
Marsden (4)	(122/1956)	Diesel

However, *Earsdon* could not take her place because of her sinking, as has already been described, and was replaced by Ridley's steam tug *Marty* (175/1906), which France Fenwick took on charter and then, in turn, chartered to Tyne Tugs. This charter was to be very short-lived since *Marty* was withdrawn for breaking-up in May 1959. The Tyne tug *Great Emperor* did not join Tyne Tugs Ltd., for she was withdrawn from service and broken-up in February 1959. With these changes the chartered tugs now took the new funnel colours of red with a black top, and the much-admired blue and white funnel design of France Fenwick disappeared from the Tyne.

The France Fenwick tugs operating at Sunderland were not affected by the new arrangement on the Tyne, and continued to operate under the control of, and wearing the colours of, their owning company. At the beginning of 1959 this fleet consisted of:-

Roker	(119/1905)	Steam
Souter	(123/1910)	Steam
President	(130/1876)	Steam
Cleadon	(148/1899)	Steam
Houghton	(133/1904)	Steam
Eppleton Hall	(166/1914)	Steam
Grangetown	(176/1943)	Diesel
Ryhope	(176/1943)	Diesel
Cornhill	(176/1943)	Diesel
Prestwick	(119/1955)	Diesel

Only two months later *President* was broken up. She was then over eighty years of age! In 1960 the Board of Directors included the chairman, K.R. Pelly, M.C., the managing director and secretary, W.R. Nicholson, A.D. Pelly, DFC., and Thomas McEwan. The principal shareholder was the William France Fenwick company. The Sunderland towage manager G.M. Martin died in May 1960 and he was succeeded by Gordon Bell. In the following year K.R. Pelly received the honour of a Knighthood.

In March 1961 a new tug was ordered from Richard Dunston at Hessle for service on the River Tyne, and *Fulwell* was sold for £1,200 for breaking up at the end of the year. She was then sixty-two years old and had been with France Fenwick since 1934. The new tug, named *Alnmouth* (173/1962), arrived in May 1962 and was placed on charter with Tyne Tugs Ltd. She had cost about £85,000 and was the first tug on the Tyne with a British-built diesel engine. At the same time *Robert Redhead* was taken off charter to Tyne Tugs and

President *being broken up at Sunderland in 1959.* *(H. Appleyard).*

Alnmouth leaving Newcastle Quay for a demonstration trip on 7th June 1962. *(The Shields Gazette).*

transferred to Sunderland. In November consideration was being given to the possibility of converting *Beamish* to diesel propulsion, but since the cost would have been about £28,000 this was thought to be too expensive. Meanwhile, she was laid up. The paddle tug *Roker,* which had worked on the River Wear since 1905, was sold for further service to Grangemouth & Forth Towing Co.Ltd. with whom she worked for a further three-and-a-half years. before being broken up. The France Fenwick office at Sunderland was moved, in November 1962, from 21 Bridge Street to Midland Bank Chambers in Fawcett Street.

Beamish was, after all, re-engined with a 1400 BHP, 16-cylinder, General Motors diesel, which had been manufactured in 1947. The work was carried out at R.B. Harrison's yard on the Tyne and completed in October 1963. In December *Souter* was advertised for sale through the agents

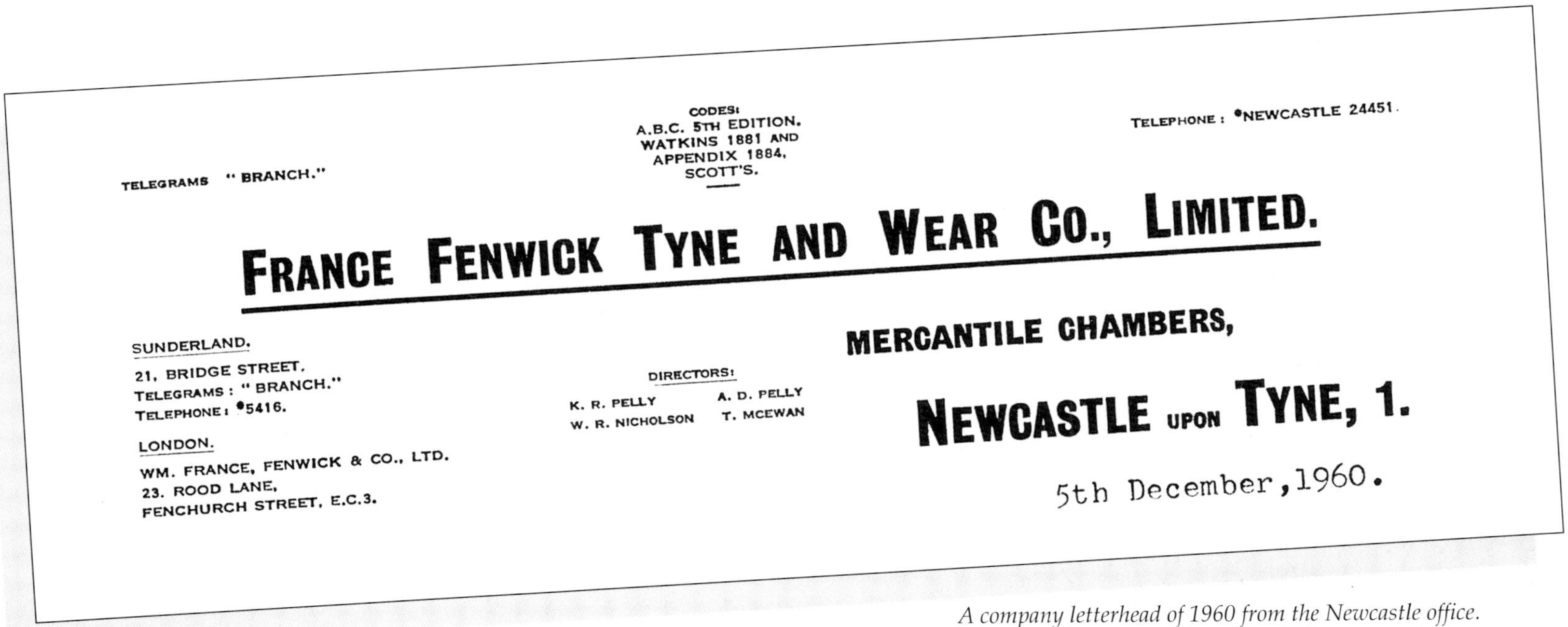

TELEGRAMS " BRANCH."

CODES:
A.B.C. 5TH EDITION.
WATKINS 1881 AND
APPENDIX 1884.
SCOTT'S.

TELEPHONE : *NEWCASTLE 24451.

FRANCE FENWICK TYNE AND WEAR CO., LIMITED.

SUNDERLAND.
21, BRIDGE STREET.
TELEGRAMS : " BRANCH."
TELEPHONE : *5416.

LONDON.
WM. FRANCE, FENWICK & CO., LTD.
23. ROOD LANE,
FENCHURCH STREET, E.C.3.

DIRECTORS:
K. R. PELLY   A. D. PELLY
W. R. NICHOLSON   T. MCEWAN

MERCANTILE CHAMBERS,

NEWCASTLE UPON TYNE, 1.

5th December,1960.

A company letterhead of 1960 from the Newcastle office. Rather less flamboyant than that of other companies!

Eppleton Hall *and* Robert Redhead *in Sunderland Docks c.1963. (World Ship Society – Teesside Branch).*

An unusual overhead view of Ryhope *with a foyboat being towed sternfirst. (World Ship Society – Teesside Branch).*

H.E. Moss & Co., and by the end of the following month she had been purchased for £1,300 by Hughes Bolckow Ltd. who proceeded to scrap her at their Battleship Wharf at Blyth. Subsequently, in 1964, there were four further changes in the tug fleet. A new tug, *Dunelm* (2) (132/1964), was built by Richard Dunston and *Kronos*, another of the standard war-time United States Army tugs, was purchased from the Bergen Steamship Co. This latter tug, after refurbishment at R.B. Harrison, was renamed *Whitburn* (2) (176/1943). Both of these acquisitions were put to work at Sunderland.

In October 1964 *Houghton* was sold for breaking-up at Dunston, and in November *Eppleton Hall* was sold to Seaham Harbour Dock Co. for further service. These changes left the company, at the end of 1964, with a fleet of eighteen tugs, ten on charter to Tyne Tugs Ltd. and eight working at Sunderland.

Although *Eppleton Hall* passes out of the scope of this story it is perhaps worth relating her subsequent history. The Seaham Harbour Dock Co. kept her until November 1967 when they sold her to Clayton & Davie Ltd. at Dunston for demolition. Work started on breaking her up; all of her woodwork had been burnt out of her, but most of the steel-

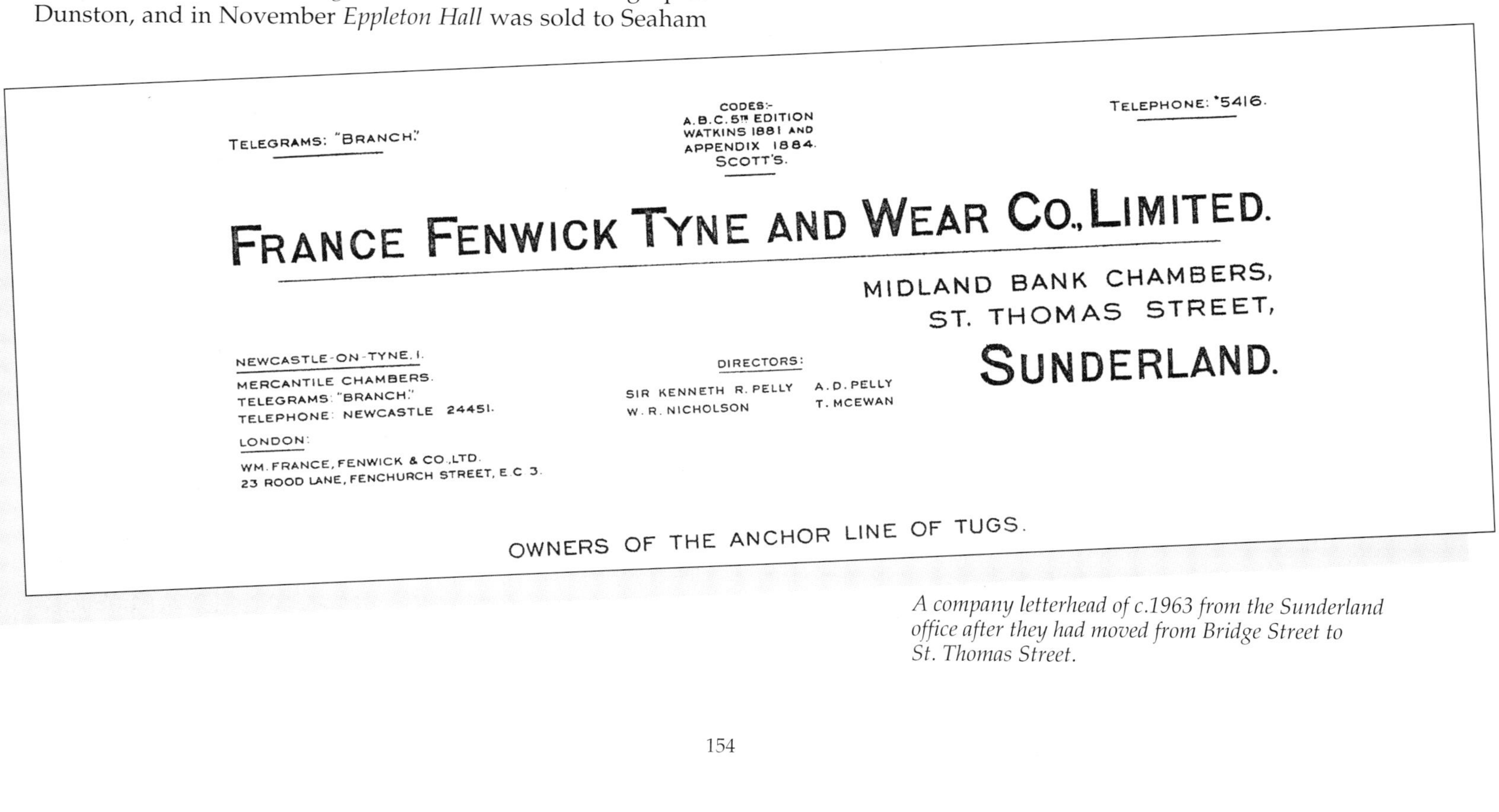

TELEGRAMS: "BRANCH."

CODES:-
A.B.C. 5TH EDITION
WATKINS 1881 AND
APPENDIX 1884.
SCOTT'S.

TELEPHONE: 5416.

FRANCE FENWICK TYNE AND WEAR CO., LIMITED.

MIDLAND BANK CHAMBERS,
ST. THOMAS STREET,
SUNDERLAND.

NEWCASTLE-ON-TYNE, 1.
MERCANTILE CHAMBERS.
TELEGRAMS: "BRANCH."
TELEPHONE: NEWCASTLE 24451.

LONDON:
WM. FRANCE, FENWICK & CO. LTD.
23 ROOD LANE, FENCHURCH STREET, E.C. 3.

DIRECTORS:
SIR KENNETH R. PELLY A. D. PELLY
W. R. NICHOLSON T. MCEWAN

OWNERS OF THE ANCHOR LINE OF TUGS.

A company letterhead of c.1963 from the Sunderland office after they had moved from Bridge Street to St. Thomas Street.

The final touches being made to Eppleton Hall *for her long journey to the United States of America in 1969. (South Shields Gazette).*

work remained, when Scott Newhall, the editor of the *'San Francisco Chronicle'* and trustee of the San Francisco Maritime Museum, came to England and bought her from the breakers for £2,500 in May 1969. His intention was to have her restored and to steam and sail her to the U.S.A. He entrusted the work to R.B. Harrison, the France Fenwick subsidiary at Bill Quay on the Tyne, with the instruction to restore her to her appearance when built in 1914. During this work she was converted from coal to oil-firing of her boilers, had an extra mast added near the stern, and was provided with a full outfit of sails. Before she left the Tyne on 18th September 1969, her skipper, Scott Newhall, was presented with a Davy Lamp and an inscribed silver salver by France Fenwick's managing director, W.R. Nicholson, in the presence of the local Mayors.

Eppleton Hall now wore the 'stars and stripes' at her stern and was registered in San Francisco. She carried a crew of eleven volunteers, three of whom were young girls. It had been intended that her first port of call should be Cherbourg, but she had to put into Dover on 24th September because of fuel pump problems and severe gales in the North Sea. Off Corunna, on 4th October, she was out of fuel and had to be re-fuelled by a German ship. Her next ports of call were

This attractive model of George V (2) *when steam powered is obviously a crowd-puller.* *(Newcastle Chronicle & Journal)*

Lisbon, Madeira and St. Vincent in the Cape Verde Islands, where she arrived on 17th November 1969. She left there on 23rd and set out on the long haul across the south Atlantic, making a landfall at Georgetown in Guyana. Christmas was spent in Trinidad, and, after passing through the Panama Canal, San Diego was reached on 14th March 1970, and finally, San Francisco on 23rd March. After an epic voyage lasting six months the paddle tug *Eppleton Hall* was put on display at the Maritime Museum where she remains to this day. It is hoped that the reader will have thought that this digression was worth while; now to return to the France Fenwick story.

On 30th June 1966 Sir Kenneth Pelly, MC., retired as chairman of the company and was succeeded by T. McEwan. This post also carried with it the chairmanship of R.B. Harrison's repair yard at Bill Quay. In March 1966 *Hendon* (2) was sold for £4,000 for breaking up at Bo'ness, and, towards the end of the year, *Wearmouth* was taken off charter to Tyne Tugs and moved to Sunderland. In November 1967 *Robert Redhead* went to the breakers after having given a total service of sixty-seven years at Liverpool, Southampton, Swansea, on the Tyne, and finally at Sunderland. Two more of similar age, *Cullercoats* and *Cleadon,* went to the same breakers in 1968.

In April 1968 William France Fenwick was taken over by Jessel Securities Ltd., a finance company located in

A company advertisement from the cover of an industrial handbook.

London. France, Fenwick Tyne & Wear, a subsidiary company of William France Fenwick, thus came under the control of Jessel. Likewise, R.B. Harrison, a subsidiary of France Fenwick. Jessel stated that they had no intention of interfering with the charter agreement involving Tyne Tugs. This charter agreement, which had become effective on 1st January 1959, had now almost run its allotted span of ten years, so, since all participants were very satisfied with the results, a new charter on similar lines was drawn up to run for a further period of ten years from 1st January 1969. Notwithstanding the take-over by Jessel, this new agreement was still in the name of France Fenwick, who were to continue trading under their old name.

Following the recent changes in the fleet, and at the commencement of the new Tyne Tugs charter period, the company owned the following tugs:-

On charter to Tyne Tugs Ltd.

George V (2)	(217/1915)	1,086 BHP	Now a completely diesel-powered fleet
Beamish	(242/1944)	1,400 BHP.	
Ashbrooke	(119/1955)	1,080 BHP. 14BP	
Alnwick	(119/1955)	1,080 BHP. 14BP	
Bamburgh	(122/1956)	1,080 BHP. 14BP	
Marsden (4)	(122/1956)	1,080 BHP. 14BP	
Alnmouth	(173/1962)	1,080 BHP. 16.5BP	

Based at Sunderland

Grangetown	(176/1943)	700 BHP	Diesel
Ryhope	(176/1943)	700 BHP	
Cornhill	(176/1943)	700 BHP	
Prestwick	(119/1955)	1,080 BHP. 14BP	
Dunelm (2)	(132/1964)	970 BHP. 14BP	
Whitburn (2)	(176/1943)	700 BHP	
Wearmouth	(182/1929)	900 IHP.	Steam

- fourteen tugs in all, with an average age of about twenty years. Only days into the new charter period *Wearmouth* went to the breakers, leaving the company for the first time with a completely diesel-powered fleet.

The Final Years. 1969 - 1977.

In January 1969 Jessel Securities put France Fenwick on the market for sale as a going concern. William Cory & Son Ltd., United Towing Co. and Tees Towing Co. Ltd. were amongst those who showed interest though nothing positive came of these overtures.

At the end of 1968 Hector Stewart, the company commodore up to the formation of Tyne Tugs Ltd. in 1959, had retired after fifty-four years' service in tugs, and having been a skipper since the early 1920s. He estimated that in that time he had handled about 50,000 ships. Strangely, although officially retired, he was to carry out one further job for the company. On 4th January 1969, after an evening out at his club with family and friends, he had retired to bed late in the day. In the early hours of the following morning he was aroused by the telephone. It was the tug office wanting to ask his advice about a vessel which was in trouble in the North Sea. Giving advice was not sufficient for him, however, and he was soon dressed and on his way, with other crew members, to man *George V* (2). All of this happened without anyone else in his house knowing that he had gone (it must have been a very good party!) or where he had gone to! By 4am. the tug was on its way to the assistance of the small Swedish motor vessel *Rigon* (376/1949), which, outward bound with coal from the river, had developed engine defects reportedly

75 miles ENE of the Tyne. Nothing was found when this position was reached at 10am. but further information was received from the casualty, giving his position as a further fifty miles away. At 4pm. *George V* (2) arrived on the scene and began to tow the vessel to the Tyne, reaching there at 1am. on the 5th, and seeing her fast at a South Shields buoy berth. Hector Stewart arrived home at 2am. on his 69th birthday, Sunday 5th January 1969, and found his house in darkness. When he opened the door and switched on the light he found all of his family and many friends filling the hall and staircase ready to greet him with 'Happy Birthday to You!'

Such a man could not rest idly at home, so he soon obtained a job as a building site watchman, but that gave him little satisfaction. By the end of 1969 he had gone back to sea as relief mate on a number of colliers trading between the Tyne and the Thames. He did this for approximately five months. *Beamish,* which had been laid-up for over a year, was sold, via an intermediary, to Rex Mellenger, Canada, and renamed *Rex Mellenger.* Lo and behold, Hector Stewart was to captain her on her transatlantic voyage. She left the Tyne on 8th June 1970 but was soon back again after blowing a gasket when off Seaham Harbour. She did not leave again until late in September. This time she managed to get as far as Southampton, where she was sold to another Canadian company who soon changed her name back to *Beamish.* With this change went Hector Stewart's chance of an ocean voyage. How disappointed he must have been! *Beamish,* now renamed *Thunder Cape,* is still working in Canada.

In the spring of 1970 *Grangetown* and *Ryhope* were sold to Brodospas, Yugoslavia, and *Marsden* (4) was taken off charter with Tyne Tugs and transferred to Sunderland. This left France Fenwick with five tugs on charter to Tyne Tugs and five working on the Wear. At this time Lawson-Batey had seven tugs on charter to Tyne Tugs. In August 1970 the Wear tugboatmen came out on strike over a pay claim. Thirty men were involved and two launches, which had already been delayed for five months due to a shipyard fitters' strike, were held up. The Wear men, who were members of the General and Municipal Workers' Union, returned to work after ten days.

In November 1973 a new tug was ordered from Richard Dunston. She was to be much larger than anything currently in the fleet, with an engine output of 3,180 BHP and costing about £410,000. She was to be a sister vessel to the tug ordered by Lawson-Batey at the same time and was to be capable of handling the largest vessels then being built in the area, as well as carrying out sea towage work. Delivery was specified as July 1975.

In January 1975 Jessel Securities, the owners of France Fenwick, were again trying to sell the tug company, having failed to do so in 1969. Various shipowners and the Rothschild organisation made bids. Because of the Tyne Tugs charter agreement any sale of the France Fenwick company had to have the agreement of Lawson-Batey. Lawson-Batey themselves stated that they were not interested in buying France Fenwick at that time due to their other financial commitments. The following month Jessel Securities were in severe financial difficulties and clearly would not be able to continue for very long. In the midst of all this doubt and worry over the survival or otherwise of their business, France Fenwick moved their Sunderland office from Midland Bank Chambers to Gladstone House, South Dock, Sunderland.

Jessel Securities went into voluntary liquidation in January 1976. By March of that year Lawson-Batey still did not wish to make an individual bid for France Fenwick but in the following month they made a joint bid with the Belgian tug company Union de Remorquage et de Sauvetage S.A. France Fenwick was thought to be worth £495,000 and an offer of £454,454 was made for the shares held by Jessel. This, however, was turned down as it was considered by the owners that a price of around £700,000 would be more appropriate. The size of the towage business of France Fenwick at this time can be judged by the number of jobs being carried out; in 1976 around 1,300 jobs at Sunderland and about 6,000 jobs, jointly with Lawson-Batey, on the Tyne.

With delivery having been promised for July 1975 the new tug on order had only progressed as far as being christened *Hendon* (3) by Mrs. Doreen Nicholson, wife of the managing director, on 22nd November 1976. Even then, because of a labour dispute, the tug was not launched until three days later.

By the early months of 1977 the idea of a joint purchase of France Fenwick by Lawson-Batey and the Letzer family's Belgian company had fallen by the wayside. Cory Ship Towage Co.Ltd. continued to show an interest in the company but in April 1977 a finance company, London & Yorkshire Trust Ltd., came on the scene. They were interested in buying the Jessel Securities' holding of France Fenwick shares. If this took place then Lawson-Batey indicated their willingness in turn to purchase the tug business, plus certain assets of France Fenwick from the finance company. An agreement was signed in May 1977 under which Lawson-Batey purchased, for a total of £555,553, the following business assets of France Fenwick:-

	£
Goodwill	1
Stocks	29,472
Plant, motor vehicles and furniture	7,880
Shares in Tyne Tugs Ltd.	1,200
Ten Tugs	517,000
	£555,553

For a very short period in September 1970 Beamish *was renamed* Rex Mellenger *when she was sold to Canadian owners. In this view she appears much larger than she actually was. (Maritime Photo Library).*

REG. NO. 149753 (ENGLAND)
CODES:
A.B.C. 5TH EDITION
WATKINS 1881 AND
APPENDIX 1884
SCOTT'S.

TELEGRAMS "BRANCH".

TELEPHONE: NEWCASTLE 24451

FRANCE FENWICK TYNE AND WEAR CO., LIMITED.

REG. OFFICE
MERCANTILE CHAMBERS
NEWCASTLE UPON TYNE.
NE1 3DP

SUNDERLAND
Gladstone House
South Dock
Telegrams: "Branch"
Telephone 75416

DIRECTORS
~~[illegible]~~
W. R. Nicholson
(Managing Director)

M.E.A.Keeling
M.A. Hudson
H.E. King
B.J. Hardy.

5th May, 1977

A letterhead from the very last days of the company's existence. At that time the Sunderland office had been at Gladstone House, South Dock for the past two years.

This arrangement did not include the purchase of R.B. Harrison's shipyard at Bill Quay, nor the tug *Hendon* (3) (266/1977), which was just completing. A year later this new tug was sold to The Alexandra Towing Co.Ltd. without change of name. The shipyard was sold to Scottish Universal Investments in September 1977.

The tugs sold to Lawson-Batey on 5th May 1977 were:-

George V (2)	(217/1915)	1,086 BHP	On charter to Tyne Tugs Ltd.
Ashbrooke	(119/1955)	1,080 BHP. 14BP	
Alnwick	(119/1955)	1,080 BHP. 14BP	
Bamburgh	(122/1956)	1,080 BHP. 14BP	
Alnmouth	(173/1962)	1,080 BHP. 16.5BP	

Cornhill	(176/1943)	700 BHP	Working at Sunderland
Prestwick	(119/1955)	1,080 BHP. 14BP	
Dunelm (2)	(132/1964)	970 BHP. 14BP	
Whitburn (2)	(176/1943)	700 BHP.	
Marsden (4)	(122/1956)	1,080 BHP	

The first five stayed on charter to Tyne Tugs. The others were soon chartered to a new company, set up by Lawson-Batey, namely Wear Tugs Ltd. This arrangement had one nice feature in that the attractive France Fenwick funnel colours were retained for the Wear Tugs fleet for a further nine years.

For those tugmen working on the Tyne, as their employers were already Tyne Tugs Ltd. no change was necessary. The men and staff at Sunderland had their employment with France Fenwick terminated and they were immediately re-employed by Wear Tugs Ltd. on their existing terms. The office at Sunderland was kept open for the control of operations in that area. Thus, in 1977, France, Fenwick Tyne & Wear Co.Ltd., after fifty-nine years of existence, passed into oblivion. When it is remembered that France Fenwick had, at one time, by far the largest fleet of tugs in the area, it is surprising that they had to disappear. On the other hand, perhaps it was not so surprising, since they were in the control of financiers based away from the area, who had no experience of, or direct interest in, shipping affairs in the north-east of England. Fifteen years later, the name is still well-remembered and often referred to in the local newspaper columns.

APPENDIX A TO CHAPTER EIGHT

France, Fenwick Tyne & Wear Co. Ltd. 1918-1977

Board of Directors

Chairmen	H.M. Stobart	1918-1947	Retired
	(Sir) K.R. Pelly	1947-1966	Retired
	T. McEwan	1966-1974	Retired
	J.E. Keville	1974-1977	
Managing Directors	J.E. Dawson	1918-1924	
		1924-1938	Joint M.D. Retired
	F. Dawson	1924-1938	Joint M.D.
		1938-1949	
		1949-1950	Joint M.D. Retired
	A.M.H. Wardlaw	1949-1950	Joint M.D.
		1950-1956	Resigned
	W.R. Nicholson	1957-1977	
Directors	K.R. Pelly	1918-1947	Became chairman
	F.W. Taylor	1918-?	
	E. Edwards	1918-1938	Died
	J. Leslie	c.1937-c.1946	
	H.F. Charlton	c.1938-c.1946	
	A.M.H. Wardlaw	1946-1949	Became Joint M.D.
	A.D. Pelly	?-1969	Resigned
	T. McEwan	1956-1966	Became Chairman
	W.R. Nicholson	1956-1957	Became M.D.
	C.R. Benzecry	c.1968-1977	Resigned
	J.E. Keville	c.1968-1974	Became Chairman
	R.C. Bowen	May 1977 only	Resigned
	R.G. Newman	May 1977 only	Resigned
	M.E.A. Keeling	1977 only	
	H.E. King	1977 only	
	M.A. Hudson	1977 only	
	B.J. Hardy	1977 only	
Secretaries	H.F. Carter	1918-1939	
	A.M.H. Wardlaw	1939-1946	Became Director
	W.R. Nicholson	1946-1977	Also Director

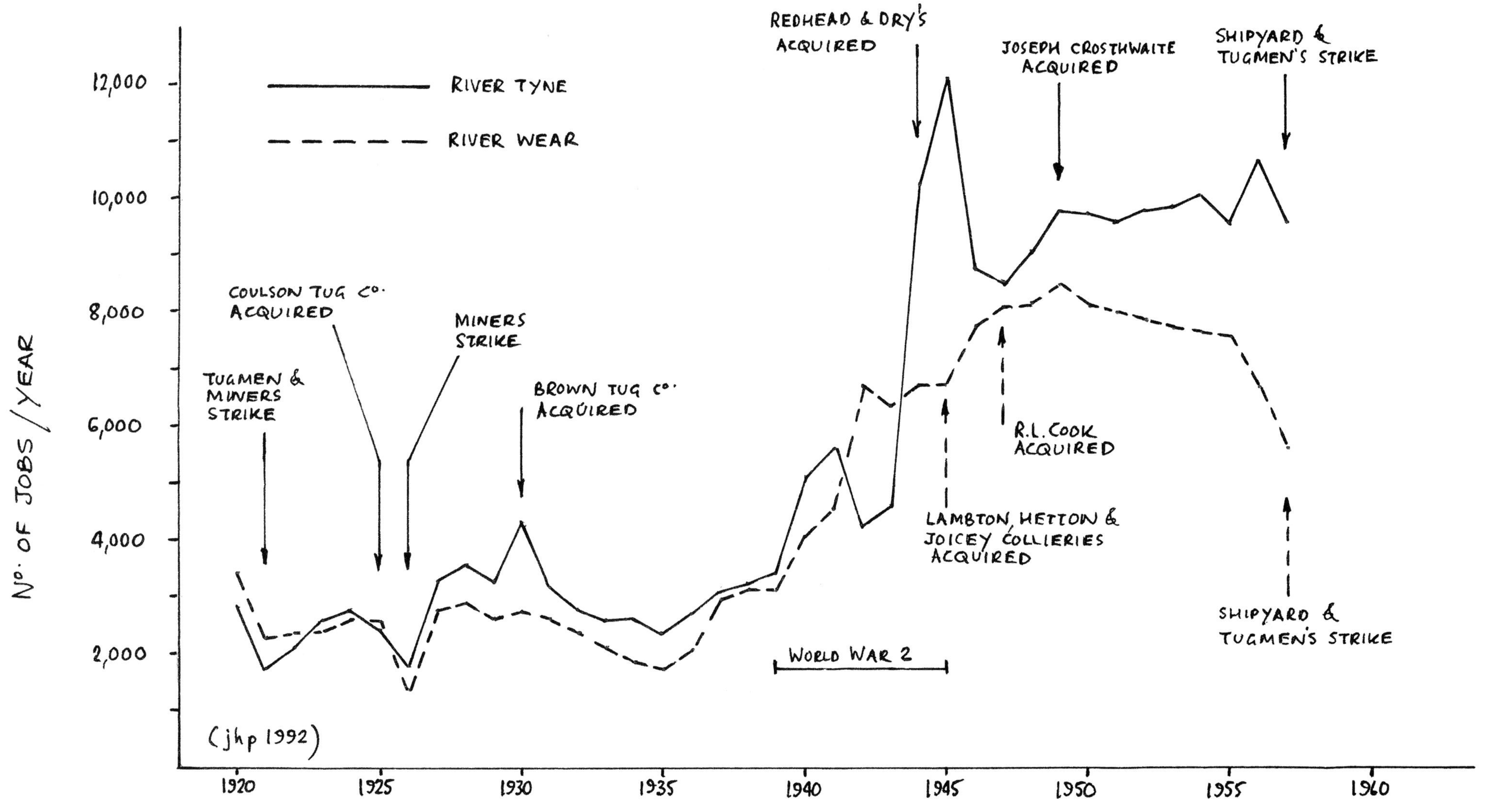

NUMBER OF JOBS CARRIED OUT BY FRANCE, FENWICK TYNE & WEAR Co. LTD 1920-1957

APPENDIX C TO CHAPTER EIGHT

REF. No.	NAME	Ex.	To
F.			
1	TRANSIT	S15	
2	ROKER	S4	
3	HENDON (1)	S8	
4	UNITED	S10	
5	SOUTER	S9	
6	SUNDERLAND	S11	
7	HARWICH	S13	CN14
8	MARSDEN (2)	S14	
9	WEXFORD		
10	IRON/SEABURN		
11	LORD WARDEN	AST13	
12	TRITON	AST15	
13	RESCUE (3)	AST16	
14	RAINBOW (2)	AST14	
15	CRICCIETH		
16	MARSDEN (3)		
17	DUNELM (1)		
18	HENDON (2)		
19	HECTOR	CN10	
20	CORSAIR (3)	CN11	
21	FLYING SCUD	CN9	
22	PANTHER	CN13	
23	CULLERCOATS		
24	WEARMOUTH		
25	PRESIDENT	BR2	
26	CLEADON		
27	FULWELL		
28	CONQUEROR (2)	RD1	
29	GEORGE V (2)	RD7	LB34
30	WASHINGTON	RD2	
31	MALTA	RD6	
32	GREAT EMPEROR (2)	RD5	
33	ROBERT REDHEAD	RD9	
34	HOUGHTON	LH8	
35	LUMLEY	LH9	
36	EPPLETON HALL	LH10	
37	STAG (5)	CK15	
38	SNOWDON	CK17	
39	EARSDON	CE8	
40	BEAMISH		
41	GRANGETOWN		
42	RYHOPE		
43	CORNHILL		LB36
44	PRESTWICK		LB37
45	ASHBROOKE		LB38
46	ALNWICK		LB39
47	BAMBURGH		LB40
48	MARSDEN (4)		LB41
49	ALNMOUTH		LB42
50	DUNELM (2)		LB43
51	WHITBURN (2)		LB35
52	HENDON (3)		
	Tugs owned at 31 Dec: 12 14 13 26 20 10 0		
	Managed tugs		
M1	EMPIRE PLANE		
M2	EMPIRE SARA		
M3	EMPIRE ANDREW		
M4	EMPIRE JANE		
M5	EMPIRE MUSTANG	RD/M1	
M6	EXPATRIATE		

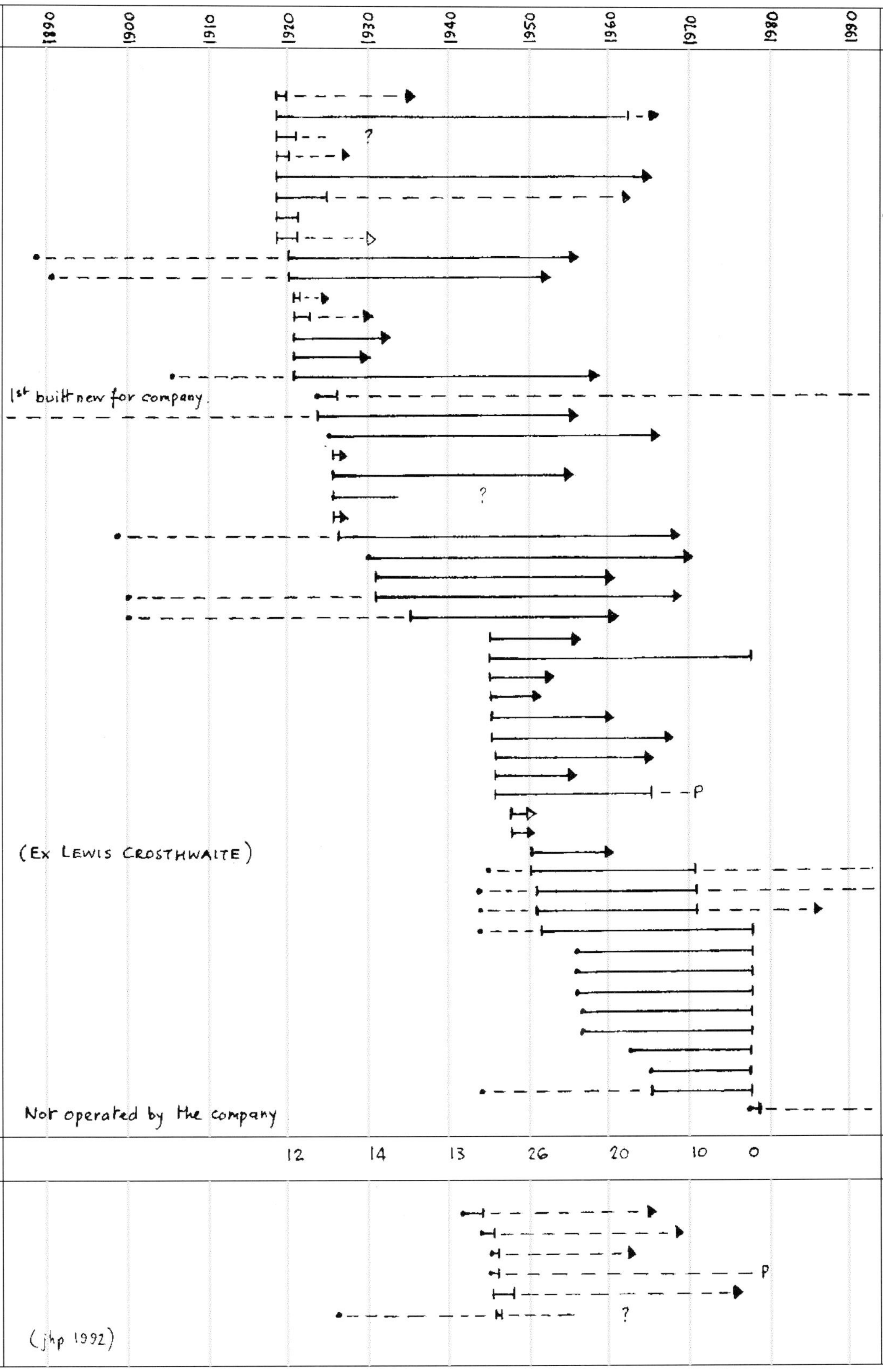

Tug Ownership Chronology.

FRANCE, FENWICK TYNE & WEAR Co. LTD. 1918 – 1977

France, Fenwick Tyne & Wear Co. Ltd. Fleet List

A model of Transfer *(ex* Transit*) to the scale of ⅜" = 1 foot in the ofices of Tyne & Wear Tugs Ltd. It has not been explained why a model of this particular tug which spent only a little over one year in the ownership of Sunderland Towage and France Fenwick was made nor why it is one that has survived to the present day.*

F 1 **TRANSIT/TRANSFER** **(1918-1919)**
See Sunderland Towage fleet list, S 15

Roker *in 1958. (Maritime Photo Library).*

F 2 **ROKER** **(1918-1962)**
See Sunderland Towage fleet list, S 4

F 3 **HENDON (1)** **(1918-1921)**
See Sunderland Towage fleet list, S 8

F 4 **UNITED** **(1918-1919)**
See Sunderland Towage fleet list, S 10

Souter *in 1957. (Maritime Photo Library).*

F 5 **SOUTER** **(1918-1964)**
See Sunderland Towage fleet list, S 9

F 6 **SUNDERLAND** **(1918-1924)**
See Sunderland Towage fleet list, S 11

F 7 **HARWICH** **(1918-c.1921)**
See Batey fleet list, B 15

F 8 **MARSDEN (2)** **(1918-1921)**
See Sunderland Towage fleet list, S 14

Wexford *in 1950s. (Maritime Photo Library).*

F 9 **WEXFORD** **(1919-1956)**
ON 70154 141 g 108.0 x 18.9 x 9.9 ft. Iron Paddle
B: J.T. Eltringham, South Shields. (YN 138)
E: L. 2 Cy. 28" x 50". 360 IHP. 10.25K. J.P. Rennoldson & Sons, South Shields.

PR: 1888 Wexford; 1922 Sunderland.

03.1888:	Completed for Wexford Harbour Commissioners.
07.1915:	Hired by Royal Navy. 04.1918: Returned to owners.
1919:	Purchased by France, Fenwick Tyne & Wear Co.Ltd.
06.1956:	Sold, for £1675, to British Iron & Steel Corporation. Allocated to Thomas Young & Sons (Shipbreakers) Ltd., Sunderland, for breaking up.

Seaburn. *(John Clarkson, Longton).*

F 10 **IRON / SEABURN** **(1919-1952)**
ON 97391 118 g 95.4 x 19.1 x 9.9 ft. Steel Paddle
B: J.T. Eltringham & Co., South Shields (YN 152)
E: L. 2 Cy. 28" x 50". 350 IHP. 10.25K. J.P. Rennoldson & Sons, South Shields.

PR: 1890 West Hartlepool; 1932 Sunderland.

1890:	Completed, as IRON, for £5,025, for The North Eastern Railway Co.
12.1919:	Purchased by France Fenwick Tyne & Wear Co.Ltd.
06.1932:	Renamed SEABURN.
04.1952:	Sold to British Iron & Steel Corporation. Allocated to C.W. Dorkin Ltd., Dunston, for breaking up.

F 11 **LORD WARDEN** **(1920 only)**
See Anchor Steam Tug fleet list, AST 13

Triton. *(Harrington collection).*

Criccieth *in 1955. (Maritime Photo Library).*

F 12 **TRITON** **(1920-1922)**
See Anchor Steam Tug fleet list, AST 15

F 13 **RESCUE (3)** **(1920-1932)**
See Anchor Steam Tug fleet list, AST 16

F 14 **RAINBOW (2)** **(1920-1929)**
See Anchor Steam Tug fleet list, AST 14

F 15 **CRICCIETH** **(1920-1958)**
ON 122803 102 g 90.4 x 17.9 x 9.8 ft. Steel Screw
B: Cox & Co. Falmouth (YN 113)
E: C. 2 Cy. 18" & 36" x 24". 530 IHP. 12K. By Builder.

PR: 1905 Swansea; 1922 Newcastle.

11.1905:	Completed, as CONQUEROR, for Swansea Steam Tug Owners.
c.1906:	Sold to Steam Tug Conqueror Ltd., Swansea
02.1915:	Hired by Royal Navy as Boom Defence Vessel
1917:	Renamed CRICCIETH
03.1920:	Returned to owners
1920:	Purchased by France, Fenwick Tyne & Wear Co.Ltd.
03.1958:	Sold, for £800, to British Iron & Steel Corporation. Allocated to C.W. Dorkin Ltd., Gateshead for breaking up.

F 16 **MARSDEN (3)** **(1923-1926)**
ON 148041 241 g 105.0 x 25.5 x 12.7 ft. Steel Screw
B: J.P. Rennoldson & Sons Ltd., South Shields. (YN 322)
E: T. 3 Cy. 16½", 27" & 44" x 30". 950 IHP. 12K. By Builder

PR: 1923 Newcastle; 1926 Buenos Aires

10.1923:	Completed for France, Fenwick Tyne & Wear Co.Ltd.
1926:	Sold to Cia Argentina de Nav. Angel Gardella Ltda., Buenos Aires
1992:	Still listed in Lloyds Register. Enquiries to Argentina to check existence received no reply!

Marsden *(3) on trials when new.* *(Harrington collection).*

Dunelm *(1). (Author's collection).*

F 17 **DUNELM (1)** **(1923-1956)**
ON 88778 165 g. 115.2 x 19.7 x 10.0 ft. Iron Paddle
B: Hepple & Co., North Shields (YN 376)
E: L. 2 Cy. 30¼" x 48". 460 IHP. By Builder.

PR: 1884 Newcastle

08.1883:	Completed, as J.C. STEVENSON, for Tyne Improvement Commission
03.1923:	Purchased by France, Fenwick Tyne & Wear Co.Ltd., renamed DUNELM
05.1956:	Sold, for £2,000, to British Iron & Steel Corporation. Allocated to C.W. Dorkin Ltd., Gateshead for breaking up

Hendon *(2) as built with short main mast. (Harrington collection).*

F 18 **HENDON (2)** **(1924-1966)**
ON 148060 241 g. 105.0 x 25.5 x 12.7 ft. Steel Screw
B: J.P. Rennoldson & Sons Ltd., South Shields. (YN 323)
E: T. 3 Cy. 16½", 27" & 44" x 30". 982 IHP. 12 K. By Builders

PR: 1924 Newcastle

03.1924:	Completed for France Fenwick Tyne & Wear Co.Ltd.
03.1940:	Hired by Royal Navy. 07.1942: Returned to owners.
01.1959:	Chartered to Tyne Tugs Ltd.
03.1966:	Sold, for £4,000, to P.W. MacLellan, Bo'ness, for breaking up.

F 19 **HECTOR** **(1925-1926)**
See White fleet list, W 3

Corsair *(3). (Maritime Photo Library).*

F 20 **CORSAIR (3)** **(1925-1955)**
See White fleet list, W 4

F 21 **FLYING SCUD** **(1925-c.1933)**
See Coulson fleet list, CN 9

F 22 **PANTHER** **(1925-1926)**
See Coulson fleet list, CN 13

Cullercoats *as built with short main mast. (Harrington collection).*

F 23 **CULLERCOATS** **(1926-1968)**
ON 149414 181 g. 106.5 x 21.1 x 11.2 ft. Steel Screw
B: J.P. Rennoldson & Sons, South Shields. (YN 184)
E: C. 2 Cy. 22" & 40" x 26". 600 IHP., 12 K. By Builder

PR: 1898 Amsterdam; c.1919 S'Gravenhage; 1926 Newcastle

03.1898:	Completed, as CYCLOP, for Amsterdam Tug & Salvage Co.
c.1919:	Sold to Bureau Wijsmuller, Scheep Transport en Sleepvaart Maats.
09.1926:	Purchased by France Fenwick Tyne & Wear Co.Ltd., renamed CULLERCOATS.
07.1956:	Re-engined by R.B. Harrison & Son Ltd., Bill Quay, using engine ex French trawler BEARN. T. 3 Cy. 12¾", 23" & 37" x 26", 700 IHP. 11.5K, made by J. Lewis & Sons, Aberdeen.
01.1959:	Chartered to Tyne Tugs Ltd.
05.1968:	Sold to Clayton & Davie Ltd., Dunston, for breaking up.

Wearmouth *as built with short main mast. (Author's collection)*

F 24 **WEARMOUTH** **(1929-1969)**
ON 160311 182 g. 95.2 x 25.1 x 11.5 ft. Steel Screw
B: Cochrane & Sons Ltd., Selkby. (YN 1036)
E: C. 2 Cy. 21" & 43" x 27". 900 IHP. 11.75 K. Earle's Shipbuilding & Engineering Co.Ltd., Hull.

PR: 1929 Sunderland.

02.1929:	Completed for France, Fenwick Tyne & Wear Co.Ltd.
01.1959:	Chartered to Tyne Tugs Ltd. until 12.1966 when transferred to Sunderland.
01.1969:	Sold to Clayton & Davie Ltd., Dunston, for breaking up.

President *in 1956. (Maritime Photo Library).*

F 25 **PRESIDENT** **(1930-1959)**
See Brown fleet list, BR 2

Cleadon. *(Harrington Collection).*

F 26 **CLEADON** **(1930-1968)**

ON 110579 148 g. 92.0 x 20.0 x 10.9 ft. Steel Screw
B: John Jones & Sons, Liverpool. (YN 146)
E: C. 2 Cy. 16" & 42½" x 27". 622 IHP. By Builder.

PR: 1899 Liverpool; 1930 Newcastle

1899:	Completed, as TOXTETH, for The Alexandra Towing Co.Ltd., Liverpool.
11.1901:	Sold to The Admiralty, renamed C.111 / TOXTETH.
10.1930:	Purchased by France, Fenwick Tyne & Wear Co.Ltd., renamed CLEADON.
1955:	Sank off Sunderland Fish Quay. Reason unknown. Later raised.
1968:	Sold to Clayton & Davie Ltd., Dunston, for breaking up.

Fulwell. *(Harrington Collection).*

F 27 **FULWELL** **(1934-1961)**

ON 110561 149 g. 92.2 x 20.7 x 10.9 ft. Steel Screw
B: John Jones & Sons, Liverpool. (YN 145)
E: C. 2 Cy. 16⅜" & 42½" x 27". 700 IHP. By Builder.

PR: 1899 Liverpool; 1934 Sunderland.

03.1899:	Completed, as NELSON, for The Alexandra Towing Co.Ltd., Liverpool.
06.1934:	Purchased by France, Fenwick Tyne & Wear Co.Ltd., renamed FULWELL.
01.1959:	Chartered to Tyne Tugs Ltd. until 09.1959 when laid up.
12.1961:	Sold, for £4,200, to British Iron & Steel Corporation. Allocated to C.W. Dorkin Ltd., Gateshead, for breaking up.

Conqueror *(2). (Local History Library, South Shields Central Library).*

F 28 **CONQUEROR (2)** **(1944-1956)**

See Redhead fleet list, RH 19

George V *(2) when steam-powered. (John Clarkson, Longton).*

George V *(2) after re-building and diesel-engined. (Company achives/ Turners).*

F 29 **GEORGE V (2)** **(1944-1977)**

See Redhead fleet list, RH 18

Washington. *(George Scott).*

F 30 **WASHINGTON** **(1944-1952)**

See Dry fleet list, D 8

Malta. *(George Scott).*

F 31 **MALTA** **(1944-1951)**

See Allen fleet list, RSA 4

Great Emperor *(2). (John Clarkson, Longton).*

F 32 **GREAT EMPEROR (2)** **(1944-1959)**
See Dry fleet list, D 12

Robert Redhead. *(A. Duncan).*

F 33 **ROBERT REDHEAD** **(1944-1967)**
See Redhead & Dry fleet list, RD 9

Houghton. *(A. Duncan).*

F 34 **HOUGHTON** **(1945-1964)**
See Lambton fleet list, LH 8

Lumley. *(George Scott).*

F 35 **LUMLEY** **(1945-1955)**
See Lambton fleet list, LH 9

Eppleton Hall. *(John Clarkson, Longton).*

F 36 **EPPLETON HALL** **(1945-1964)**
See Lambton fleet list, LH 10

F 37 **STAG (5)** **(1947-1950)**
See Cook fleet list, CK 15

F 38 **SNOWDON** **(1947-1950)**
See Dry fleet list, D 11

Earsdon *in 1955. (Maritime Photo Library).*

F 39 **EARSDON** **(1949-1959)**
See Crosthwaite fleet list, CE 8, as LEWIS CROSTHWAITE

Beamish *newly painted before attending the Coronation Review of the fleet in June 1953. (Turners Photography Ltd.)*

F 40 **BEAMISH** **(1949-1970)**
ON 180286 242 g. 106.7 x 26.7 x 11.6 ft. Steel Screw
B: J.S. Watson (Gainsborough) Ltd., Gainsborough. (YN 1545)
E: T. 3 Cy. 16", 26" & 43" x 30". 850 IHP. 11.5 K. C.D. Holmes & Co.Ltd., Hull.

PR: 1946 Bristol; 1949 Newcastle; 1972 Thunder Bay; 1989 Nanticoke

09.1944:	Completed, as EMPIRE PAUL, for Ministry of War Transport. Southampton, Isle of Wight and South of England Royal Mail Steam Packet Co.Ltd., appointed Managers.
1946:	Sold, for £24,000 to R. & J.H. Rea Ltd., London, renamed QUEENSGARTH.
04.1949:	Purchased by France, Fenwick Tyne & Wear Co. Ltd., renamed BEAMISH.
08.1951:	Chartered to Southampton, Isle of Wight and South of England Royal Mail Steam Packet Co.Ltd. for 3 months.
06.1953:	Represented Towage Industry at Coronation Naval Review at Spithead.
01.1959:	Chartered to Tyne Tugs Ltd.
10.1963:	Re-engined by R.B. Harrison & Sons Ltd., Bill Quay. Converted to diesel propulsion using General Motors Corporation engine made 1947. Oil. 2 SA. 16 Cy. 8.5" x 10". 1400 BHP. 12 K.
04.1970:	Sold to Pounds Marine Shipping Ltd., Southsea.
09.1970:	Sold to Rex Mellenger, Canada, renamed REX MELLENGER.
10.1970:	Sold to Industrial Insulators (1961) Ltd., Canada, renamed BEAMISH.
09.1972:	Sold to Western Engineering Services Ltd., Canada, renamed THUNDER CAPE.
1989:	Sold to Misner Offshore Services (Canada) Ltd.
1992:	Still in service.

Grangetown *in 1950. (Company Archives/ Turners).*

F 41 **GRANGETOWN** **(1950-1970)**

ON 181152 176 g. 82.5 x 24.2 x 10.0 ft. Steel Screw
B: American Shipbuilding Co., Buffalo, N.Y. (YN 918)
E: Oil. 2 SA. 8 Cy. 8½" x 10". 700 BHP. 14 K. General Motors Corporation, Cleveland.

PR: 1943 ?; 1950 Sunderland; 1970 Split

10.1943:	Completed, as DPC 95, for Defence Plant Corporation, U.S.A.
1943:	Transferred to United States Army, renamed ST 764.
10.1950:	Purchased by France, Fenwick Tyne & Wear Co. Ltd., renamed GRANGETOWN.
1970:	Sold to 'Brodospas', Yugoslavia, rrenamed POLARIS.
1992:	Still in service.

Ryhope *in 1950. (Company Archives/ Turners).*

F 42 **RYHOPE** **(1950-1970)**

ON 181151 176 g 82.5 x 24.2 x 10.0 ft. Steel Screw
B: Decatur Iron & Steel Co., Decatur, Alabama. (YN 74)
E: Oil. 2 SA. 8 Cy. 8½" x 10". 700 BHP. 14 K. General Motors Corporation, Cleveland.

PR: 1943 —; 1950 Sunderland; 1970 Split.

11.1943:	Completed, as DPC 74, for Defence Plant Corporation, U.S.A.
1943:	Transferred to United States Army, renamed ST 761.
10.1950:	Purchased by France, Fenwick Tyne & Wear Co.Ltd., renamed RYHOPE.
1970:	Sold to 'Brodospas'. Yugoslavia, renamed KASTOR.
1986:	Broken up by 'Brodospas' at Split.

Cornhill. *(Author's collection)*

F 43 **CORNHILL** **(1951-1977)**

ON 169227 176 g 82.5 x 24.2 x 10.0 ft. Steel Screw
B: American Shipbuilding Co., Buffalo, N.Y. (YN 922)
E: Oil. 2 SA, 8 Cy. 8½" x 10". 700 BHP. 14 K. General Motors Corporation, Detroit.

PR: 1943 ?; 1951 Newcastle.

11.1943:	Completed, as DPC 99, for Defence Plant Corporation, U.S.A.
1943:	Transferred to United States Army, renamed ST 767.
04.1951:	Purchased by France, Fenwick Tyne & Wear Co.Ltd., renamed CORNHILL.
05.1977:	Transferred, for £18,000, to Lawson-Batey Tugs Ltd. and chartered to Wear Tugs Ltd. 04.1978.
07.1980:	Sold, for £25,000, to W.G.S. Crouch & Sons Ltd., Greenhithe, renamed HANNAH SPEARING.
11.1985:	Sold to Marine & General Engineering, Sunbury-on-Thames.
02.1988:	Sold to John Gaskill, Skipton.
1992:	Still in service. No owner quoted in Lloyds Register.

Prestwick. *(Author's collection)*

F 44 **PRESTWICK** **(1955-1977)**

ON 186840 119 g 93'4" x 25'4" x 8'4" Steel Screw
B: P.K. Harris & Sons Ltd., Appledore. (YN 56)
E: Oil. 2 SA. 12 Cy. 8½" x 10". (Made 1944). 1080 BHP. 14 BP. ll K. General Motors Corporation, La Grange, Illinois.

PR: 1955 Newcastle.

03.1955:	Completed for France, Fenwick Tyne & Wear Co.Ltd.
05.1977:	Transferred, for £40,000, to Lawson-Batey Tugs Ltd.
04.1978:	Chartered to Wear Tugs Ltd.
04.1986:	Chartered to Tyne & Wear Tugs Ltd.
12.1987:	Sold, for £20,000, to Medway Port Services Ltd., Strood, and Corveda Marine Ltd., Stanmore.
1992:	Still in service. Owner Medway Port Services only.

Ashbrooke *when on charter to Tyne Tugs Ltd. (Author's collection).*

F 45 **ASHBROOKE** **(1955-1977)**
ON 186845 119 g 93'4" x 25'3" x 8'4" Steel Screw
B: P.K. Harris & Sons Ltd., Appledore. (YN 57).
E: Oil. 2 SA. 12 Cy. 8½" x 10". (Made 1944) 1080 BHP. 14 BP. 11 K.
General Motors Corporation, La Grange, Illinois.

PR: 1955 Newcastle; 1982 Colombo.

07.1955:	Completed for France, Fenwick Tyne & Wear Co.Ltd.
01.1959:	Chartered to Tyne Tugs Ltd.
05.1977:	Transferred, for £40,000, to Lawson-Batey Tugs Ltd. Remained on charter.
05.1982:	Sold, for £15,000, to Orchid Maritime Co.Ltd., Colombo, Sri Lanka.
1984:	Renamed GIGAS and later that year reverted to ASHBROOKE.
1992:	Listed in Lloyds Register but without owner or port of registry.

Alnwick *with name on bridge front. (Maritime Photo Library).*

F 46 **ALNWICK** **(1955-1977)**
ON 186847 119 g 93'4" x 25'3" x 8'6" Steel Screw
B: P.K. Harris & Sons Ltd., Appledore. (YN 58).
E: Oil. 2 SA. 12 Cy. 8½" x 10" (Made 1941). 1080 BHP. 14 BP. 11 K.
General Motors Corporation, Cleveland, Ohio.

PR: 1955 Newcastle; 1987 Gibraltar.

10.1955:	Completed for France Fenwick Tyne & Wear Co.Ltd.
01.1959:	Chartered to Tyne Tugs Ltd.
05.1977:	Transferred, for £40,000, to Lawson-Batey Tugs Ltd. Remained on charter.
05.1981:	Sank whilst moored at Sunderland.
06.1981:	Sold to, and salvaged by, Eurosalve Ltd., Folkestone.
08.1983:	Sold to W.G.S. Crouch & Sons, Greenhithe.
05.1986:	Sold to Estuary Tugs Ltd., Sark.
03.1987:	Sold to Lavington International Ltd., Gibraltar; renamed NORE CREST.
1992:	Still in service.

Bamburgh. *(Company Archives/ Shields Gazette).*

F 47 **BAMBURGH** **(1956-1977)**
ON 186854 122 g 93'4" x 25'3" x 8'4" Steel Screw
B: P.K. Harris & Sons Ltd., Appledore. (YN 59).
E: Oil. 2 SA. 12 Cy. 8½" x 10" (Made 1944). 1080 BHP. 14 BP.
General Motors Corporation, Cleveland, Ohio.

PR: 1956 Newcastle; 1982 Colombo; 1985 Piraeus.

03.1956:	Completed for France, Fenwick Tyne & Wear Co.Ltd.
01.1959:	Chartered to Tyne Tugs Ltd.
05.1977:	Transferred, for £45,000, to Lawson-Batey Tugs Ltd. Remained on charter.
05.1982:	Sold, for £15,000, to Orchid Maritime Co.Ltd., Colombo, Sri Lanka.
1984:	Renamed ATLAS and later that year reverted to BAMBURGH.
1985:	Sold to Mani Maritime Co., Piraeus; renamed SOSTIS.
1992:	Still in service.

Marsden *(4). (Company archives).*

F 48 **MARSDEN (4)** **(1956-1977)**
ON 186859 122 g 93'4" x 25'3" x 8'10¼" Steel Screw
B: P.K. Harris & Sons Ltd., Appledore. (YN 60)
E: Oil. 2 SA. 12 Cy. 8½" x 10" (Made 1944). 1080 BHP. 14 BP.
General Motors Corporation Cleveland, Ohio.

PR: 1956 Newcastle.

07.1956:	Completed for France Fenwick Tyne & Wear Co.Ltd.
01.1959:	Chartered to Tyne Tugs Ltd. until 05.1970 when transferred to River Wear.
05.1977:	Transferred, for £45,000, to Lawson-Batey Tugs Ltd.
04.1978:	Chartered to Wear Tugs Ltd.
04.1986:	Chartered to Tyne & Wear Tugs Ltd.
12.1987:	Sold, for £20,000, to Medway Port Services, Strood, and Corveda Marine Ltd., Stanmore.
03.1989:	Owners now Medway Port Services alone.
1992:	Still in service.

Alnmouth *when on charter to Tyne Tugs Ltd. in 1962. (Maritime Photo Library).*

F 49 ALNMOUTH (1962-1977)
ON 186925 173 g 99′6″ x 25′3″ x 10′8¼″ Steel Screw Kort nozzle
B: Richard Dunston (Hessle) Ltd., Hessle. (YN S.784)
E: Oil. 4 SA. 6 Cy. 12½″ x 14½″. 1080 BHP. 16.5 BP. 11 K.
Ruston & Hornsby Ltd., Lincoln.

PR: 1962 Newcastle; 1987 Piraeus.

05.1962: Completed, for c.£85,000, for France, Fenwick Tyne & Wear Co.Ltd. Chartered to Tyne Tugs Ltd. on delivery.
05.1977: Transferred, for £134,005, to Lawson-Batey Tugs Ltd.
04.1986: Chartered to Tyne & Wear Tugs Ltd.
02.1987: Sold, for £32,500, to Thiela Shipping Co., Greece, renamed THIELA.
1990: Owners now Megaiohari Hellenic Tugboats.
1992: Still in service.

Dunelm *(2). (Author's collection).*

F 50 DUNELM (2) (1964-1977)
ON 305311 132 g 91′9″ x 25′8″ x 10′3½″ Steel Screw
B: Richard Dunston (Hessle) Ltd., Hessle. (YN S.812)
E: Oil. 4 SA. 8 Cy. 10¼″ x 14¼″. 970 BHP. 11K. 14 BP. Ruston & Hornsby Ltd., Lincoln.

PR: 1965 Sunderland; 1988 Fowey.

11.1964: Completed for France Fenwick Tyne & Wear Co.Ltd.
05.1977: Transferred, for £130,000, to Lawson-Batey Tugs Ltd.
04.1978: Chartered to Wear Tugs Ltd.
04.1986: Chartered to Tyne & Wear Tugs Ltd.
03.1988: Sold, for £55,000, to Fowey Harbour Commissioners, renamed PENDENNICK.
1992: Still in service.

Whitburn *(2). (Author's collection).*

F 51 WHITBURN (2) (1964-1977)
ON 305312 176 g 86′7″ x 24′7″ x 10′0″ Steel Screw
B: American Shipbuilding Co., Buffalo, N.Y. (YN 923)
E: Oil. 2 SA. 8 Cy. 8½″ x 10″. 700 BHP. General Motors Corporation, Detroit.

PR: 1943 —; 1949 Aarhus; 1950 Bergen; 1965 Sunderland; 1977 Port Said.

11.1943: Completed, as DPC 100, for Defence Plant Corporation, U.S.A.
1943: Transferred to United States Army, renamed ST 768
1949: Sold to A/S Danena, Aarhus, renamed KRONOS.
1950: Sold to Bergen Steamship Co.
1964: Purchased by France, Fenwick Tyne & Wear Co.Ltd., renamed WHITBURN.
05.1977: Transferred, for £18,000, to Lawson-Batey Tugs Ltd.
10.1977: Sold, for £36,000, to Fouad Hassan Hamza & Sons, Port Said, renamed ABOU SHANAB.
1992: Still in service.

F 52 HENDON (3) (————)
ON 363073 266 g. 32.8 x 9.68 x 4.16 m. Steel Screw
B: Richard Dunston (Hessle) Ltd., Hessle. (YN 907)
E: Oil. 4 SA. 6 Cy. 380 x 457 mm. 3180 BHP. 45 BP. 13 K. Mirrlees Blackstone Ltd., Stockport.

PR: 1977 Newcastle.

05.1977: Completed for France, Fenwick Tyne & Wear Co.Ltd. but not operated by them nor transferred to Lawson-Batey Tugs Ltd. with other France Fenwick assets.
05.1978: Sold to The Alexandra Towing Co.Ltd.
1992: Still in service.

F / M1 EMPIRE PLANE (1941-1943)
ON 168777 129 g. 92.5 x 20.5 x 8.4 ft. Steel Screw
B: Richard Dunston Ltd., Thorne.
E: T. 3 Cy. 12″, 20″ & 32″ x 22″. 500 IHP. McKie & Baxter Ltd., Paisley.

PR: 1941 Goole.

12.1941: Completed for Ministry of War Transport. France, Fenwick Tyne & Wear Co.Ltd. appointed managers.
1943: Managers now Coast Lines Ltd., London.
02.1945: Sank at Greenock, refloated 02.1945.
1946: Owners now Ministry of Transport.
1947: In Royal Navy Service.
1958: Sold to Shipping & Freighting Co.Ltd., London, renamed RIVER ESKIMO.
1964: Broken up at Southampton.

Empire Sara *as* Presto. *(Author's collection).*

F / M2 **EMPIRE SARA** **(1943-1944)**
ON 169307 276 g. 105.2 x 26.6 x 12.2 ft. Steel Screw
B: Cochrane & Sons Ltd., Selby. (YN 1267)
E: T. 3 Cy. 15", 25" & 42" x 27". 850 IHP. Amos & Smith Ltd., Hull.

PR: 1943 Hull.

08.1943:	Completed for Ministry of War Transport. France, Fenwick Tyne & Wear Co.Ltd. appointed managers.
09.1944:	Managers now The Alexandra Towing Co. Ltd., Liverpool.
05.1946:	Sold, for £22,500, to Ellerman's Wilson Line Ltd., Hull, renamed PRESTO.
03.1968:	Transferred to United Towing Co.Ltd., Hull.
06.1968:	Sold to Hughes Bolckow Ltd., Blyth, for breaking up.

F / M3 **EMPIRE ANDREW** **(1944-1945)**
ON 169095 138 g 94.1 x 21.3 x 9.7 ft. Steel Screw
B: Richard Dunston Ltd., Thorne. (YN T.386)
E: T. 3 Cy. 12", 20" & 32" x 22". 500 IHP. Worsley Mesnes Ironworks Ltd., Wigan.

PR: 1944 Goole; 1948 Gdynia.

02.1944:	Completed for Ministry of War Transport. France, Fenwick Tyne & Wear Co.Ltd. appointed managers.
1945:	Managers now Elliott Steam Tug Co., London.
1946:	Owners now Ministry of Transport.
1948:	Sold to Zegluga Polska S.A., Poland, renamed DZIK.
1950:	Transferred to Zarzad Portu Gdansk & Gdynia, Poland.
1953:	Transferred to Polish Government.
c.1962:	Broken up.

Empire Jane *as* Taioma. *(World Ship Photo Library).*

F / M4 **EMPIRE JANE** **(1944-1945)**
ON 165268 232 g. 105.2 x 27.1 x 11.7 ft. Steel Screw
B: A. Hall & Co.Ltd., Aberdeen. (YN 699).
E: T. 3 Cy. 16", 25" & 42" x 27". 850 IHP. By Builder.

PR: 1944 Aberdeen; 1947 Wellington, N.Z.

05.1944:	Completed for Ministry of War Transport. France, Fenwick Tyne & Wear Co.Ltd. appointed managers.
08.1945:	Managers now Townsend Bros. Ferries Ltd.
1946:	Owners now Ministry of Transport.
1947:	Sold to Union Steam Ship Co. of New Zealand Ltd., renamed TAIOMA.
1975:	Sold to Development Services Ltd., Singapore.
1978:	Sold to Robert Arthur Owens, New Zealand.
1978:	Preserved by Tauranga District Museum, New Zealand.

Empire Mustang. *(Local History Library, South Shields Central Library).*

F / M5 **EMPIRE MUSTANG** **(1944-1947)**
See Redhead & Dry fleet list, RD/M1

F / M6 **EXPATRIATE** **(1945-1946)**
ON 169203 134 g. 89.9 x 21.0 x 10.8 ft. Steel Screw
B: G. Seebeck A.G., Wesermunde. (YN 458)
E: T. 3 Cy. 320, 520 & 600 mm. By Builders.

PR: 1925 Bremen; 1945 Newcastle; 1946 ??

07.1925:	Completed for Unterweser Reederei A.G., as EVERSAND.
09.1945:	Taken over by Ministry of War Transport as war reparations, renamed EXPARIATE. Managed by France, Fenwick Tyne & Wear Co.Ltd.
02.1946:	Allocated to the U.S.A. by the Tripartite Commission.
c.1947:	Transferred to United States Maritime Commission.
c.1952:	Transferred to United States Department of Commerce.
1954:	Not in Lloyds Register. Fate unknown.

CHAPTER NINE

Sunderland Towage Co. Ltd. 1900 - 1918

Sunderland Towage Co.Ltd. 1900 - 1918.

In the latter part of the nineteenth century the Sunderland collier company of Taylor & Sanderson owned a number of tugs to assist their vessels in and around Sunderland harbour. These tugs were managed by Fenwick, Stobart & Co.Ltd. One such was the wooden paddle tug *Annandale* (66/1869) registered in the ownership of John W. Taylor c. 1890, and later John E. Dawson c.1896. The Taylor & Sanderson interest in the tugs declined to such an extent that on 1st August 1900 a new company was incorporated to own and manage the tugs involved in these Sunderland operations. The new business was, not surprisingly, named Sunderland Towage Co.Ltd., and they became the owners of *Annandale*.

The initial capital of the new company was £2,400 in £1 shares. These were taken up by Fenwick, Stobart & Co.Ltd. (176), J.W. Taylor (130), J.W. Taylor Jnr. (113), F.W. Taylor (201), W. Sanderson (108), J.E. Dawson (131) T.R. Blumer (50), W. Blumer (50), J. Raine (100), C.H. Marr Jnr. (150), S. Gunn (100), H.K.S. Marshall (111), J.E. Walker (133), S. Marshall (89), W.J. Branfoot (100), A.E. Usher (112), W. Fenwick (22), J. Moon (44), T. Speeding (88), N. Priestman (88), F. Dawson (11), and nineteen others (293). The three directors of the company were J.W. Taylor Jnr. (chairman), F.W. Taylor and J.E. Dawson (managing). The registered office of the company was at 34 Sunniside, Sunderland until 1908 when it was transferred to 21 Bridge Street, Sunderland. The secretary was David Joseph who, on retiring in 1902, was succeeded by Edwin Edwards.

The funnel colours of the Taylor & Sanderson collier fleet were black with a band made up of alternating pale blue and white vertical stripes. This band was said to represent the blue and white armbands worn by police constables. Just why this should be used by Taylor & Sanderson is not known. A variation of this colour scheme, still with the blue and white stripes but with the lower part of the funnel buff, was used for the tugs of Sunderland Towage Co.Ltd. The Taylor & Sanderson house flag design was blue and white vertical stripes. Whether this was used for the tugs is not known, but it does seem likely, since this colour scheme was later retained by France, Fenwick Tyne & Wear Co.Ltd.

At about the time of the establishment of the limited liability company Sunderland Towage purchased the iron paddle tug *Renown* (83/1876) from M. Dent of Jarrow and the wooden paddle tug *Australia* (97/1885), latterly operated as a trawler, was acquired from North Shields owners. Continuing the build-up of their fleet the company, in

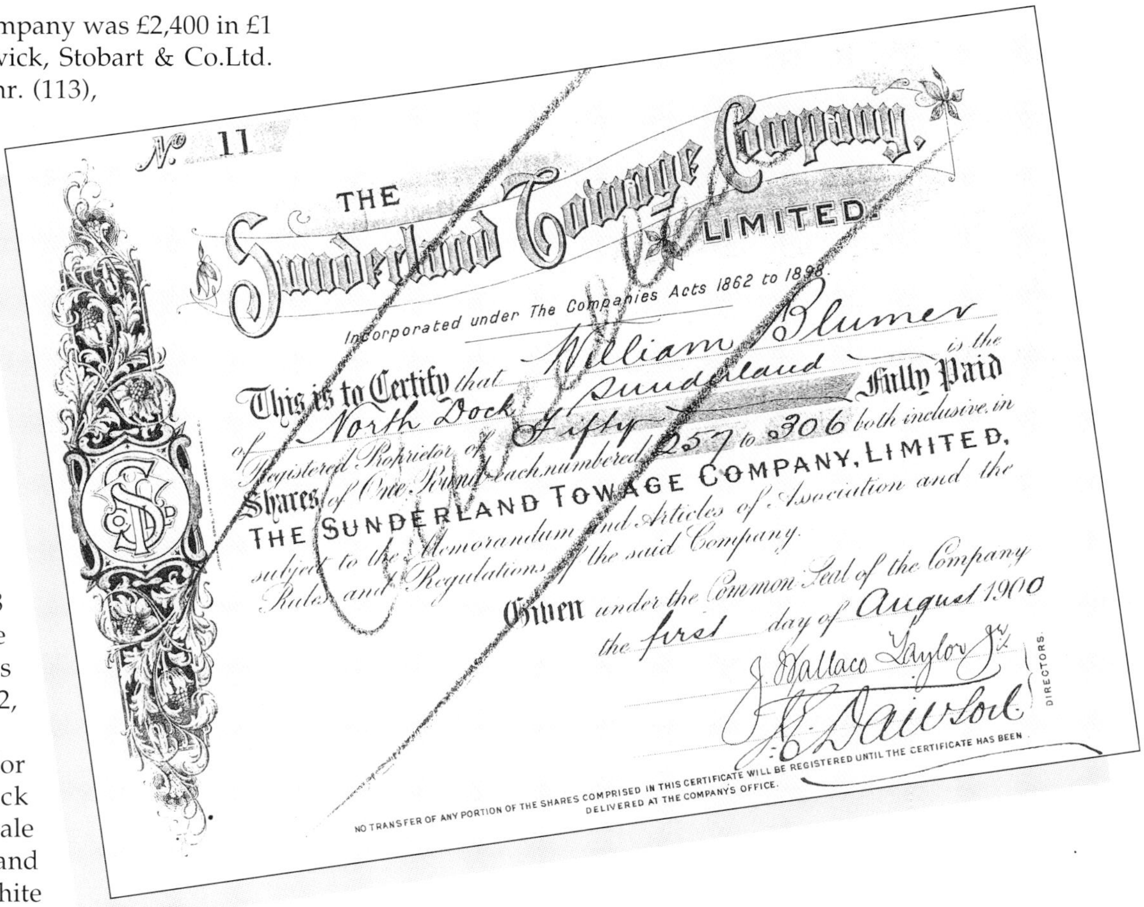

No. 11

THE Sunderland Towage Company LIMITED.

Incorporated under The Companies Acts 1862 to 1898.

This is to Certify that William Blumer of North Dock Sunderland is the Registered Proprietor of Fifty Fully Paid Shares of One Pound each numbered 257 to 306 both inclusive in THE SUNDERLAND TOWAGE COMPANY, LIMITED, subject to the Memorandum and Articles of Association and the Rules and Regulations of the said Company.

Given under the Common Seal of the Company the first day of August 1900

DIRECTORS

NO TRANSFER OF ANY PORTION OF THE SHARES COMPRISED IN THIS CERTIFICATE WILL BE REGISTERED UNTIL THE CERTIFICATE HAS BEEN DELIVERED AT THE COMPANYS OFFICE.

William Blumer's share certificate No.11 from the first batch of 2,400 shares when Sunderland Towage Co. was incorporated on 1st August 1900. The certificate is signed by the chairman and the managing director. (Author's collection).

February and March 1905, took delivery of the new steel paddle tugs *Roker* (119/1905) and *Whitburn* (1) (119/1905) from J.P. Rennoldson & Sons of South Shields. Again, in 1906, two more new paddle tugs were built and named *Marsden* (1) (131/1906) and *Hendon* (1) (130/1906). One strange thing about these two tugs is that, although well over 100 tons gross, they were never recorded in Lloyds Register but did appear in the Mercantile Navy List. Also, though operated by Sunderland Towage, *Roker* and *Whitburn* (1) were officially registered in the name of J.E. Dawson the managing director and not the company. With the arrival of the new tugs *Annandale* was sold to William Cross for £180. *Rapid* (1) (90/1868) was purchased from J. Batey & Son Ltd. about 1906 and in that year *Renown* was sold to Liverpool owners - though she never reached her new home, being wrecked on the way.

Thus Sunderland Towage, at the end of 1906, operated a fleet of six tugs:-

Name	Gross tonnage/Year	Hull Construction	Means of Propulsion
Australia	(97/1885)	Wood	Paddle
Roker	(119/1905)	Steel	Paddle
Whitburn (1)	(119/1905)	Steel	Paddle
Rapid (1)	(90/1868)	Wood	Paddle
Marsden (1)	(131/1906)	Steel	Paddle
Hendon (1)	(130/1906)	Steel	Paddle

- a very modern fleet by the standards of other north-east coast tug companies of the time. *Australia* was soon sold to Jarrow Tug & Lighter Co.Ltd. for £1,100 in August 1907.

In November 1910 the company took delivery of a new steel screw tug - their first with this means of propulsion. She had a compound steam engine developing 400 IHP and was named *Souter* (123/1910). Having in recent years bought five new tugs, it is surprising to find that in 1911 the thirty-eight year old wooden paddle tug *United* (87/1873) was purchased after having had seven previous owners. She cost only £250 so must have been something of a bargain, since she worked on the north-east coast for another sixteen years! *Rapid* (1) was sold in the same year for further local service. In 1913 *Whitburn* (1) was sold to Grangemouth & Forth Towage Co.Ltd. for whom she was to work for the next thirty-nine years. Just after the beginning of the First World War the company took delivery of their second screw tug. Somewhat larger than *Souter*, at 95 feet long, she was *Sunderland* (172/1914) and had been built by Rennoldson at South Shields. Late in 1915 she was hired by the Royal Navy and did not return to her owners, then France Fenwick, until March 1919. Whilst on hire *Sunderland* went to Archangel in North Russia. After the explosion of an ammunition ship in that port all members of the tug's crew were given a decoration for saving life. Both *Souter* and *Sunderland* had been acquired with the purpose of servicing the increasing amount of coastal towage which the company were now obtaining.

In April 1915 the iron screw tug *Symbol* (119/1881) was purchased from South Shields owners and by September she was taken on hire by the Royal Navy for a period of eight months, as were several of the company's other tugs. *Marsden* (1) went to the Dardanelles, and was driven ashore in a gale in October 1915 at Suvla Bay, and subsequently was destroyed by shell-fire from the Turkish Army. The company received £6,500 in compensation from the Admiralty. *Hendon* (1) was hired from June 1915 and was not returned to her owners until March 1920; *Souter* was hired for two-and-a-half months in 1915/1916. To make up in some way for the loss of availability of those tugs on hire the wooden paddle tug *Harwich* (125/1877) was purchased from J. Batey & Son Ltd. Also, in September 1917, delivery was taken of the new screw tug *Marsden* (2) (195/1917), but in the month following she, too, went on hire to the Royal Navy for two years. Whereas earlier tugs supplied by Rennoldson had taken about six months to build, *Marsden* (2) had been on the stocks for over two years due to the builders' work for the Admiralty and shortage of materials and labour. Sunderland Towage had to pay an increase of 28% over the original contract price due to inflation.

In 1918 the iron paddle tug *Transit* (169/1884) was purchased from the Hon. Mrs. Jessie Platt of Stornoway though she was not formally registered with the company. This vessel had originally been owned in Sunderland as the trawler *Frank Buckland*. When sold to the Platt family in 1891 she was renamed *Transit* and, up to the time of her purchase by Sunderland Towage Co.Ltd., had been used by Mrs. Platt as a yacht in the Hebrides during the shooting season. In fact she was so included in Lloyds Register of Yachts. A large model of *Transit* still exists today in the offices of Tyne & Wear Tugs Ltd.

The wreck of Marsden *(1) at Suvla Bay, October 1915.* *(Harrington collection).*

Marsden *(2) and* Souter *towing the remains of* Brussels *(1,380/1902) which had been sunk as a blockship at Zeebrugge by the Germans in October 1918. She was later raised and brought to England. It is thought that this photograph shows her leaving the Humber estuary for the Tyne. (Company archives).*

During the First World War J.E. Dawson was appointed chairman of the War Emergency Committee of the River Wear Commissioners. Under his leadership this committee did valuable work in restoring financial equilibrium to the port of Sunderland's affairs which, due to the war, had become very precarious. The Commission was reorganised and an alliance with the Corporation of Sunderland established. Since its formation in 1900 Sunderland Towage Co.Ltd. had built seven new tugs out of the fifteen it had owned; a notably different position from most of the other local companies. It should also be noted that the new tugs had all been named after localities in the Sunderland area.

In March 1918 William France, Fenwick & Co.Ltd. purchased a substantial proportion of the five thousand shares of Sunderland Towage Co.Ltd. and the remaining shareholders transferred their holdings into a new company, named France, Fenwick Tyne & Wear Co.Ltd., which was established as a subsidiary of William France Fenwick. The eight tugs transferred to the new company were:-

Roker	(119/1905)	Steel	Paddle	
Hendon (1)	(130/1906)	Steel	Paddle	On hire to RN
Souter	(123/1910)	Steel	Screw	
United	(87/1873)	Wood	Paddle	
Sunderland	(172/1914)	Steel	Screw	On hire to RN
Harwich	(125/1877)	Wood	Paddle	
Marsden (2)	(195/1917)	Steel	Screw	On hire to RN
Transit	(169/1884)	Iron	Paddle	

Symbol (119/1881) was not transferred to France, Fenwick Tyne & Wear Co.Ltd., soon being sold to the government and thence to Middlesbrough owners.

The capital of France, Fenwick Tyne & Wear Co.Ltd. was initially £5,000 in £1 shares. These were owned by William France, Fenwick & Co. Ltd. (1039), J.E. Dawson (624), E. Edwards (105), T.W. Greenwell (100), The Gunn family (125), C.H. Marr Jnr. (150), H.K.S. Marshall (201), S. Marshall (89), J. Raine (125), T. Speeding (110), Florence Taylor (125), F.W. Taylor (452), J.W. Taylor Jnr. (398), J.E. Walker (133) and twenty-four others (1224). Most of these shareholders, it should be noted, had been members since the company's formation in 1900.

So it was that the name Sunderland Towage Co.Ltd. disappeared from the scene. Reconstituted as France, Fenwick Tyne & Wear Co.Ltd. it was to become the largest tug-owning company on the north-east coast of England.

APPENDIX A TO CHAPTER NINE

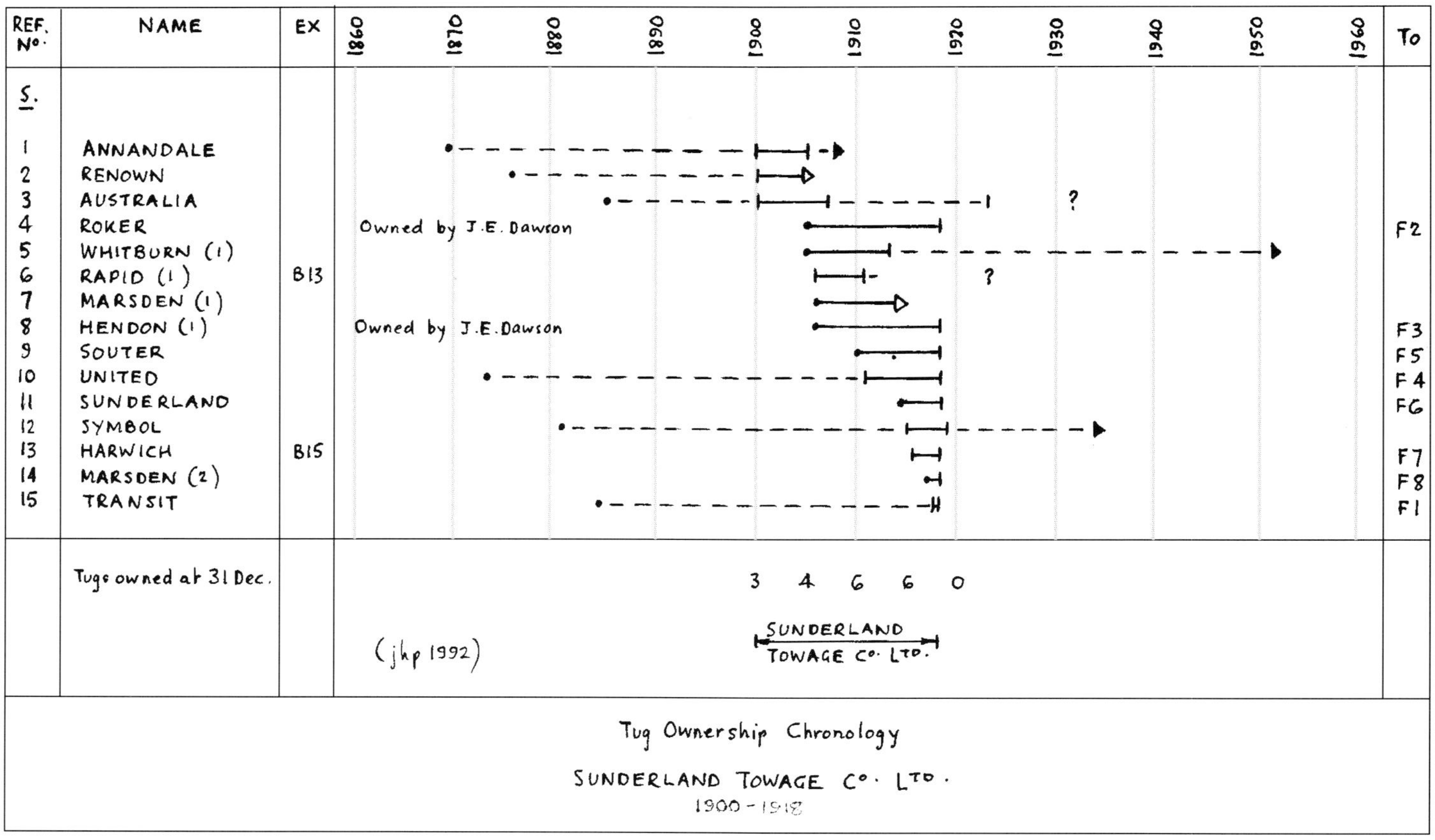

Sunderland Towage Co. Ltd. Fleet List

S 1 **ANNANDALE** **(1900-1905)**
ON 60527 66 g. 79.7 x 17.0 x 8.9 ft. Wood Paddle
B: Richard Stobbs, North Shields.
E: ? x 48". 30 NHP. 12 K. Scott.

PR: 1869 North Shields.

07.1869: Completed for W. Reed, J. Reed and G. Pigg, North Shields.
c.1890: Purchased by John W. Taylor, Sunderland.
c.1896: Transferred to John E. Dawson, Sunderland.
08.1900: Transferred to Sunderland Towage Co.Ltd.
06.1905: Sold, for £180, to William Cross.
c.1908: Sold. Possibly broken up.

S 2 **RENOWN** **(1900-1906)**
ON 72532 83 g. 87.2 x 17.6 x 8.8 ft. Iron Paddle
B: J.T. Eltringham, South Shields.
E: L. 1 Cy. 31½" x 40". 33 NHP. Hepple & Co., South Shields.

PR: 1876 Cardiff; 1900 Sunderland.

08.1876: Completed for C.O. Rundle, Cardiff.
1884: Sold to M. Dent, Jarrow.
08.1900: Purchased, for £1,250, by Sunderland Towage Co.Ltd.
11.1906: Sold, for £250, to G. Hooton, Liverpool, but went ashore at Monifieth on delivery voyage. Believed total loss.

Australia. *(Michael Morris).*

S 3 **AUSTRALIA** **(1900-1907)**
ON 89811 97 g. 90.5 x 18.4 x 9.7 ft. Wood Paddle
B: J.P. Rennoldson & Sons, South Shields.
E: L. 1 Cy. 33" x 50". 36 NHP. By Builder.

PR: 1885 North Shields.

1885: Completed for John Dodds et al, North Shields.
08.1900: Purchased by Sunderland Towage Co.Ltd.
08.1907: Sold, for £1,100, to Jarrow Tug & Lighter Co.Ltd.
1921: Sold to Peter Walker, North Shields.
1923: Registry closed.

S 4 **ROKER** **(1905-1918)**
ON 119208 119 g. 95.0 x 19.0 x 10.0 ft. Steel Paddle
B: J.P. Rennoldson & Sons, South Shields. (YN 234)
E: L. 2 Cy. 28" x 50". 350 IHP. 10 K. By Builder.

PR: 1905 Sunderland.

02.1905: Completed for J.E. Dawson, Sunderland.
05.1918: Transferred to France, Fenwick Tyne & Wear Co.Ltd.
06.1962: Sold to Grangemouth & Forth Towing Co.Ltd.
04.1966: Sold to J.A. White & Co. for breaking up at St. Davids, Fife.

Whitburn *(1). (John Clarkson, Longton).*

S 5 **WHITBURN (1)** **(1905-1913)**
ON 119211 119 g. 95.0 x 19.0 x 10.0 ft. Steel Paddle
B: J.P. Rennoldson & Sons, South Shields. (YN 236)
E: L. 2 Cy. 28" x 50". 350 IHP. 10 K. By Builder.

PR: 1905 Sunderland; 1913 Grangemouth.

03.1905: Completed, for £4,650, for Sunderland Towage Co.Ltd.
12.1913: Sold, for £3,750, to Grangemouth & Forth Towing Co.Ltd.
07.1914: Hired by Royal Navy. 04.1919: Returned to owners.
1940: Requisitioned by Royal Navy. 1945: Returned to owners.
1958: Sold for breaking up.

S 6 **RAPID (1)** **(c.1906-1911)**
See Batey fleet list, B 13.

Marsden *(1) when new. (H. Appleyard).*

S 7 **MARSDEN (1)** **(1906-1915)**
ON 123943 131 g. 95.0 x 19.1 x 10.0 ft. Steel Paddle
B: J.P. Rennoldson & Sons, South Shields. (YN 244?)
E: L. 2 Cy. 28" x 50". 350 IHP. 10 K. By Builder.

PR: 1906 Sunderland.

09.1906: Completed, for £4,975, for Sunderland Towage Co.Ltd.
06.1915: Hired by Royal Navy.
10.1915: Wrecked at Suvla Bay, Gallipoli.

S 8 **HENDON (1)** **(1906-1918)**
ON 123945 130 g 95.0 x 19.1 x 9.9 ft. Steel Paddle
B: J.P. Rennoldson & Sons, South Shields. (YN 245?)
E: L. 2 Cy. 28" x 50". 350 IHP. 10 K. By Builder.

PR: 1906 Sunderland.

09.1906: Completed for J.E. Dawson, Sunderland.
06.1915: Hired by Royal Navy.
05.1918: Transferred to France Fenwick Tyne & Wear Co.Ltd.
03.1920: Returned to owners by Royal Navy.
08.1921: Sold to Board of Trade. No further information.

Souter. *(John Clarkson, Longton).*

S 9 SOUTER (1910-1918)

ON 132054 123 g. 85.0 x 20.1 x 10.1 ft. Steel Screw
B: J.P. Rennoldson & Sons, South Shields. (YN 269)
E: C. 2 Cy. 16" & 34" x 22". 400 IHP. 11 K. By Builder.

PR: 1910 Sunderland.

11.1910: Completed, for £4,800, for Sunderland Towage Co.Ltd.
10.1915: Hired by Royal Navy. 01.1916: Returned to owners.
05.1918: Transferred to France, Fenwick Tyne & Wear Co.Ltd.
01.1964: Sold, for £1,300, to Hughes Bolckow Ltd., Blyth, for breaking up.

United *on the Humber. (Humberside Museum).*

S 10 UNITED (1911-1918)

ON 68899 87 g. 88.9 x 18.3 x 9.4 ft. Wood Paddle
B: James Jackson, South Shields.
E: L. 1 Cy. 34" x 48". 38 NHP. Wear Engine Works, Sunderland.

PR: 1873 Sunderland; 1876 Grimsby; 1894 Yarmouth; 1905 Hull.

10.1873: Completed for John Morgan et al, Sunderland.
05.1874: Sold to Richard McSheen, Grimsby.
1877: Sold to Richard Watnough, Grimsby.
1886: Sold to W. Cooper, Grimsby.
1892: Sold to G.E.J. Moody, Cleethorpes.
1894: Sold to Hewett Co., London.
1904: Sold to City Towing Co., Hull.
03.1911: Purchased, for £250, by Sunderland Towage Co.Ltd.
05.1918: Transferred to France, Fenwick Tyne & Wear Co.Ltd.
1919: Sold to Michael McKeown, Newcastle.
c.1927: Broken up.

Sunderland *on the Mersey in 1924. (John Clarkson, Longton).*

S 11 SUNDERLAND (1914-1918)

ON 132080 172 g. 95.0 x 22.0 x 10.8 ft. Steel Screw
B: J.P. Rennoldson & Sons Ltd., South Shields. (YN 293).
E: C. 2 Cy. 19" & 40" x 26". 96 NHP. By Builder.

PR: 1914 Sunderland; 1924 Liverpool.

11.1914: Completed, for £6,400, for Sunderland Towage Co.Ltd.
12.1915: Hired by Royal Navy.
05.1918: Transferred to France, Fenwick Tyne & Wear Co.Ltd.
03.1919: Returned to owners by Royal Navy.
03.1924: Sold to Mersey Docks & Harbour Board, renamed ASSISTANT.
—.1946: Sold to J.H. Lamey Ltd., Liverpool, renamed J.H. LAMEY.
04.1962: Sold to Routledge Bros., Garston, for breaking up.

S 12 SYMBOL (1915-1919)

ON 134960 119 g. 90.6 x 18.3 x 11.8 ft. Iron Screw
B: L. Tilkin & Co., Liége.
E: C. 2 Cy. 20" & 35" x 22". 350 IHP. By Builder.

PR: 1881 Antwerp; 1914 South Shields; 1919 South Shields; 1919 Middlesbrough.

1881: Completed, as SAMSON, for Soc.Anon. de Remorquage Helice, Antwerp.
04.1913: Sold to L. Brown, South Shields; renamed SYMBOL.
04.1915: Purchased, for £2,300, by Sunderland Towage Co.Ltd.
09.1915: Hired by Royal Navy. 06.1916: Returned to owners.
12.1919: Sold to G. Alder, Middlesbrough, after 3 months ownership by H.M.Government.
1923: Re-engined by Richardsons, Westgarth & Co.Ltd., Middlesbrough. C. 2 Cy. 18" & 36" x 22". 450 IHP. Renamed GRAYTHORPE.
01.1930: Sold to A.M. Crosthwaite and transferred to Middlesbrough Towage Co.Ltd.
08.1934: Sold for breaking up.

S 13 HARWICH (1916-1918)

See Batey fleet list, B 15.

Marsden *(2). (John Clarkson, Longton).*

S 14 MARSDEN (2) (1917-1918)

ON 137265 195 g. 100.2 x 24.0 x 11.2 ft. Steel Screw
B: J.P. Rennoldson & Sons Ltd., South Shields. (YN 303).
E: C. 2 Cy. 20" & 42" x 27". 700 IHP. 11½ K. By Builder.

PR: 1917 Sunderland; 1921 Swansea.

09.1917: Completed, for £12,500, for Sunderland Towage Co.Ltd.
10.1917: Hired by Royal Navy.
05.1918: Transferred to France, Fenwick Tyne & Wear Co.Ltd.
11.1919: Returned to owners by Royal Navy.
04.1921: Sold to British Tanker Co.Ltd., renamed MUMBLES.
02.1931; Wrecked at Oxwich Point, Swansea Bay.

Transit *as* Brigadier.

S 15 TRANSIT (1918 only)

ON 90314 169 g. 116.0 x 19.6 x 10.4 ft. Iron Paddle
B: J.T. Eltringham, South Shields. (YN 122).
E: L. 1 Cy. 38" x 56". 75 NHP. J.P. Rennoldson & Sons, South Shields.

PR: 1885 Sunderland; 1919 Glasgow.

12.1884: Completed, as FRANK BUCKLAND, for J. Siddell et al, Sunderland.
04.1891: Sold to J.A. Platt, Marchington, renamed TRANSIT.
07.1907: Transferred to Hon. Mrs. Jessie Platt, Stornoway.
1918: Purchased by Sunderland Towage Co.Ltd.
05.1918: Transferred to France Fenwick Tyne & Wear Co.Ltd.
08.1918: Hired by Royal Navy, renamed TRANSFER.
05.1919: Returned to owners.
06.1919: Sold to Steel & Bennie Ltd., Glasgow, renamed BRIGADIER.
11.1936: Sold, for £405, to T.W. Ward Ltd., Preston, for breaking up.

• CHAPTER TEN •

Anchor Steam Tug Co. Ltd.

The Anderson/Chater Partnership. c.1867 - 1903.

The precise date of the origin of the partnership between John Anderson and Robert Chater of North Shields has not been discovered, but it is known that in 1867 they purchased the tug *Engineer* (47/1853). They did not retain her for long, however, since by 1871 she had disappeared from the Mercantile Navy List. In the meantime, in 1869, the wooden paddle tug *Defiance* (74/1854) was purchased from Greenock owners, each partner owning thirty-two of the sixty-four shares. She, too, only stayed with the partnership for a short time, going to other North Shields owners in 1872. *Imperial Prince* (63/1858) was purchased from George Stephenson of North Shields in 1873 and *Teazer* joined the fleet in 1875. This latter tug was renamed *Stag* (1) (69/1863) on arrival on the Tyne and was the first of three to be so named by the company.

The first new tug and the first with an iron hull for the partnership was delivered by J.T. Eltringham of South Shields in 1876. She was *Admiral* (102/1876) and powered by a single cylinder side-lever engine constructed by T.R. Scott & Co. of North Shields. Very surprisingly she was sold after only two years to G. Drover of Cowes, but returned to the north-east coast for Sunderland owners four years later. Her skipper in 1876 was T. Chater, no doubt a relative of her part owner. *Imperial Prince* was sold in 1876 and the following year *Stag* (1) went to Sunderland. These disposals were partly balanced by the purchase of *Defence* (69/1862) in August 1877 and further still by *Fiery Dragon* (72/1869) from G. Drover of Cowes in September 1878. As was mentioned above *Admiral* went to Drover at the same time so it is clear that some sort of exchange arrangement with a financial payment to the partnership was negotiated. It seems possible that Drover may have been a shipbroker. *Defence* suffered serious damage when in collision with an unlit and unidentified French fishing smack in September 1878 off the Longstone light. She managed to return to port using only her starboard paddle wheel.

The company's second new tug, *Stag* (2) (90/1878) was delivered in February 1878 with the ownership shared this time between J. Anderson (24/64), R. Chater (24/64) and Mrs. Dorothy Anderson (16/64). The latter person's share was transferred to J. Anderson in October 1882. This latest acquisition brought the Anderson/Chater fleet strength up to three tugs for the first time. In 1882 *Defence* was sold to owners in south-west Scotland and the following year *Fiery Dragon* was sold to Duff & Appleby of Gateshead. The newly-built wooden paddle tug *Brothers* (3) (96/1883) was delivered by J.R. Maxwell of South Shields in 1883, followed by the larger *Stag* (3) (123/1885) two years later, from the same builders. At this time the earlier *Stag* (2) had been sold to Blyth Harbour Commissioners to whom she gave a further twenty-nine years service. The twenty-one-year old *Flying Childers* (84/1864) was purchased in 1885. She had been built originally for the Clyde Shipping Co. and had later been owned at Middlesbrough and Grimsby before coming to the Tyne. Another ex-Clyde Shipping Co. tug, *Flying Arrow* (123/1875), was purchased in 1886. On this occasion Mrs. D. Anderson was the major shareholder. On her death in 1889 the ownership was transferred to Thomas Anderson and, later, G.R. Anderson who were possibly sons of John and Dorothy Anderson. With a length of 108.6 feet *Flying Arrow* was the longest tug which the company owned, although her gross tonnage was to be exceeded slightly by a later acquisition.

In 1890 *Flying Childers* was disposed of to owners at Great Yarmouth leaving a fleet of three in number, until 1895 when the double-engined iron paddle tug *Lord Warden* (118/1868) was purchased via J.T. Eltringham & Co. from the Thames where she had worked since built new. The year 1897 saw *Brothers* (3) sold to R.L. Cook of Sunderland, by whom she was renamed *Rescue* (2) and, in 1899, *Rainbow* (2) (125/1864) another iron paddler, was purchased from the Great Eastern Railway Co. In 1900 *Stag* (3) joined *Brothers* (3) under the ownership of R.L. Cook, and in May 1903 *Flying Arrow* was sold to G.T. Pearson at Hartlepool.

Anchor Steam Tug Co.Ltd. 1903 - 1920

On 25th May 1903 the John Anderson/Robert Chater partnership was converted into a limited liability company with the title Anchor Steam Tug Co.Ltd. It is believed that James Westoll Ltd., shipowners of Sunderland, had a stake in the new company. The company's offices were at 8 New Quay, North Shields, and at Westoll's premises at 13 John Street, Sunderland. The tugs owned at the time of the incorporation of the new company were:-

Name	Gross Tonnage/Year	Hull construction	Means of propulsion
Lord Warden	(118/1868)	Iron	Paddle
Rainbow (2)	(125/1864)	Iron	Paddle

The funnel colours of the company were white with a black top and a blue oblique anchor on the upper part of the white section. The house flag was a white pennant with the blue oblique anchor superimposed.

Immediately following the setting up of the new company the iron screw tug *Triton* (75/1883) and the iron paddle tug *Rescue* (3) (142/1869) were acquired. On Whit Monday, 16th May 1910, the company's first screw tug *Triton* came to grief in unusual circumstances. She was on the annual pleasure cruise with the North Shields "New Quay Water Rats", - another name for the tug companies' water clerks and other riverside workers, - when she came into collision in poor visibility with the 'Tribal' class destroyer HMS *Viking* which was on sea trials after having been built at Palmers shipyard at Jarrow. *Triton* sank on the edge of the Black Middens, Tynemouth, but was later salvaged and returned to service. Several of the thirty passengers sustained injuries but there was no loss of life. What happened to the piano and other musical instruments which were on board for the special entertainment of the party has not been discovered. HMS *Viking* was unusual also, maybe unique, in that she had six funnels. Perhaps the sight of the piano or the six funnels had caused the respective captains to lose their concentration!

During wartime it was common practice for vessels, whilst in port, to receive sealed orders which were not to be opened until they were three miles out to sea. On an occasion in the First World War this happened to one of the Anchor Steam Tug Company's tugs whilst she was lying at Hebburn. She rapidly cast off and steamed the nine miles to sea, whereupon the sealed orders were opened, revealing the instruction to proceed to Dunston, which was six miles upstream from her berth at Hebburn!

In July 1920 the Anchor Steam Tug Co.Ltd. was taken over by France, Fenwick Tyne & Wear Co.Ltd. of Sunderland with the intention of giving the latter company a ready-made presence on the River Tyne. The four tugs which were transferred were:-

Lord Warden	(118/1868)	Iron	Paddle
Rainbow (2)	(125/1864)	Iron	Paddle
Triton	(75/1883)	Iron	Screw
Rescue (3)	(142/1869)	Iron	Paddle

all of which had been with the company since 1903. Thus ended the independent existence of a company which had its origins in the Anderson/Chater partnership some fifty-three years earlier.

APPENDIX A TO CHAPTER TEN

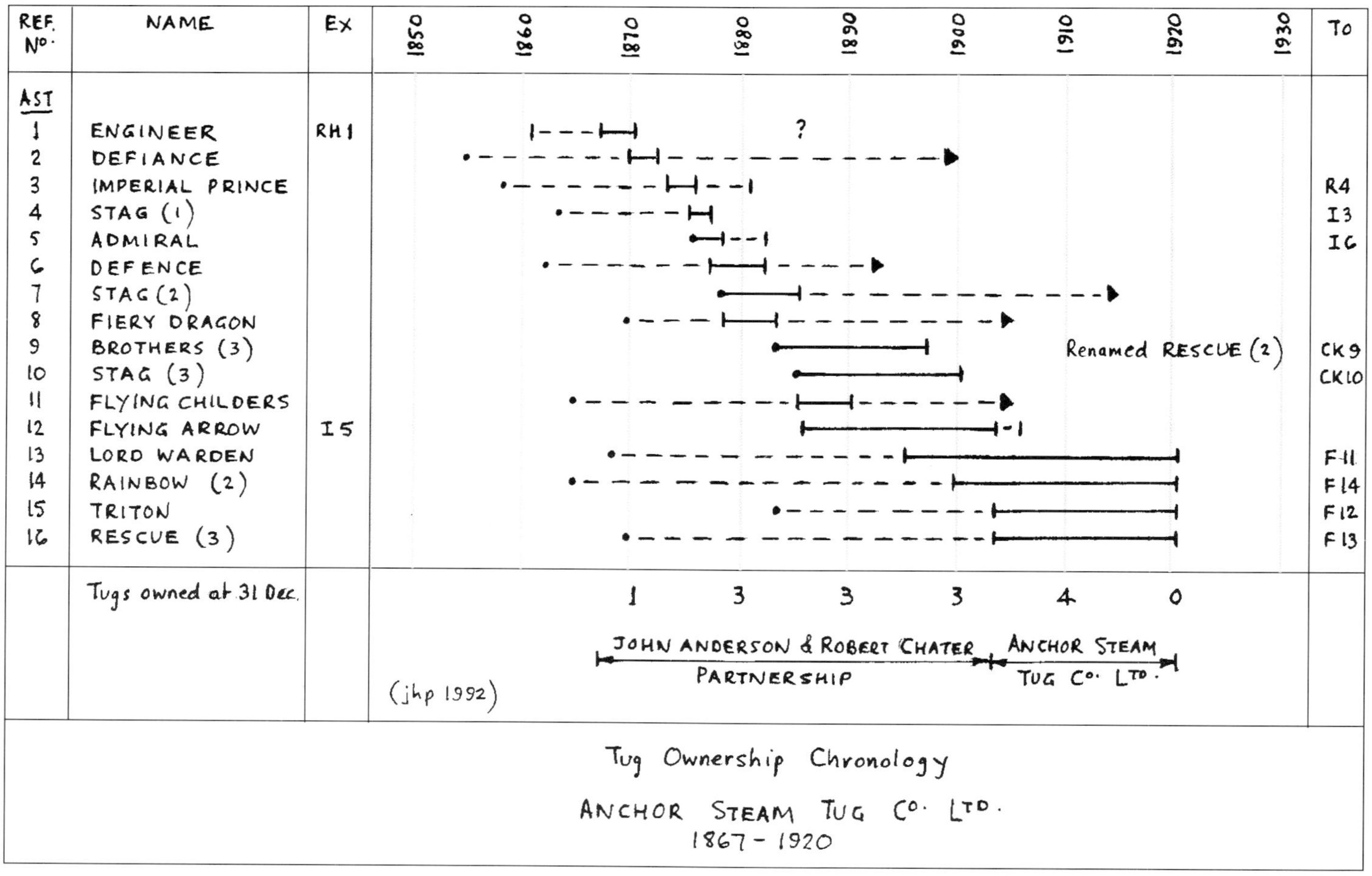

Anchor Steam Tug Co. Ltd. Fleet List

AST 1 — **ENGINEER** **(1867-c.1870)**
See Redhead fleet list, RH 1

AST 2 — **DEFIANCE** **(1869-1872)**
ON 7486 — 74 g. — 83.4 x 17.1 x 9,3 ft. — Wood — Paddle
B: Andrew Woodhouse, South Shields.
E: L. 1 Cy. 35″ x 46″. 50 NHP. J.D. Marshall, South Shields.

PR: 1854 ?; 1866 Greenock; 1869 North Shields.

1854:	Completed, owners not known.
1866:	Owned by James Park, Greenock.
1869:	Purchased by John Anderson and Robert Chater, North Shields.
1872:	Sold to John Turnbull, North Shields.
1876:	Sold to E.H. Lishman.
1878:	Sold to R. Moore, Newcastle.
1884:	Sold to W. Purdy, North Shields.
1891:	Sold to T. Dowson, South Shields.
1900:	Broken up.

AST 3 — **IMPERIAL PRINCE** **(1873-1876)**
ON 28668 — 63g. — 80.8 x 15.6 x 9.0 ft. — Wood — Paddle
B: James Dowey, North Shields.
E: 25 NHP. J.P. Rennoldson, South Shields.

PR: 1858 Shields.

1858:	Completed for G. Chisholm, Shields.
c.1869:	Sold to George Stephenson, North Shields.
1873:	Purchased by J. Anderson and R. Chater, North Shields.
1876:	Sold to E.H. Lishman.
1877:	Sold to R. Moore, Newcastle.
1881:	Purchased by James Ridley, Bill Quay.
1888:	Transferred to John Ridley, Bill Quay.
1895:	Not in register.

AST 4 — **STAG (1)** **(1875-1877)**
ON 45614 — 69 g. — 79.3 x 16.9 x 9.1 ft. — Wood — Paddle
B: J.D. Burdis, North Shields.
E: L. 1 Cy. 29½″ x 48″. 28 NHP. William Scott, North Shields.

PR: 1863 Montrose; 1876 North Shields; 1877 Sunderland.

1863:	Completed, as TEAZER, for Montrose owners.
1865:	Owners amalgamated with others and became The Montrose Steam Tug Co.Ltd.
1875:	Purchased by J. Anderson and R. Chater, North Shields.
01.1876:	Renamed STAG.
07.1877:	Purchased by J.H.W. Culliford and A. Irwing, Sunderland.
07.1884:	Owners now J.H.W. Culliford, A. Irwing and J.L. Clark, Sunderland.
07.1890:	Owner now A. Irwing only.
05.1897:	Vessel broken up.

AST 5 — **ADMIRAL** **(1876-1878)**
ON 70421 — 102 g. — 96.5 x 18.1 x 9.4 ft. — Iron — Paddle
B: J.T. Eltringham, South Shields.
E: L. 1 Cy. 35″ x 51″. 40 NHP. T.R. Scott & Co., North Shields.

PR: 1876 North Shields; 1878 Liverpool; 1882 Sunderland; 1884 Inverness; 1888 Bo'ness.

1876:	Completed for J. Anderson and R. Chater, North Shields.
1878:	Sold to G. Drover, Cowes.
1882:	Sold to J.H.W. Culliford and A. Irwing, Sunderland.
04.1883:	Sold to Inverness Steam Trawling Co.Ltd.
09.1888:	Sold to J.S. & D.S. Wilson.
01.1890:	Sold to Elizabeth and J.S. Wilson.
09.1895:	Sold to J.S. Wilson, Charlestown, Fife.
1917:	Purchased by Lawson Steam Tug-boat Co.Ltd.
07.1918:	Captured by German submarine 2 miles N. of Flamborough Head and sunk by gunfire. No lives lost.

AST 6 — **DEFENCE** **(1877-1882)**
ON 43631 — 69 g. — 81.8 x 17.3 x 8.9 ft. — Wood — Paddle
B: John Turnbull, North Shields.
E: 1 Cy. 30 NHP. Marshall Bros., South Shields.

PR: 1862 South Shields; 1877 North Shields; 1882 Ardrossan; 1886 Belfast.

07.1862:	Completed for John Hall et al, South Shields.
c.1870:	Sold to W.H. Burn, South Shields.
08.1877:	Purchased by J. Anderson and R. Chater, North Shields.
1882:	Sold to J. Murray, Saltcoates.
1886:	Sold to T. McCracken, Belfast.
1891:	Sold to J. Hull, Belfast.
1893:	Broken up.

AST 7 — **STAG (2)** **(1878-1885)**
ON 76654 — 90 g. — 92.0 x 17.9 x 9.8 ft. — Wood — Paddle
B: Wouldhave & Johnston, North Shields.
E: L. 1 Cy. 33⅛″ x 48″. 36 NHP. T.R. Scott & Co., North Shields.

PR: 1878 North Shields; 1910 Blyth.

02.1878:	Completed for J. Anderson, R. Chater and Dorothy Anderson, North Shields.
10.1882:	Owners now J. Anderson and R. Chater only.
03.1885:	Sold to Blyth Harbour Commissioners.
12.1914:	Sold to Hughes Bolckow & Co.Ltd., Blyth, for breaking up.

AST 8 — **FIERY DRAGON** **(1878-1883)**
ON 62552 — 72 g. — 82.4 x 16.8 x 9.0 ft. — Wood — Paddle
B: Readhead, Softley & Co., South Shields.
E: L. 1 Cy. 30″ x 48″. 30 NHP. By Builder.

PR: 1870 Sunderland; 1872 Portsmouth; 1878 North Shields.

12.1869:	Completed for Marjiam & Co., Sunderland.
1872:	Sold to J. White, Portsmouth.
1874:	Sold to G. Drover, Cowes.
09.1878:	Purchased by J. Anderson and R. Chater, North Shields.
05.1883:	Sold to R. Duff and R. Appleby, Gateshead.
c.1905:	Vessel broken up.

AST 9 — **BROTHERS (3)** **(1883-1897)**
ON 84880 — 96 g. — 92.0 x 18.4 x 9.7 ft. — Wood — Paddle
B: J.R. Maxwell, South Shields.
E: L. 1 Cy. 34″ x 50″. 40 NHP. Baird & Barnsley, North Shields.

PR: 1883 North Shields; 1897 Sunderland.

1883:	Completed, as BROTHERS, for J. Anderson and R. Chater, North Shields.
09.1897:	Purchased by Robert Lumley Cook, Sunderland, renamed RESCUE.
09.1908:	Transferred to Mrs. Elizabeth S. Cook, Sunderland.
10.1912:	Transferred to Robert Cook Baister, Sunderland.
07.1915:	Broken up.

AST 10 — **STAG (3)** **(1885-1900)**
ON 89815 — 123 g. — 100.0 x 19.5 x 10.5 ft. — Wood — Paddle
B: J.R. Maxwell, South Shields.
E: L. 1 Cy. 36¾″ x 52″. 45 NHP. Baird & Barnsley, North Shields.

PR: 1886 North Shields; 1900 Sunderland.

04.1885:	Completed for J. Anderson and R. Chater, North Shields.
11.1900:	Purchased by Robert Lumley Cook, Sunderland.
09.1908:	Transferred to Mrs. Elizabeth S. Cook, Sunderland.
10.1912:	Transferred to Robert Cook Baister, Sunderland.
12.1921:	Transferred to Robert Lumley Cook Baister, Sunderland.
10.1922:	Sold to T.C. Potts, Sunderland, for breaking up.

AST 11 — **FLYING CHILDERS** **(1885-1890)**
ON 50351. — 84 g. — 88.4 x 17.9 x 9.2 ft. — Wood — Paddle
B: Andrew Woodhouse, South Shields.
E: L. 1 Cy. 33½″ x 48″. 40 NHP. J.P. Rennoldson, South Shields.

PR: 1864 Glasgow; 1868 Glasgow; 1873 Middlesbrough; 1875 Grimsby; 1885 North Shields; 1890 Yarmouth; 1896 Hull.

1864:	Completed for Clyde Shipping Co., (J.P. Kidston).
04.1873:	Sold to John Robinson and Thomas Robson, Middlesbrough.
06.1875:	Sold to J.H. Manton, Grimsby.
10.1885:	Purchased by J. Anderson and R. Chater, North Shields.
10.1890:	Sold to E.A. Durrant, Yarmouth.,
1896:	Sold to J. Guy, Hull.
1898:	Sold to Flying Childers Steam Tug Co., Kings Lynn.
1902:	Sold to Flying Childers Steam Tug Co., Hull.
12.1905:	Vessel broken up by M. McMillan.

AST 12 — **FLYING ARROW** **(1886-1903)**
See Irwing fleet list, I 5

AST 13 **LORD WARDEN** **(1895-1920)**
ON 60856. 118 g. 100.4 x 18.3 x 10.1 ft. Iron Paddle
B: Readhead, Softley & Co., South Shields. (YN 36).
E: L. 2 Cy. 28" x 48". 55 NHP. By Builder.

PR: 1868 London; 1895 North Shields; 1920 Middlesbrough.

1868:	Completed for J. Martin, Gravesend.
1874:	Sold to W.J. Homewood, London.
1879:	Sold to T.A.F. Banks, Gravesend.
1886:	Sold to F. Pattison, Gravesend.
1895:	Purchased by J. Anderson, North Shields, via J.T. Eltringham & Co., South Shields.
05.1903:	Transferred to Anchor Steam Tug Co.Ltd., North Shields.
07.1920:	Transferred to France, Fenwick Tyne & Wear Co.Ltd.
09.1920:	Sold to C.W. Robinson, Middlesbrough.
09.1920:	Sold to Charles Duncan & Sons Ltd., Middlesbrough.
01.1924:	Sold for breaking up.

AST 14 **RAINBOW (2)** **(1899-1920)**
ON 50083. 125 g. 90.0 x 18.5 x 9.9 ft. Iron Paddle
B: J. Ash & Co., Cubitt Town.
E: L. 2 Cy. 30" x 42". 60 NHP. J. Stewart, London.

PR: 1864 London; 1885 Lowestoft; 1899 North Shields.

1864:	Completed for Great Eastern Railway Co., London.
1899:	Purchased by J. Anderson, North Shields.
05.1903:	Transferred to Anchor Steam Tug Co.Ltd.
07.1920:	Transferred to France, Fenwick Tyne & Wear Co.Ltd.
1929:	Sold for breaking up at Sunderland.

AST 15 **TRITON** **(1903-1920)**
ON 86130 75 g. 84.7 x 16.3 x 9.2 ft. Iron Screw
B: Cox & Co., Falmouth.
E: C. 2 Cy. 16" & 30" x 21". 40 NHP. By Builder.

PR: 1884 Falmouth; 1899 North Shields.

1883:	Completed for The Falmouth Fishery Co.Ltd., Falmouth.
1893:	Transferred to The Falmouth Towage Co.Ltd., Falmouth.
1899:	Sold to William Black, Sunderland.
1903:	Purchased by Anchor Steam Tug Co.Ltd., North Shields.
07.1920:	Transferred to France, Fenwick Tyne & Wear Co.Ltd.
1922:	Sold to John Davies, Cardiff.
05.1930:	Broken up.

Rescue *(3) in Winlo's wooden floating dock. (Local History Library, South Shields Central Library).*

AST 16 **RESCUE (3)** **(1903-1920)**
ON 63510 142 g. 107.2 x 18.5 x 9.8 ft. Iron Paddle
B: T. Hepple & Co., Low Walker.
E: L. 2 Cy. 28½" x 54". 60 NHP. By Builder.

PR: 1869 London; 1891 Middlesbrough; 1897 North Shields.

1869:	Completed for W.J. Thomas et al, Gravesend.
c.1872:	Sold to Dr. M. Brownfield, Poplar.
c.1879:	Sold to W. Renney, London.
c.1881:	Sold to G.R. Lardner, Gravesend.
c.1885:	Sold to H. Green, Blackwall.
c.1886:	Sold to J. Mitchell, Gravesend.
11.1891:	Sold to T.F. Ward, Middlesbrough.
10.1892:	Sold to Tees Tug & Lighter Co.Ltd., Middlesbrough.
10.1897:	Sold to William Black, Sunderland.
1903:	Purchased by Anchor Steam Tug Co.Ltd., North Shields.
07.1920:	Transferred to France, Fenwick Tyne & Wear Co.Ltd.
1932:	Broken up.

• CHAPTER ELEVEN •

The Coulson Companies

John Coulson. c.1866 - 1918

The precise date of John Coulson's first involvement in the towage business is not known, but what is known is that in 1866 he bought two small wooden paddle tugs, *Active* (46/1836) and *Gosforth* (52/1857) both of which were sold in the following year. There is then a gap in the information available about John Coulson until 1873, when he, with H. Gilroy, purchased *Rainbow* (1) (37/1854). In 1875 the partners then acquired *Cumbrian* (66/1859) from owners in Boston. *Rainbow* (1) was sold in 1885 and *Cumbrian* in 1887, both for further operations on the River Tyne. *Miss Roberts* (54/1870) was purchased in 1886 from Aberdeen owners and stayed with John Coulson for eleven years. In June 1892 *Star of Hope* (79/1877) was acquired from J.H. Newton of Percy Main. Although built on the Tyne she had been initially owned abroad before returning to the north-east coast in 1880.

Around the turn of the century John Coulson's address was quoted as Mile End Road, South Shields, and shortly afterwards 19 Salmon Street, in the same town. Whether these were business or private addresses or both is not known. *Black Watch* (84/1884) was purchased from Thomas Barlow of North Shields in 1901 and remained with the Coulson business until broken-up in 1907. So far, all the seven tugs owned by John Coulson had wooden hulls, but in 1909 he obtained his first iron-hulled tug, *Prince of Wales* (88/1886). She had already had two owners in Cardiff and one each in Grangemouth and Arbroath. After three years, in 1912, she disappeared from the registers and was probably broken up. In 1917, *Flying Scud* (74/1877) was added to the fleet following a very brief spell on hire to the Royal Navy.

The Coulson Tug Co.Ltd. 1918 - 1925

In August 1918 John Coulson took over the John White towing concern and absorbed the two tugs owned by that firm. The combined company was named the Coulson Tug Co.Ltd. John White Sr. had commenced tug ownership in 1852 when he took delivery of the newly-built *Corsair* (1) (81/1852). Following the sale of this tug in 1859 there is a gap in the information discovered regarding the White business until, in 1877, John White Jr. took delivery of another new tug, *Corsair* (2) (85/1877), which, after transfer of ownership to female members of the family, was sold to Blyth in 1891. *Hector* (101/1882) was another new tug acquired in 1882, ownership of which also for a time went to the female side of the family. She was in the news on 27th July 1913 when her crew rescued a man washed overboard from the submarine *C.27* whilst leaving the Tyne. *Hector's* skipper, James Weddle, was presented with a silver tobacco box by Lt. G.S. White, the rescued man, and the crew received gifts of money. *Corsair* (3) (153/1892), a large paddle tug with an iron hull was purchased in 1892. It will have been noted that all four of the 'White' tugs had been built new for the business. At the time this was quite unusual for Tyne-based owners, since most companies had to make do with second-hand tonnage. The funnel colours of the 'White' tugs were white with a deep

Dunera towing Glentworth (5,677/1920) at Blyth in the 1920s. *(Company archives).*

black top, superimposed on which were four white vertical panels. As a result of these latter items the tugs were nicknamed the 'Tombstone Fleet'. It is not known what the funnel colours of the Coulson tugs were prior to 1918, but after the merger the new limited liability company continued to use the White company's colours.

Following the merger the combined fleet consisted of:-

Name	Gross Tonnage/Year	Hull Construction	Means of Propulsion	
Star of Hope	(79/1877)	Wood	Paddle	Ex Coulson
Flying Scud	(74/1877)	Iron	Paddle	
Hector	(101/1882)	Wood	Paddle	Ex White
Corsair (3)	(153/1892)	Iron	Paddle	

with an average age of thirty-six years.

Star of Hope was disposed of in 1920 and in April 1921 the steel screw-propelled tug *Archdale* was purchased from the British Government and renamed *John Coulson* (119/1896). In the same year another screw tug *Panther* (150/1884) was obtained from The Canadian Pacific Railway Co.Ltd. She had begun her existence as *Elliott & Jeffery*, the only tug known to this writer which was named after her owners, builders and engine builders. About the same time the wooden paddle tug *Harwich* (125/1877) was purchased from France, Fenwick Tyne & Wear Co.Ltd. and in 1922 the iron screw tug *Dunera* (91/1892) was brought from the Thames where she had worked since being built. These last two tugs were set to work at Blyth in competition with the Blyth Tug Co.Ltd. In June 1923 *John Coulson* was sold to owners at Belfast, thus reducing the fleet to six in number.

In 1925 the Coulson Tug Co.Ltd. was taken over by France, Fenwick Tyne & Wear Co.Ltd. The four tugs transferred were:-

Flying Scud	(74/1877)	Iron	Paddle
Hector	(101/1882)	Wood	Paddle
Corsair (3)	(153/1892)	Iron	Paddle
Panther	(150/1884)	Iron	Screw

The other two tugs *Harwich* (125/1877) and *Dunera* (91/1892) were sold to Blyth Tug Co.Ltd. in March 1926 for £1,500 and £2,000 respectively, thus ending the sixty years existence of the Coulson enterprise.

APPENDIX A TO CHAPTER ELEVEN

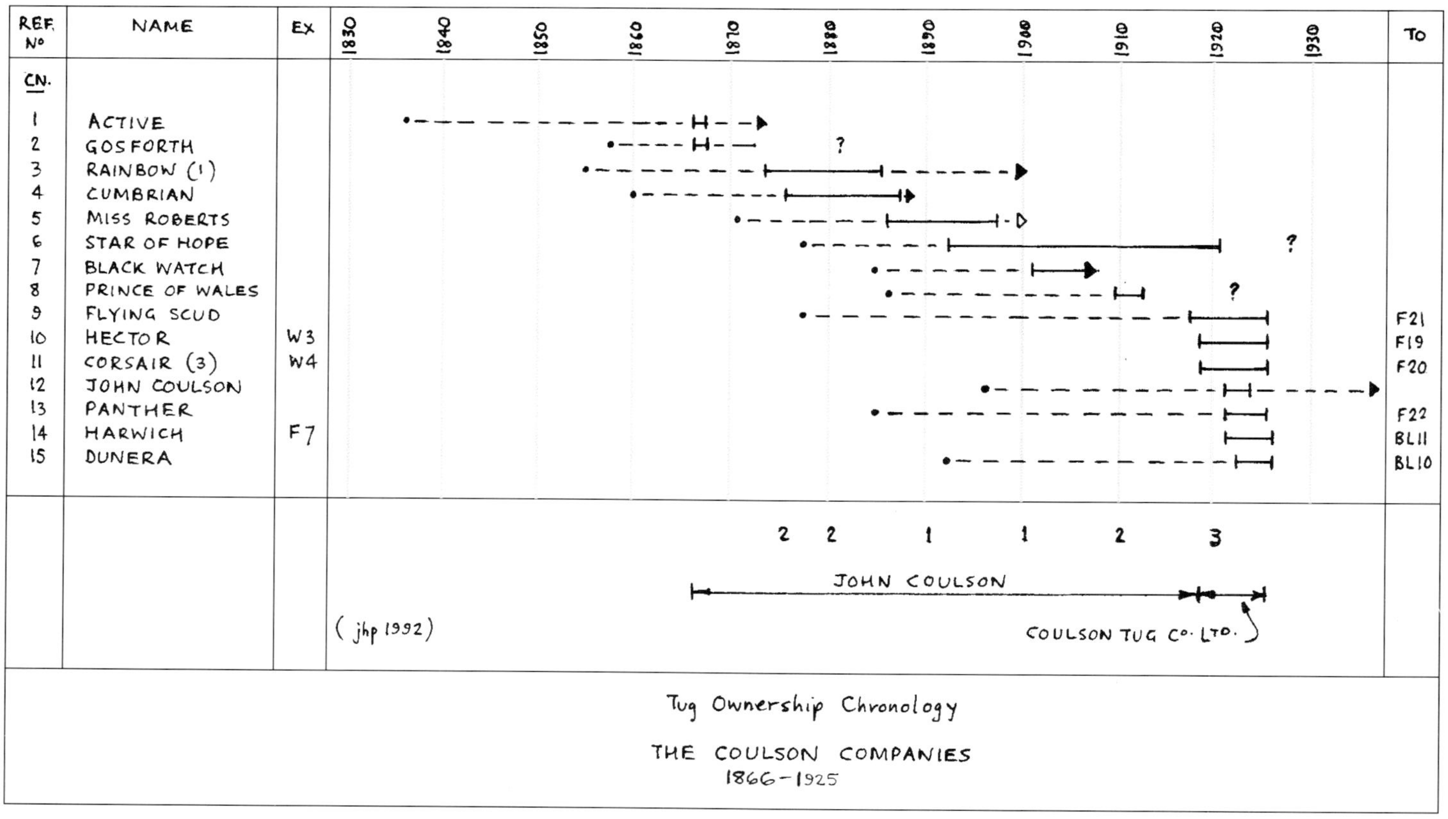

APPENDIX B TO CHAPTER ELEVEN

The Coulson Companies Fleet List

CN 1 **ACTIVE** **(1866-1867)**
ON 15258 46 g. 66.1 x 15.2 x 7.7 ft. Wood Paddle
B: At South Shields.
E: L. 1 Cy. 25 NHP.

PR: 1836 ?; 1853 Shields; 1869 Middlesbrough.

1836: Completed. Owner not known.
1853: Sold to G. Brown, South Shields.
1858: Sold to William and John Armstrong, South Shields.
1866: Purchased by John Coulson, North Shields.
1867: Sold to E.R.D. Halder and P.S. Wheatley, Middlesbrough.
1869: Sold to Alex. Petrie et al, Middlesbrough.
12.1873: Broken up.

CN 2 **GOSFORTH** **(1866-1867)**
ON 15999 52 g. 75.8 x 14.9 x 7.0 ft. Wood Paddle
B: At North Shields.
E: 26 NHP.

PR: 1857 Shields; 1867 Guernsey.

1857: Completed for W. Snowball et al, Shields.
1860: Sold to G. Knox et al, Shields.
1866: Purchased by John Coulson, North Shields.
1867: Sold to A. Collenette, Guernsey.
c.1872: Sold to W.W. Bird, Guernsey.
1873: Not in register.

CN 3 **RAINBOW (1)** **(1873-1885)**
ON 2165 37 g. 70.7 x 14.1 x 6.8 ft. Wood Paddle
B: At North Shields.
E: 20 NHP.

PR: 1854 Shields; 1863 Newcastle.

1854: Completed for B. Dain, North Shields.
1858: Sold to T. Hepple and another, North Shields.
1863: Sold to J. Hopper Jnr., Newcastle.
1873: Purchased by John Coulson and H. Gilroy jointly.
1885: Sold to R. Jull, St. Anthony's.
1900: Broken up.

CN 4 **CUMBRIAN** **(1875-1887)**
ON 26873 66 g. 83.6 x 15.9 x 8.8 ft. Wood Paddle
B: At Low Walker.
E: 30 NHP.

PR: 1859 Newcastle; 1859 Boston.

1859: Completed for Robert Bell.
07.1859: Sold to W. Wilkinson, Boston.
1860: Sold to J.H. Thomas et al, Boston.
1875: Purchased by John Coulson and H. Gilroy jointly.
10.1887: Sank at her moorings, later raised.
1887: Sold to J. Barker, North Shields.
1888: Broken up.

CN 5 **MISS ROBERTS** **(1886-1897)**
ON 60542 54 g. 76.7 x 15.8 x 8.2 ft. Wood Paddle
B: Heckels Bros., North Shields.
E: L. 1 Cy. 26″ x 43″. 22 NHP. By Builder.

PR: 1870 North Shields; 1883 Aberdeen; 1886 South Shields.

03.1870: Completed for Hugh Roberts et al.
1875: Sold to G. Smith et al, North Shields.
02.1883: Sold to C. Noble Jnr., Aberdeen.
1886: Purchased by John Coulson.
06.1897: Sold to T. Hutchinson, Monkwearmouth.
12.1900: Sank 4½ miles east of Hartlepool.

CN 6 **STAR OF HOPE** **(1892-1920)**
ON 79242 79 g. 84.7 x 17.9 x 9.2 ft. Wood Paddle
B: Wouldhave & Johnston, North Shields.
E: L. 1 Cy. 30½ x 48″. 30 NHP. Baird & Barnsley, North Shields.

PR: 1877 ?; 1880 North Shields.

1877: Completed, as WAKENDAM No.9, for foreign owners.
11.1880: Sold to J.H. Newton,Percy Main.
06.1892: Purchased by John Coulson, South Shields.
1918: Transferred to The Coulson Tug Co. Ltd.
1920: Registry closed.

CN 7 **BLACK WATCH** **(1901-1907)**
ON 89792 84g 89.0 x 18.0 x 9.6 ft. Wood Paddle
B: J.R. Maxwell, South Shields.
E: L. 1 Cy. 31″ x 48″. 32 NHP. Baird & Barnsley, North Shields.

PR: 1884 North Shields.

1884: Completed for Thomas Barlow et al, North Shields.
01.1901: Purchased by John Coulson.
10.1907: Registry closed, vessel broken up.

CN 8 **PRINCE OF WALES** **(1909-1912)**
ON 89206 88 g. 88.5 x 17.7 x 8.8 ft. Iron Paddle
B: At North Shields.
E:

PR: 1886 Cardiff.

1886: Completed for R. Morgan, Cardiff.
1897: Sold to W.H. Tucker, Cardiff.
1899: Sold to W. Taylor, Grangemouth.
1901: Sold to W.K. McDonald, Arbroath.
1909: Purchased by J. Coulson, South Shields.
1912: Not in register.

CN 9 **FLYING SCUD** **(1917-1925)**
ON 70512 74 g. 83.4 x 17.1 x 8.4 ft. Iron Paddle
B: Jones, Neath.
E: L. 1 Cy. 31½″ x 50″. 35 NHP. William Scott, North Shields.

PR: 1877 Swansea; 1895 Peterhead.

11.1877: Completed for R. Rosser, Swansea.
1895: Sold to William H. Leask, Peterhead.
10.1917: Hired by Royal Navy for less than one month.
1917: Purchased by John Coulson
1918: Transferred to The Coulson Tug Co. Ltd.
1925: Transferred to France, Fencwick Tyne & Wear Co. Ltd.
c.1933: Fate not known.

CN 10 **HECTOR** **(1918-1925)**
See White fleet list, W 3.

CN 11 **CORSAIR (3)** **(1918-1925)**
See White fleet list, W 4.

CN 12 **JOHN COULSON** **(1921-1923)**
ON 105277 119g. 95.8 x 18.2 x 10.1 ft. Steel Screw
B: Allsup & Co. Ltd., Preston.
E: C. 2 Cy. 17½″ & 35¼″ x 24″. 320 IHP. 10K. W. Allsup & Sons, Preston.

PR: 1896 Plymouth; 1913 Middlesbrough; 1921 South Shields.

1896: Completed, as VIXEN, for T.W. Fox, Plymouth.
1908: Sold to G. Robinson & Co., Middlesbrough.
05.1909: Sold to The Tees Tug Co. Ltd.
11.1914. Renamed ARCHDALE.
09.1915: Requisitioned by Royal Navy. 12.1915: Returned to owners.
03.1919: Sold to Dept. of Director-General of Movements & Railways.
04.1921: Purchased by Coulson Tug Co. Ltd., renamed JOHN COULSON.
06.1923: Sold to Mrs. Jane E. Strong, Belfast.
09.1931: Transferred to James E. Strong, Belfast.
1937: Sold for breaking up at Glasgow.

CN 13 **PANTHER** **(1921-1925)**
ON 89188 150 g. 105.2 x 19.8 x 10.7 ft. Iron Screw
B: Elliott & Jeffery, Cardiff.
E: C. 2 Cy. 19″ & 35″ x 20″. 70 NHP. By Builder.

PR: 1884 Cardiff; 1901 Liverpool.

1884: Completed, as ELLIOTT & JEFFERY, for Elliot & Jeffery, Cardiff.
1901: Sold to Mersey Towing Co. Ltd. (Elder Dempster & Co.), renamed PANTHER.
1903: Managers now Canadian Pacific Railway Co. Ltd.
1921: Purchased by Coulson Tug Co. Ltd.
1925: Transferred to France, Fenwick Tyne & Wear Co. Ltd.
1926: Broken up.

CN 14 **HARWICK** **(c1921-1926)**
See Batey fleet list, B 15.

CN 15 **DUNERA** **(1922-1926)**
ON 99078 91 g. 86.0 x 19.1 x 9.9 ft. Iron Screw
B: Hepple & Co., North Shields.
E: C. 2 Cy. 17" & 32" x 24". 80 NHP. By Builder.

PR: 1892 London

05.1892:	Completed for Gravesend United Tug Co.
1922:	Purchased by Coulson Tug Co,. Ltd.
03.1926:	Purchased, for £2,000, by Blyth Tug Co. Ltd.
1937:	Renamed DUNERA II.
12.1937:	Sold, for £300, to Clayton & Davie Ltd., Dunston for breaking up.

APPENDIX C TO CHAPTER ELEVEN

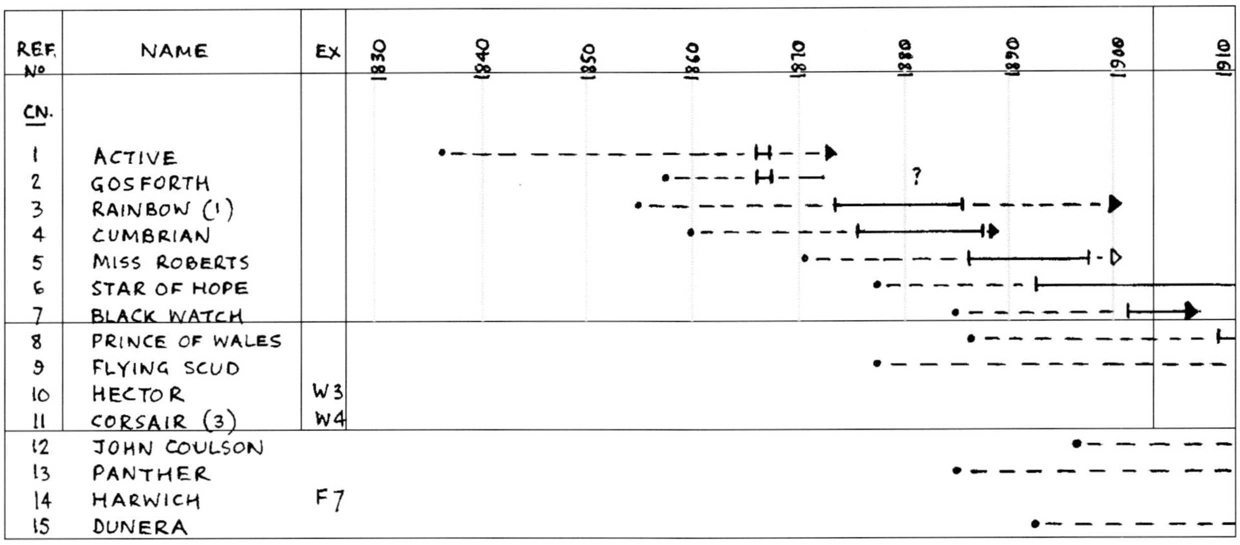

APPENDIX D TO CHAPTER ELEVEN

John White Fleet List

W 1 **CORSAIR (1)** **(1852-1859)**
ON 16185 81 g. 76.6 x 15.0 x 8.7 ft. Wood Paddle
B: On the River Tyne.
E: L. 1 Cy. 35 NHP.

PR: 1852 Shields; 1859 Leith.

1852:	Completed for J. White, South Shields.
1859:	Sold to A. Boyne, Leith.
11.1867:	Wrecked.

W 2 **CORSAIR (2)** **(1877-1891)**
ON 69877 85 g. 89.6 x 18.5 x 9.5 ft. Wood Paddle
B: J.P. Rennoldson, South Shields.
E: L. 1 Cy. 33" x 50". 40 NHP. By Builder.

PR: 1878 South Shields; 1899 Blyth.

1877:	Completed for John White Jnr., South Shields.
04.1884:	Transferred to Edith S. White and Ann Lee White.
05.1891:	Sold, for £1,400, to R.L. Manners, Blyth.
08.1921:	Transferred to Blyth Steam Tug Co.
07.1923:	Transferred to Blyth Tug Co.Ltd.
08.1923:	Sold, for £200, to Mrs. M.E. Robson, St. Anthony's.
1927:	Vessel broken up.

W 3 **HECTOR** **(1882-1918)**
ON 81977 101 g. 94.3 x 18.3 x 9.8 ft. Wood Paddle
B: J.R. Maxwell, South Shields.
E: L. 1 Cy. 34" x 51". 38 NHP. Baird & Barnsley, North Shields.

PR: 1882 South Shields.

10.1882:	Completed for J. White, Newcastle.
1888:	Transferred to Miss G. White and Juliana White.
07.1913:	Transferred to John White.
08.1918:	Purchased by Coulson Tug Co.Ltd.
1925:	Transferred to France, Fenwick Tyne & Wear Co.Ltd.
04.1926:	Broken up.

W 4 **CORSAIR (3)** **(1892-1918)**
ON 99465 153 g. 105.0 x 19.1 x 9.8 ft. Iron Paddle
B: J.P. Rennoldson & Sons, South Shields. (YN 141).
E: L. 2 Cy. 28" x 50". 350 IHP. 11 K. By Builder.

PR: 1892 South Shields.

09.1892:	Completed for John White, Newcastle.
08.1918:	Purchased by Coulson Tug Co.Ltd., (John Carr Coulson, Manager).
09.1925:	Transferred to France, Fenwick Tyne & Wear Co.Ltd.
11.1955:	Sold, for £1,700, to British Iron & Steel Corporation. Allocated to J.J. King & Co.Ltd., Gateshead, for breaking up.

Corsair *(3) on a pleasure outing. It is noteworthy that the sexes appear to be severely segregated! The ladies forward and the men aft. (Local History Library, South Shields Central Library).*

• CHAPTER TWELVE •

The Brown Companies

George Brown. 1888 - 1923.

Although the tugs mentioned in this chapter were registered in the name of George Brown or the succeeding limited company, they were in fact really owned by the South Shields shipbuilders John Readhead & Sons, which itself became a limited company in 1909. By this means the shipyard had, under its control, a small number of tugs always available for launches, fitting-out movements and trials. Of course, these tugs were used for other jobs when not required by Readhead.

The iron paddle tug *Contest* (87/1866) was purchased in 1888. Although built by Readhead & Softley she had spent her early years working on the River Thames for T.W. Elliott and had been re-engined by Readhead in 1875. The next acquisition by George Brown, in 1891, had also been built by Readhead but on this occasion had been completed for the Limerick Towing Co., Ireland. She was *President* (130/1876), a large double-engined iron paddle tug of 420 IHP which measured exactly 100 feet long. The third tug purchased, again built and engined by Readhead, arrived back on the Tyne after twenty years' service in the London area. *Vigilant* (115/1874) as she was named, was an iron paddle tug of similar length to *President* and a sister tug to *Ulysses* of the Lawson fleet, and *Warrior* belonging to Joseph Crosthwaite. In 1896 the wooden paddle tug *Great Britain* (1) (100/1874) was purchased from J. Winlo & Co., a South Shields shiprepairer, who had owned her for seventeen years. Prior to that she had been owned by T.A. Dry and partners.

Danger was always present in the crowded waters of the Tyne and its entrance channel, as is indicated by the fact that the members of the crew of *Contest* were awarded medals by the Tyne Foyboatmen's Provident Society for saving the lives of two foyboatmen on 7th August 1904 when their boat was hit by a steamer's propeller.

In 1905 *Great Britain* (1) was withdrawn from service and broken up at Shields, leaving the business with three iron-hulled paddle tugs all of which were to have very long lives. In 1915 Robert Lincoln was manager of George Brown's tugs. He had started in the tug business about 1900 as a boy working for George Brown when he spent his time chasing up and down the river with orders for the tugs. Elsewhere in this volume this job is referred to as 'water clerk'. Later he became the company secretary. In July 1920 George Brown purchased the steel screw tug *Knight Prendor* from the Prendeville organisation at Liverpool. She was a large tug, 109 feet long, and having a compound steam engine giving 520 IHP. On arrival in the Tyne she was renamed *Great Britain* (2) (184/1898).

The Brown Tug Co.Ltd. 1923 - 1930.

In July 1923 George Brown converted his business into a limited liability company entitled The Brown Tug Co.Ltd. The company's funnel colours were black with narrow white over red bands and the house-flag included a large letter 'B' on probably a red background with a thin white border near the edge of the flag. It is likely that the funnel marks and the flag had been used since the early days of George Brown's

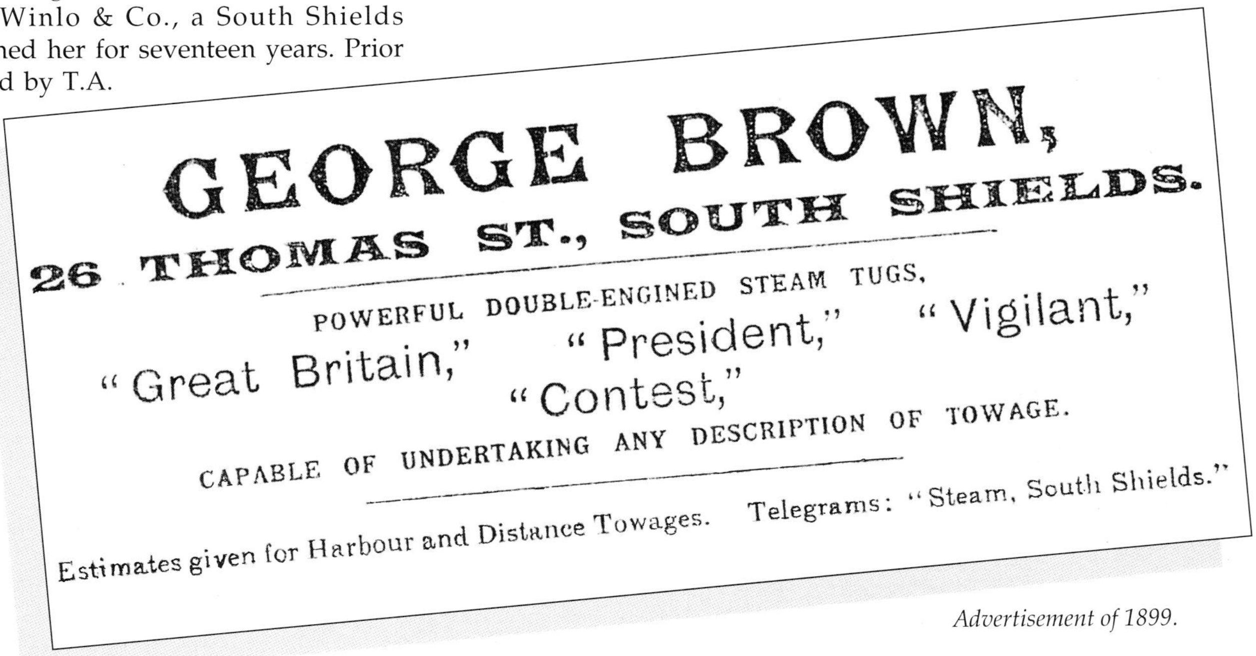

Advertisement of 1899.

President *(left) and* Hendon *(2) towing* San Gaspar *c.1930.* *(Harrington collection).*

interest in tugs. At this time the company office was in Dean Street, South Shields. George Brown died on 9th April 1925 at the age of eighty-nine, having been in the towage business for 37 years. What he did in the first thirty-six years of his working life has not been discovered. Presumably Robert Lincoln then ran the business until 1930.

Rather surprisingly, *Great Britain* (2), 'only' 30 years old, was sold for £405 for breaking-up in 1928 and two years later *Contest* and *Vigilant* suffered the same fate. These latter two were sixty-four and fifty-six years old respectively. In 1930 the company was taken over by France, Fenwick Tyne & Wear Co.Ltd. who transferred the one remaining tug, *President,* to their already large fleet. It is interesting to note that this tug survived until 1959 when she was broken up, eighty-three years after being built!

APPENDIX A TO CHAPTER TWELVE

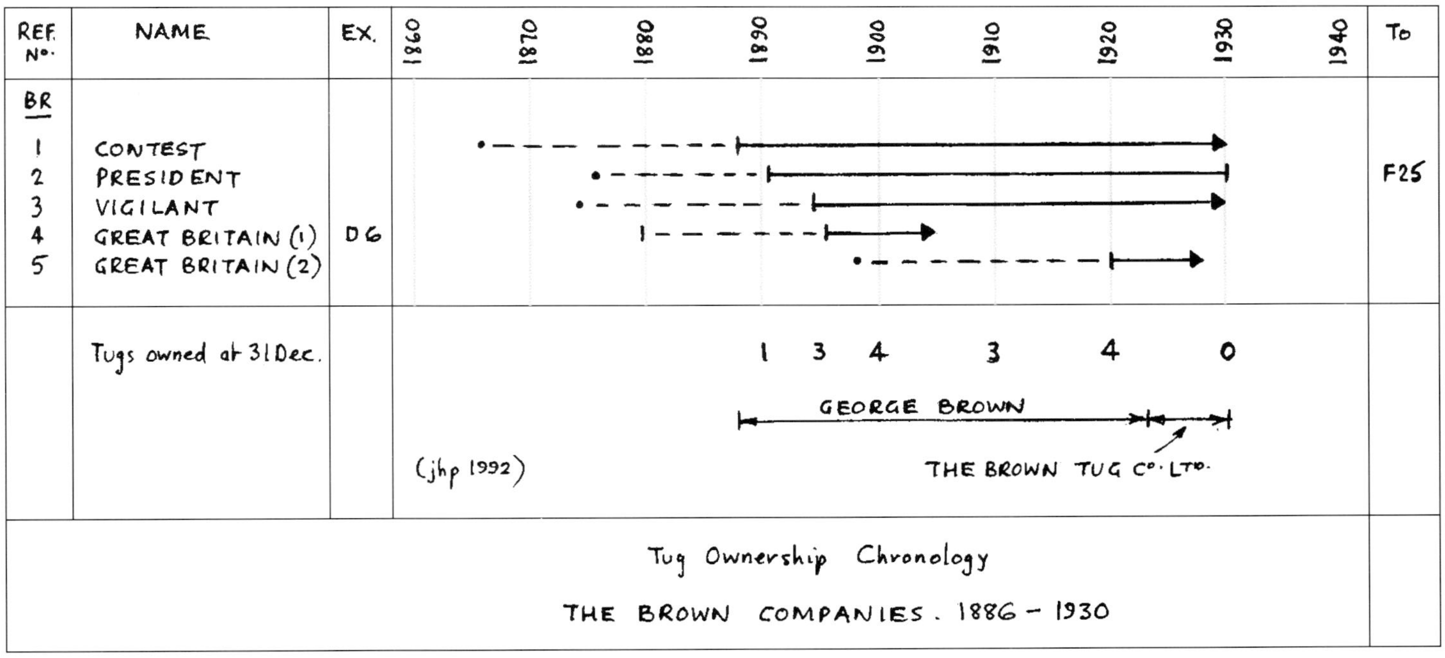

APPENDIX B TO CHAPTER TWELVE

The Brown Companies Fleet List

Contest *(Local History Library, South Shields Central Library).*

BR 1 **CONTEST** **(1888-1930)**
ON 54668 87 g. 88.4 x 17.2 x 8.9 ft. Iron Paddle
B: Readhead & Softley, South Shields.
E: Details not known.

PR: 1866 London; 1875 London; 1883 North Shields; 1888 South Shields.

1866:	Completed for T.W. Elliott, London.
1875:	Re-engined. L. 2 Cy. 25″ x 45″. 60 NHP. by J. Readhead & Co.
1882:	Sold to G. Drover, Isle of Wight.
12.1883:	Sold to J.T. Davison, North Shields.
02.1886:	Sold to H. Fenwick and T. Lawson.
1886:	Sold to J.R. Thompson & R. Young, South Shields.
1888:	Purchased by George Brown et al, South Shields.
07.1923:	Owners restyled The Brown Tug Co.Ltd.
04.1930:	Broken up on River Tyne.

President *(Local History Library, South Shields Central Library).*

BR 2 **PRESIDENT** **(1891-1930)**
ON 86652 130 g. 100.0 x 20.0 x 9.9 ft. Iron Paddle
B: John Readhead & Co., South Shields. (YN 124)
E: L. 2 Cy. 30″ x 52″. 420 IHP. By Builder.

PR: 1876 Limerick; 1884 Limerick; 1891 South Shields.

08.1876:	Completed for Limerick Towing Co., Limerick.
1891:	Purchased by George Brown et al, South Shields.
07.1923:	Owners restyled Brown Tug Co.Ltd.
1930:	Transferred to France, Fenwick Tyne & Wear Co.Ltd.
02.1959:	Sold, for £600, to British Iron & Steel Corporation. Allocated to Thomas Young & Sons (Shipbreakers) Ltd., Sunderland, for breaking up.

Vigilant *(Local History Library, South Shields Central Library).*

BR 3 **VIGILANT** **(1894-1930)**
ON 70588 115 g. 100.3 x 18.1 x 9.4 ft. Iron Paddle
B: John Readhead & Co., South Shields. (YN 102).
E: L. 2 Cy. 26½″ x 48″. 70 NHP. By Builders.

PR: 1874 London; 1894 South Shields.

1874:	Completed for Dr. M. Brownfield, Poplar.
02.1894:	Purchased by George Brown et al, South Shields.
07.1923:	Owners became Brown Tug Co.Ltd., South Shields.
1930:	Broken up on River Tyne.

Great Britian *(1) left with other unidentified paddle tugs (Local History Library, South Shields Central Library).*

BR 4 **GREAT BRITAIN (1)** **(1896-1905)**
See Dry fleet list, D 6.

BR 5 **GREAT BRITAIN (2)** **(1920-1928)**
ON 108586 184 g. 109.0 x 21.1 x 10.4 ft. Steel Screw
B: Cumming & Ellis, Inverkeithing. (YN 30).
E: C. 2 Cy. 21″ & 42″ x 24″. 520 IHP. Hutson & Son, Glasgow.

PR: 1898 Granton; 1908 Swansea; 1915 Liverpool; 1920 South Shields.

1898:	Completed, as MOUNT VISO, for Muirhead's Trawlers Ltd., Newhaven.
c.1905:	Sold to H.O. Rice, Swansea.
1908:	Sold to Thomas H. Jones (later & Co.).
c.1915:	Sold to Knight Prendor Tug Co.Ltd., Liverpool, (J. Prendeville, Manager) renamed KNIGHT PRENDOR.
03.1918:	Hired by Royal Navy, renamed SIR THOMAS.
01.1919:	Returned to owners, renamed KNIGHT PRENDOR.
07.1920:	Purchased by G. Brown, renamed GREAT BRITAIN.
07.1923:	Owners became Brown Tug Co.Ltd.
06.1928:	Sold, for £405, to T.W. Ward, Inverkeithing, for breaking up.

• CHAPTER THIRTEEN •

The Irwing Companies

Anthony Irwing. c.1867 - 1911.

One of the many collier companies trading from Sunderland in the latter half of the nineteenth century was Culliford & Clark Ltd., the partners being James Henry Wood Culliford and J.L. Clark. It would appear that the tugowner Anthony Irwing was contracted to handle the colliers of Culliford & Clark in and about the port of Sunderland from the late 1860s. The tugs involved were sometimes registered in the name of Anthony Irwing, sometimes J.H.W. Culliford and at other times Culliford and Clark and even all three together. Nevertheless the various tugs owned over the years have been treated here as one fleet under the name of Irwing, who was responsible for their management even if not the outright owner of them all. That they were one fleet contracted to the collier company is borne out by the fact that the tugs carried a Culliford & Clark funnel design. This was a complicated arrangement of the letters 'C' and 'C' interlinked back to back and joined with a horizontal figure-of-eight all within the outline of a shield. This design was in white on an otherwise black funnel.

About 1867 Anthony Irwing purchased the wooden paddle tug *Tartar* (61/1853) from Middlesbrough owners via the North Shields shipbuilder Richard Stobbs. After three years she was sold to other Sunderland owners. It is not known what tugs, if any, Irwing owned in the period 1870-1876 either individually or in partnership with others, though it seems likely that he would have been involved in the towage business in some way. About 1876 *Pilot* (67/1860), a very similarly-sized tug to *Tartar,* was acquired by Irwing following sixteen years working on the Tyne. In July 1877 *Stag* (1) (69/1863) was purchased from Anderson and Chater of North Shields and registered in the ownership of J.H.W. Culliford and A. Irwing. Later she was to be registered in the ownership of Anthony Irwing (32 shares), Culliford (16 shares) and Clark (16 shares). The large iron paddle tug *Florence* (150/1879) was built new for Culliford and others in 1879 by J.T. Eltringham at South Shields. At about the same time *Pilot* was disposed of to other Sunderland owners.

In 1881 Culliford purchased the iron paddle tug *Flying Arrow* (123/1875) from Clyde Shipping Co. and for several years now there was to be a steady increase in the size of the tug fleet. *Admiral* (102/1876) was purchased by Culliford and Irwing in 1882, and a second *Pilot* (91/1871) by Culliford in the same year but she was soon renamed *Silksworth* and transferred to the ownership of Anthony Irwing. *Admiral* was sold in April 1883 for service as a trawler at Inverness. At this period it was quite common for paddle tugs to be used for fishing purposes and most were capable of carrying out either duty as and when appropriate. Another new tug was purchased in April 1885, this time having been built by J.P. Rennoldson & Sons at South Shields. She was *General Gordon* (97/1885) and was registered in the name of Anthony Irwing and partners. *Flying Arrow* was sold to the Anderson family of North Shields in March 1886 and J.W.H. Culliford purchased *Norfolk Hero* (71/1861) in 1887 from a local owner and transferred her ownership to Anthony Irwing and associates. These changes left the business with a fleet of five tugs at the end of 1887, namely:-

Vessel name	Gross tonnage/year	Hull construction	Means of propulsion
Stag (1)	(69/1863)	Wood	Paddle
Florence	(150/1879)	Iron	Paddle
Silksworth	(91/1871)	Wood	Paddle
General Gordon	(97/1885)	Wood	Paddle
Norfolk Hero	(71/1861)	Wood	Paddle

- with an average age of fifteen years.

Surprisingly, for she was only five years old, *General Gordon* was sold to Blyth Steam Tug Co. in April 1890. Her replacement in the same year was *Lord Derby* (98/1876) purchased by Anthony Irwing for £1,425 from French owners. *Pactolus* (93/1870) came to Sunderland from Ireland in October 1892 having been bought by Anthony Irwing who, four months later, shared the ownership with Susannah Irwing. In September 1896 Anthony Irwing purchased the wooden paddle tug *William Wouldhave* from G. Smith and partners of North Shields and renamed her *Stag* (4) (86/1884). The earlier *Stag* (1) was broken up in 1897. *Florence* was sold and converted into a towing lighter at Middlesbrough. *Norfolk Hero,* now thirty-nine years old, was sold in 1900 and broken up the following year.

Conqueror (1) (144/1861), a large iron paddle tug, was purchased by Anthony Irwing in 1902 from the North Eastern Railway Co. She had spent all of her working life to date at West Hartlepool where the railway company were owners of the docks. Another large tug *Agamemnon* (129/1872) joined the fleet in 1904 in the name of Anthony Irwing. *Pactolus* went to the breakers in 1905 and *Silksworth* followed in 1906. *Granville* (160/1876) was purchased in 1906 by James Irwing, a son of Anthony, from Dover Harbour Board for whom she had worked since being built. She only stayed with the Irwing business for about a year, being sold to the neighbouring Seaham Harbour Dock Co. in October 1907. Also in 1907 *Lord Derby* was sold to other Sunderland owners. At the end of 1910 the ownership of *Conqueror* (1) and *Agamemnon* was transferred to James Irwing and a large number of family associates.

Anthony Irwing & Son. 1911 - 1926

Anthony Irwing having died on 10th September 1906 at the age of eighty, the partnership of Anthony and James Irwing, from at least 1911, was trading under the style of Anthony Irwing & Son, though their tugs remained registered in the names of individuals. Their business address was 1 Barrack Street, Sunderland overlooking the docks and harbour entrance. Another North Eastern Railway Co. tug from West Hartlepool joined the Irwing fleet in November 1911. She was the steel paddle tug *N.E.R. No.20*, ex *William Gray*, and was renamed *Anthony Irwing* (119/1887) on arrival on the River Wear. Her registered ownership was the same as for *Agamemnon*. At the end of 1911 the fleet consisted of:-

Stag (4)	(86/1884)	Wood	Paddle
Conqueror (1)	(144/1861)	Iron	Paddle
Agamemnon	(129/1872)	Iron	Paddle
Anthony Irwing	(119/1887)	Steel	Paddle

- with an average age of thirty-five years.

Stag (4) came to grief in July 1914 when she sank in Hendon Dock, Sunderland, and had to be declared a total loss. The Great Yarmouth tug *King Edward VII* (138/1901) was purchased in October 1920.

Irwing Steam Tug Co.Ltd. 1926 - 1934

In 1926 the partnership was converted into a limited liability company with James Irwing as the designated manager. *Agamemnon* was sold for breaking-up in 1928, and *Conqueror* (1) also went to the breakers in 1934. In July 1934 the company was taken over by France, Fenwick Tyne & Wear Co.Ltd. but neither of the two Irwing tugs existing at that time was transferred in the deal, France Fenwick presumably being content to have the business previously carried out by the Irwing company for their own tugs. *Anthony Irwing* was broken up at Grangemouth and *King Edward VII* was sold for £425 to The Ridley Steam Tug Co.Ltd. with whom she was to stay for the next fifteen years. Thus came about the demise of the Irwing Steam Tug Co.Ltd. after it and its forbears had been in business for sixty-seven years.

APPENDIX A TO CHAPTER THIRTEEN

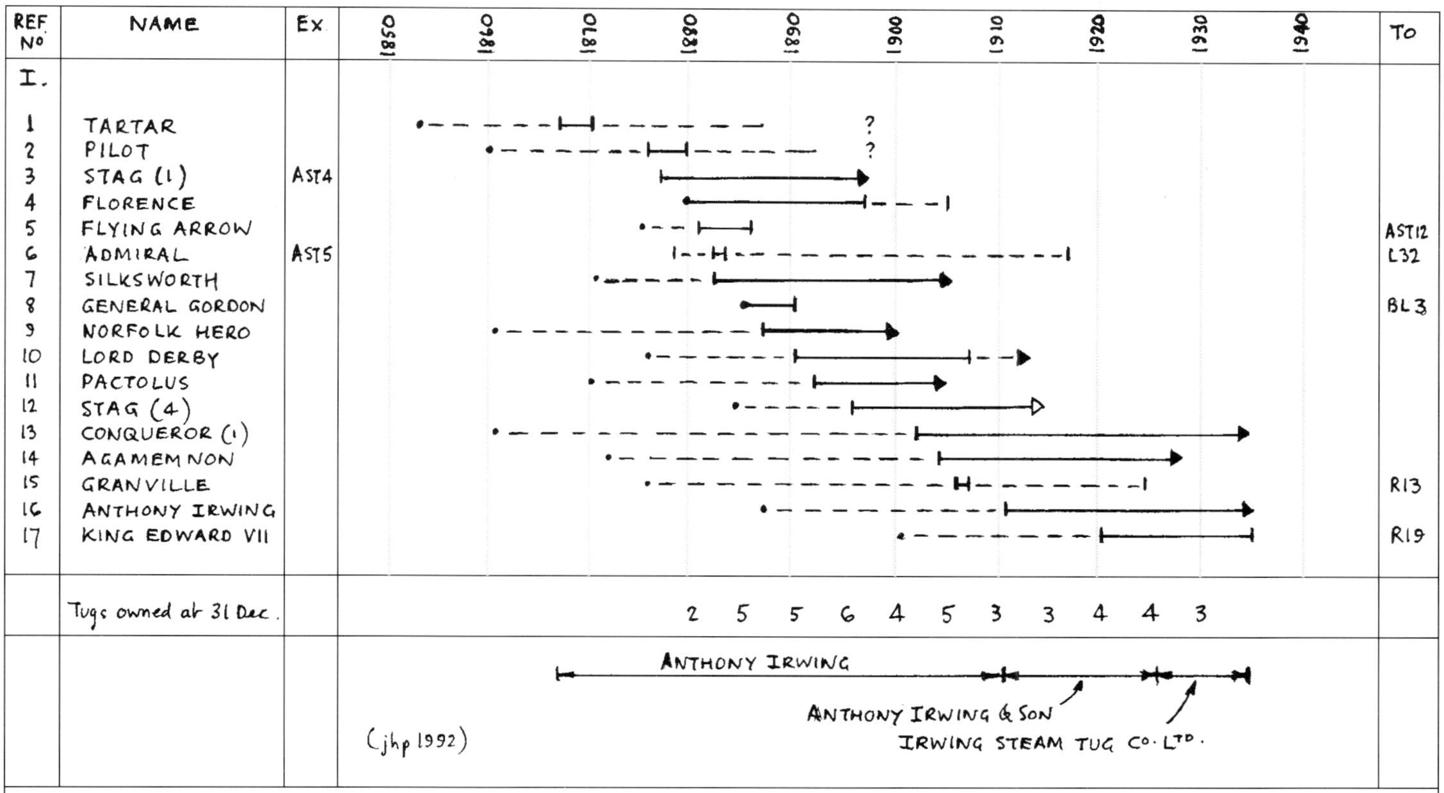

Tug Ownership Chronology

THE IRWING COMPANIES c 1867 - 1934

APPENDIX B TO CHAPTER THIRTEEN

The Irwing Companies Fleet List

I 1 **TARTAR** **(c.1867-c.1870)**
ON 7754 61 g. 79.2 x 16.5 x 8.3 ft. Wood Paddle
B: At North Shields.
E: L. 1 Cy. 30 NHP.

PR: 1853 Leith; 1856 Stockton; 1861 Middlesbrough; 1867 Sunderland.

1853:	Completed. Owner not known.
03.1856:	Sold to Isaac Wilson et al, Middlesbrough.
05.1867:	Sold to R. Stobbs and W. Scott, North Shields.
c.1867:	Purchased by Anthony Irwing, Sunderland.
c.1870:	Sold to J.G. Garrick, Sunderland.
c.1872:	Sold to J. Smurthwaite, Sunderland.
c.1876:	Sold to J.W. Snowdon, Sunderland.
1888:	Not in register.

I 2 **PILOT** **(c.1876-c.1879)**
ON 28315 67 g. 82.4 x 16.3 x 8.9 ft. Wood Paddle
B: At North Shields.
E: 30 NHP.

PR: 1860 Newcastle.

1860:	Completed for W.T. Turnbull and M. Wheldon, Newcastle.
c.1865:	Sold to T.B. Calleman et al, Newcastle.
c.1876:	Purchased by Anthony Irwing, Sunderland.
c.1879:	Sold to J.W. Snowdon.
1893:	Not in register.

I 3 **STAG (1)** **(1877-1897)**
See Anchor Steam Tug fleet list, AST 4

I 4 **FLORENCE** **(1879-1897)**
ON 81451 150 g. 113.8 x 19.2 x 8.9 ft. Iron Paddle
B: J.T. Eltringham, South Shields. (YN 77).
E: L. 1 Cy. 37" x 54". 65 NHP. J.P. Rennoldson & Sons, South Shields.

PR: 1879 Sunderland.

05.1879:	Completed for J.H.W. Culliford et al, Sunderland.
1897:	Sold to C. Douglas, Sunderland.
08.1898:	Sold to Sarah Johnson, Sunderland.
04.1899:	Sold to T. Moore, Sunderland.
02.1904:	Sold to T.P. Rayson, Sunderland.
05.1905:	Sold and converted to a towing lighter at Middlesbrough.

I 5 **FLYING ARROW** **(1881-1886)**
ON 71699 123 g. 108.6 x 18.2 x 9.3 ft. Iron Paddle
B: J.T. Eltringham, South Shields. (YN 39).
E: L. 1 Cy. 36" x 54". 60 NHP. J.P. Rennoldson, South Shields.

PR: 1875 Glasgow; 1881 Sunderland; 1886 North Shields.

1875:	Completed for Clyde Shipping Co. (G.J. Kidston).
1881:	Purchased by J.H.W. Culliford and J. Siddell, Sunderland.
03.1886:	Purchased by Dorothy Anderson, North Shields.
08.1889:	Transferred to Thomas Anderson, and later G.R. Anderson, North Shields.
05.1903:	Sold to G.T. Pearson, Hartlepool.
11.1903:	Sold to E. Derwent, Middlesbrough.
1905:	Sold to J.P. Bell, Stockton.
1906:	Sold to E. Thomas and reduced to a lighter. Registry closed.

I 6 **ADMIRAL** **(1882-1883)**
See Anchor Steam Tug fleet list, AST 5

Silksworth at launch of Good Hope in 1903. (George Scott).

I 7 **SILKSWORTH** **(1882-1906)**
ON 63003. 91 g. 89.3 x 17.8 x 9.4 ft. Wood Paddle
B: Readhead & Softley, South Shields.
E: L. 1 Cy. 32½" x 50". 35 NHP. By Builder.

PR: 1871 South Shields; 1882 Sunderland.

1871:	Completed, as PILOT, for G. Smith, South Shields.
1879:	Sold to R.A. Harrison, South Shields.
1882:	Purchased by J.H.W. Culliford, Sunderland.
1883:	Transferred to Anthony Irwing, Sunderland, renamed SILKSWORTH.
04.1906:	Vessel broken up.

I 8 **GENERAL GORDON** **(1885-1890)**
ON 90322 97 g. 92.0 x 18.4 x 9.8 ft. Wood Paddle
B: J.P. Rennoldson & Sons, South Shields.
E: L. 1 Cy. 32¼" x 50". 34 NHP. By Builder.

PR: 1885 Sunderland; 1890 North Shields; 1913 Blyth.

04.1885:	Completed for Anthony Irwing et al, Sunderland.
04.1890:	Sold to Blyth Steam Tug Co. (G. Dinsdale et al).
07.1923:	Owners restyled Blyth Tug Co.Ltd.
02.1930:	Sold, for £140, to R.H. Kirkley, Hylton, Sunderland, for breaking up.

I 9 **NORFOLK HERO** **(1887-1900)**
ON 29705 71 g. 85.5 x 16.6 x 8.8 ft. Wood Paddle
B: At Low Walker.
E: 35 NHP.

PR: 1861 ?; 1862 Dublin; 1882 Belfast; 1883 Sunderland.

1861:	Completed for J. Wilson et al, Shields.
1862:	Sold to H. Sheridan et al, and later C. Palgrave et al, Dublin.
1865:	Sold to M. Murphy, Dublin. 1874: Sold to R. Tedcastle, Dublin.
1882:	Sold to R,M. Greevy, Belfast.
1883:	Sold to T.M. Scott, Sunderland.
08.1887:	Purchased by J.H.W. Culliford et al, Sunderland.
1888:	Transferred to Anthony Irwing et al, Sunderland.
1900:	Sold.
1901:	Broken up.

I 10 **LORD DERBY** **(1890-1907)**
ON 72533 98 g. 96.0 x 17.8 x 9.3 ft. Iron Paddle
B: J.T. Eltringham, South Shields.
E: L. 1 Cy. 34½" x 54". 50 NHP. Hepple & Co., South Shields.

PR: 1876 Cardiff; 1889 (France); 1890 Sunderland.

09.1876:	Completed for R. Lee, Cardiff.
1888:	Sold to Mrs. S. Thomas, Swansea.
05.1889:	Sold to M. Delarouche, Paris.
07.1890:	Purchased, for £1,425, by Anthony Irwing, Sunderland.
09.1906:	Transferred to Susannah Irwing et al.
07.1907:	Transferred to James Irwing, then sold to Joseph Carney, Sunderland.
03.1913:	Broken up.

Pactolus at launch of E.O. Saltmarsh in 1903. (George Scott).

I 11 PACTOLUS (1892-1905)
ON 60561 93 g. 87.2 x 17.9 x 9.6 ft. Wood Paddle
B: C.W. Dodgin & Co., North Shields.
E: L. 1 Cy. 32½" x 48". 38 NHP. T.R. Scott & Co., North Shields.

PR: 1870 North Shields; 1875 Dublin; 1876 Belfast; 1885 North Shields; 1889 North Shields.

09.1870:	Completed for T.J. Newton et al, North Shields.
02.1874:	Sold to J.F.F. Common, North Shields.
1875:	Sold to J. Somers, Dublin.
1876:	Sold to H. Andrews, Belfast.
10.1892:	Purchased by Anthony Irwing, Sunderland.
02.1893:	Ownership now jointly between Anthony Irwing and Susannah Irwing.
02.1905:	Vessel broken up.

I 12 STAG (4) (1896-1914)
ON 84898 86 g. 89.8 x 18.1 x 9.3 ft. Wood Paddle
B: William Wouldhave, North Shields.
E: O.B. 1 Cy. 31½" x 48". 32 NHP. R.B. Findley & Sons, North Shields.

PR: 1884 North Shields.

1884:	Completed, as WILLIAM WOULDHAVE, for G. Smith et al, North Shields.
09.1896:	Purchased by Anthony Irwing, Sunderland, and renamed STAG.
1911:	Transferred to James Irwing.
07.1914:	Sank in Hendon Dock, Sunderland, - total loss.

Conqueror (1) at Hartlepool when owned by NER Co. (B.G. Spaldin).

I 13 CONQUEROR (1) (1902-1934)
ON 28289 144 g. 107.0 x 20.4 x 10.2 ft. Iron Paddle
B: Pile Spence & Co., West Hartlepool. (YN 37)
E: 2 Cy. 30" x 54". 60 NHP. W. Huntley & Co.

PR: 1861 West Hartlepool; 1867 West Hartlepool.

07.1861:	Completed for West Hartlepool Harbour & Railway Co. (W.C. Ward Jackson).
1867:	Transferred to North Eastern Railway Co.
1902:	Purchased by Anthony Irwing, Sunderland.
12.1910:	Transferred to James Irwing and a large number of family and associates.
08.1926:	Transferred to Irwing Steam Tug Co.Ltd., Sunderland.
1934:	Broken up at Grangemouth.

Agamemnon. *(John Clarkson, Longton).*

I 14 AGAMEMNON (1904-1928)
ON 65673 129 g. 106.3 x 18.4 x 10.0 ft. Iron Paddle
B: Readhead, Softley & Co., South Shields. (YN 75).
E: L. 2 Cy. 28" x 48". 55 NHP. By Builder.

PR: 1872 London; 1875 Memel; 1885 London.

1872:	Completed for Gravesend Pilots, T.A.F. Banks et al.
1875:	Sold to owners at Memel.
1885:	Sold to James Watkins, London. Later Watkins & Co.
c.1891:	Sold to George Petrie, Middlesbrough.
1904:	Purchased by Anthony Irwing, Sunderland.
12.1910:	Transferred to James Irwing and a large number of family and associates.
08.1926:	Transferred to Irwing Steam Tug Co.Ltd., Sunderland.
1928:	Sold for breaking up.

Granville. *(John Clarkson, Longton).*

I 15 GRANVILLE (1906-1907)
ON 68127 160 g. 101.0 x 19.1 x 10.4 ft. Iron Paddle
B: Westwood, Baillie & Co., Blackwall.
E: L. 2 Cy. 30" x 54". 400 IHP. 10 K. J. Stewart, London.

PR: 1876 Dover; 1907 Sunderland.

04.1876:	Completed for Dover Harbour Board.
1906:	Purchased by James Irwing et al, Sunderland.
10.1907:	Sold to Seaham Harbour Dock Co.
02.1924:	Purchased, for £775, by The 'Ridley' Steam Tug Co.Ltd.
10.1930:	Chartered to Blyth Tug Co.Ltd.
04.1952:	Sold, for £1,600, to British Iron & Steel Corporation. Allocated to Clayton & Davie Ltd., Dunston, for breaking up.

Anthony Irwing. *(Author's collection).*

I 16 **ANTHONY IRWING** **(1911-1935)**

ON 92921 119 g. 95.0 x 19.2 x 9.8 ft. Steel Paddle
B: J.T. Eltringham, South Shields. (YN 136).
E: L. 2 Cy. 28" x 50". 350 IHP. 10¼ K. J.P. Rennoldson & Sons, South Shields.

PR: 1888 West Hartlepool.

1887:	Completed, as WILLIAM GRAY, for North Eastern Railway Co.
08.1911:	Renamed N.E.R. No.20.
11.1911:	Purchased by James Irwing and a large number of family and associates, renamed ANTHONY IRWING.
08.1926:	Transferred to Irwing Steam Tug Co.Ltd., Sunderland.
1935:	Broken up at Grangemouth.

I 17 **KING EDWARD VII** **(1920-1935)**

ON 111083 138 g. 100.0 x 19.1 x 9.2 ft. Steel Paddle
B: J.P. Rennoldson & Sons, South Shields. (YN 209).
E: L. 2 Cy. 25" x 45". 260 IHP. 10 K. By Builders.

PR: 1901 Great Yarmouth; 1920 Sunderland.

1901:	Completed for Nicholson Towage Co.Ltd., Great Yarmouth.
10.1918:	Hired by Royal Navy. 08.1919: Returned to owners.
10.1920:	Purchased by James Irwing et al, Sunderland.
08.1926:	Transferred to Irwing Steam Tug Co. Ltd., Sunderland.
07.1935:	Purchased, for £425, by The 'Ridley'Steam Tug Co.Ltd.
10.1950:	Sold, for £275, to British Iron & Steel Corporation. Allocated to Clayton & Davie Ltd., Dunston, for breaking up.

• CHAPTER FOURTEEN •

The Redhead Companies

Robert Redhead and Associates. c.1855 - c.1874

Robert Redhead is known to have been in the towage business from about 1855, as it is recorded that, along with others, he had a share in the ownership of the tug *Engineer* (47/1853). Also in 1855 the tug *Tiger* (?/1838) was listed as being owned by two people of the name Redhead and it is quite likely that these were relatives of Robert. Before proceeding very far into the Redhead story it is worth giving the warning that there has often been confusion, particularly in the press, between the names Redhead and Readhead. Redhead is the name of the tugowners whose history is related in this chapter. Readhead is the name of the South Shields' shipbuilders who, over the years, traded successively as Readhead & Softley, John Readhead & Co., John Readhead & Sons, and John Readhead & Sons Ltd. and built several of the tugs listed in this volume. Just to complicate matters further, the Readhead shipyard had interests in the towage business through George Brown and, later, the Brown Tug Co.Ltd.

In 1857 Robert Redhead took a part interest in the new wooden paddle tug *Wards* (54/1857) which had been built at Gateshead. Presumably the name *Wards* had some connection with the owners, but what this was is not known. At this time Robert Redhead was in partnership with a chemist, William Gray, an association which was to endure for some seventeen years. Their base was at 2 Broad Chare, Newcastle, just off the Quayside. In the following year *Little Western* (36/1844) was purchased, only to be sold after less than a year with the Redhead partners. *Engineer* was sold in 1861 and another small tug/launch was purchased twelve months later. She was *Collingwood* (29/1842) and had already had several owners at Newcastle and Sunderland. In 1863 the wooden paddle tug *Brothers* (1) (57/1854) was bought from North Shields owners, but, true to form, Robert Redhead kept her for just a short while. *Royal Charlie* (48/1861) joined the fleet in 1864 and *Tiger* was disposed of to Sunderland owners. After *Brothers* (1) went to Port Glasgow in 1866 Redhead and Gray were left with a fleet of three small tugs with a gross tonnage ranging from 29 to 54 tons.

In 1867 *Collingwood* left the fleet and was replaced by the newly-built *Sensation* (91/1867). She was a much larger paddle tug than hitherto owned by Robert Redhead and William Gray and measured some ninety feet in length. Having sold a tug to the Clyde in 1866, the tug *Pearl* (72/1858) moved in the opposite direction in 1868. *Royal Charlie* was sold about 1871 to other Newcastle owners, and *Pearl* was sold to foreign owners in 1873 after five years on the Tyne. In the early 1870s William Gray relinquished his partnership with Robert Redhead, and, in his place, came Robert's son.

POWERFUL STEAM TUGS.

Quotations Given for Towing to Coast or Continent.

DOUBLE-ENGINED:

"Daring," "Ben Ledi," "Dundee," "Sensation."

ROBT. REDHEAD, Steam Tug Owner,

BROAD CHARE, NEWCASTLE-ON-TYNE.

Telegraphic Address: "DARING, NEWCASTLE-ON-TYNE."

Advertisement of 1899.

Robert Redhead & Son. c.1874 - 1916

From about 1874, Robert Redhead's son, Mark Henderson Redhead, began to take a part share in the ownership of his father's tugs. This was usually, but not always, on a 50/50 basis with each owning 32/64 of the value of the tugs. Although perhaps not formally registered as a company the business is recorded as trading under the title of Robert Redhead & Son. At that time the partnership was operating two tugs:-

Name	Gross Tonnage/Year	Hull construction	Means of propulsion
Wards	(54/1857)	Wood	Paddle
Sensation	(91/1867)	Wood	Paddle

Wards was very soon sold to owners in Hull and was replaced by the single-engined wooden paddle tug *Liverpool* (116/1870), purchased from Ireland. With a length of almost 106 feet she was quite a large tug for river operations. Just over a year later she was sold to a Harwich owner; perhaps she did not meet the Redhead's requirements in some way, or maybe they just made a good profit on the transaction.

In 1874 Robert Redhead purchased 48/64 shares in the wooden paddle tug *Dundee* (89/1865) from J. Ormston. Later Ormston relinquished his share in the vessel and it was jointly owned by Robert and Mark Henderson Redhead. In 1876 a new tug was purchased. She was *Commodore* (158/1876) - the first Redhead vessel to have an iron hull. Again, at 118 feet long, she was a large tug, but surprisingly she had only a single side-lever engine. Her time in Redhead ownership was also remarkably short, since she was sold to Limerick owners in 1878. The twelve-year-old wooden paddle tug *Bulldog* (67/1867) was acquired in 1879, only to be sold to Sunderland owners in 1883.

At the age of twenty-five James Patton entered the employ of Robert Redhead & Son in 1886, as skipper of *Sensation*. As will be noted later, he was to give the company and its successors and local Marine Insurance Associations very long service. In 1887 the iron paddle tug *Daring* (135/1876) was purchased from the River Thames and was the first of the Redhead paddle tugs to be fitted with a double engine. In 1878 she had survived a collision with a Belgian Government Mail Packet when her boiler was cut through. One presumes that the boiler was not under pressure at the time otherwise that would have been the end of her! The iron paddle tug *Ben Ledi* (114/1878) was added to the fleet in 1897, having been purchased from Ayr Harbour Trustees. At the turn of the century the Redhead fleet thus comprised four tugs:-

Sensation	(91/1867)	Wood	Paddle
Dundee	(89/1865)	Wood	Paddle
Daring	(135/1876)	Iron	Paddle
Ben Ledi	(114/1878)	Iron	Paddle

- with an average age of over twenty-eight years.

After being under Redhead ownership for thirty-four years *Sensation* foundered off Hartlepool on 10th November 1901. Her skipper, James Patton, then came ashore to superintend operations for the company. In 1903 James' son Alexander started work with Redhead at the age of about 19 years. In January of the same year Robert Redhead died, and the ownership of the tug fleet was then shared jointly between Mark Henderson Redhead and Susannah H. White. What the relationship was between the latter and Robert Redhead has not been discovered.

Dundee sank in deep water following a collision off the Tyne on 17th May 1903 whilst towing a lighter from Middlesbrough to the Tyne. Her crew was saved by the tug *Corsair*. This was not the first time she had been sunk, for in 1880 she had suffered a similar fate in the river but on that occasion she had soon been raised. Thus departed the last of the wooden-hulled paddle tugs in the Redhead fleet. 'Fleet' is perhaps a rather extravagant description, since the partnership then owned only two tugs! This situation, however, was corrected in August 1903 when *Thames* (116/1866) was purchased for £700 from the Great Western Railway Co. at Milford Haven. During the early years of the 20th century the company's office was at 63 Quayside, Newcastle.

In September 1910 the company received delivery of their first steel screw tug which was built by Armstrong, Whitworth. She was named *George V* (1) (188/1910) to commemorate the accession to the throne of King George V in May 1910. The new tug, which cost £6,500, was exactly 100 feet long and was powered by a triple-expansion steam engine developing 750 IHP. She must have made a striking impression on all who saw her on the river, since she represented such a big step forward from the paddle tugs previously owned by Redhead. Not surprisingly, *Thames* was sold to R.S. Allen a month after the delivery of *George V* (1). The new tug was soon put to good use. One interesting tow occurred in 1913 when she left Vigo on 14th January with the disabled British steamer *Glenmorven* (2,812/1901) en route for Jarrow. During bad weather 150 miles off Ferrol the crew of the steamer abandoned ship and were landed by the Dutch tug *Poolzee* (304/1900) which was also in attendance. *George V* (1), skippered by G.V. Nicholson, plodded on alone through the storm and arrived with the dead ship at Falmouth on 27th January. What was surprising was that, having owned their powerful new tug *George V* (1) for only three and a half years, the company sold her in March 1914 to the Russian Government who renamed her *Salvage Steamer No.2*.

On the outbreak of the First World War in August 1914 *Ben Ledi* was immediately taken on hire by the Royal Navy, but whether she worked away from the Tyne is not known. After having lost the use of *George V* (1) the company did not delay in rectifying the position. They ordered a new, similar but slightly larger, tug with the same name. This one, however, was built and engined by J.P. Rennoldson & Sons Ltd. at South Shields for £9,500, and had an 850 IHP steam engine which gave a free running speed of 12.5 knots. *George V* (2) (224/1915) was delivered in January 1915, but four months later she too was hired by the Royal Navy until November 1915. Later she was hired again between 1917 and 1919. *George V* (2) was to become one of the best known and loved tugs on the River Tyne for no less than sixty-three years. In 1992 she still survives with other owners away from the Tyne.

Robert Redhead & Son Ltd. 1916 - 1929.

Mark Henderson Redhead died on 2nd March 1915, as a result of which the partnership was converted into a limited liability company in 1916 under the title of Robert Redhead & Son Ltd.. James Patton became managing director of the new company which then owned the following tugs:-

Daring	(135/1876)	Iron	Paddle
Ben Ledi	(114/1878)	Iron	Paddle
George V (2)	(224/1915)	Steel	Screw

It is known that, at least from this time, the funnel colours of the Redhead company were yellow with a black top and that their house flag design consisted of red/white/red stripes with a red shield on the white section. It is probable that the funnel colours had been used from a much earlier date.

On 19th December 1916 *Daring* foundered south of St. Mary's Island, Northumberland, whilst assisting a British submarine in foggy conditions. Although a total loss it was fortunate that all her crew of five were saved. To replace this loss the iron paddle tug *Conqueror* (2) (177/1884) was purchased from V.A. Cappon of Dundee. Although already thirty-three years old she was to remain on the Tyne for a further thirty-nine years! In 1920, *George V* (2). with tugs of The John Dry Steam Tugs Ltd. fleet, towed the German dreadnought *Nassau* (18,873 displacement/1909) from the Firth of Forth for demolition at Hughes Bolckow's yard at Derwenthaugh. When passing Newcastle problems arose when the battleship temporarily became wedged between a pier of the High Level Bridge and the staging of the Swing Bridge. This tow was possibly the beginning of co-operation between the Redhead and Dry companies.

An event which was organised by the Redhead company, but which involved tugs of several other companies, was the towage of a new Tyne-built floating dock to Southampton between 17th and 24th April 1924. James Patton supervised the operation which involved the Redhead tug *George V* (2) and much was made of the event for publicity purposes. Although at least six head tugs were involved there is, to the author's knowledge, an advertisement photograph which shows *George V* (2) towing this huge dock on its own with all the other tugs removed from the picture! No doubt it was classed as a major event because James Patton was presented with a solid silver model of the World's Largest Floating Dock 'to commemorate his achievement in supervising the towage of the Southampton floating dock in the record time of four days'. The model was made to a scale of 40 feet to 1 inch by the Northern Goldsmiths Co. One wonders where it is now!

During 1929 discussions had been taking place with The John Dry Steam Tug Co.Ltd. regarding co-operation between them and the Redhead company, covering their operations on the River Tyne. An agreement was reached which resulted in Robert Redhead & Son Ltd. taking over the assets and business of The John Dry Steam Tug Co.Ltd. on 14th September 1929. The new company so formed was named Redhead & Dry's Tugs Ltd. The Redhead tugs contributed to the combined company were:-

Ben Ledi	(114/1878)	Iron	Paddle
George V (2)	(224/1915)	Steel	Screw
Conqueror (2)	(177/1884)	Iron	Paddle

- with an average age of over thirty-six years.

This amalgamation ended the seventy-four years of separate existence of the Redhead partnerships and company, during which some nineteen different tugs had been owned, only two of which had been screw-propelled. The Redhead office was, at the time of the merger, at 27 Quayside, Newcastle and this address continued in use as the headquarters of the combined company.

The advertisement which contrives to show George V *(2) towing the Southampton floating dock on her own.*

In 1924 towage of the Southampton floating dock was shared by several companies. Here we see, from left to right, Redhead's George V (2), *France Fenwick's* Hendon (2), *Stelp & Leighton's concrete-hulled* Creteboom, *France Fenwick's* Marsden (2) *partly hidden, and Dry's* Great Emperor (2). George V (2) *provides a rare sight of the Redhead house flag.* *(Harrington collection).*

MODEL OF THE WORLD'S LARGEST FLOATING DOCK.

Presented to Mr J. Patton, Managing Director, Messrs R. Redhead and Sons, Ltd., Tug Owners, Quayside, Newcastle-upon-Tyne

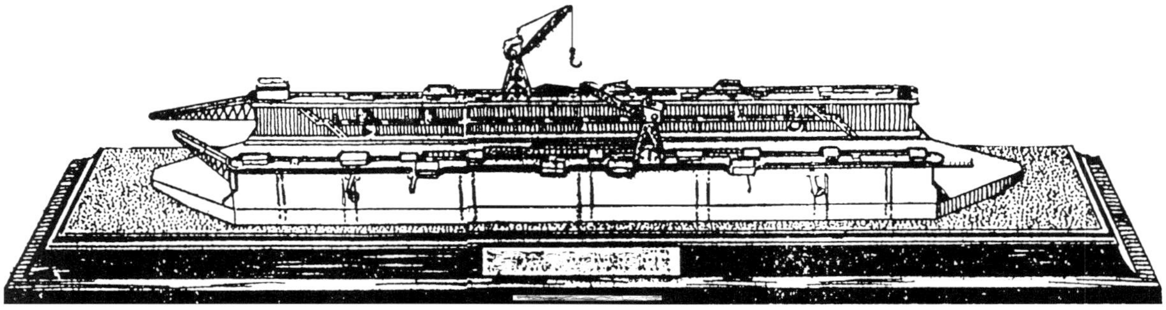

The Solid Silver Model, illustrated above, of the World's Largest Floating Dock, has been presented to Mr J. Patton, Managing Director of Messrs R. Redhead and Sons, Ltd., Tug Owners, Newcastle-upon-Tyne.

It has been specially made by the Northern Goldsmiths Company, Blackett Street, Newcastle-upon-Tyne, and is constructed to a scale of approximately 40 feet to 1 inch. It is entirely hand made, and exhibits a surprising amount of detail in the form of Movable Cranes, Gangways, Keel and Bilge Blocks, Shore Blocks, Power Houses, etc., etc.

Special attention has been given to the finish. The Model being Oxidized and Lacquered and mounted on a fine Dark Oak Plinth, the whole presenting a nautical and impressive appearance.

On the Oxidized Silver Plate appears the following inscription:–

> Presented to Mr J. Patton, Managing Director of Messrs Redhead and Sons Ltd., Tug Owners, Quayside, Newcastle-upon-Tyne, by Mrs S. H. White, to commemorate his great achievement in supervising the towing of the Southampton Floating Dock (built by Messrs Sir W. G. Armstrong, Whitworth and Co., Ltd.) from the Tyne to Southampton in the record time of 4 days–17th to 21st April, 1924.

It is a matter of congratulation that we have on Tyneside a firm of Silversmiths who are in the position to undertake the execution of this class of work.

The Northern Goldsmith Company possess exceptional facilities for the production of Models of every description, Chains of Office, Caskets, Hunting Trophies, etc., etc., and unique articles for presentation.

An illustration of the silver model of the Southampton floating dock.

APPENDIX A TO CHAPTER FOURTEEN

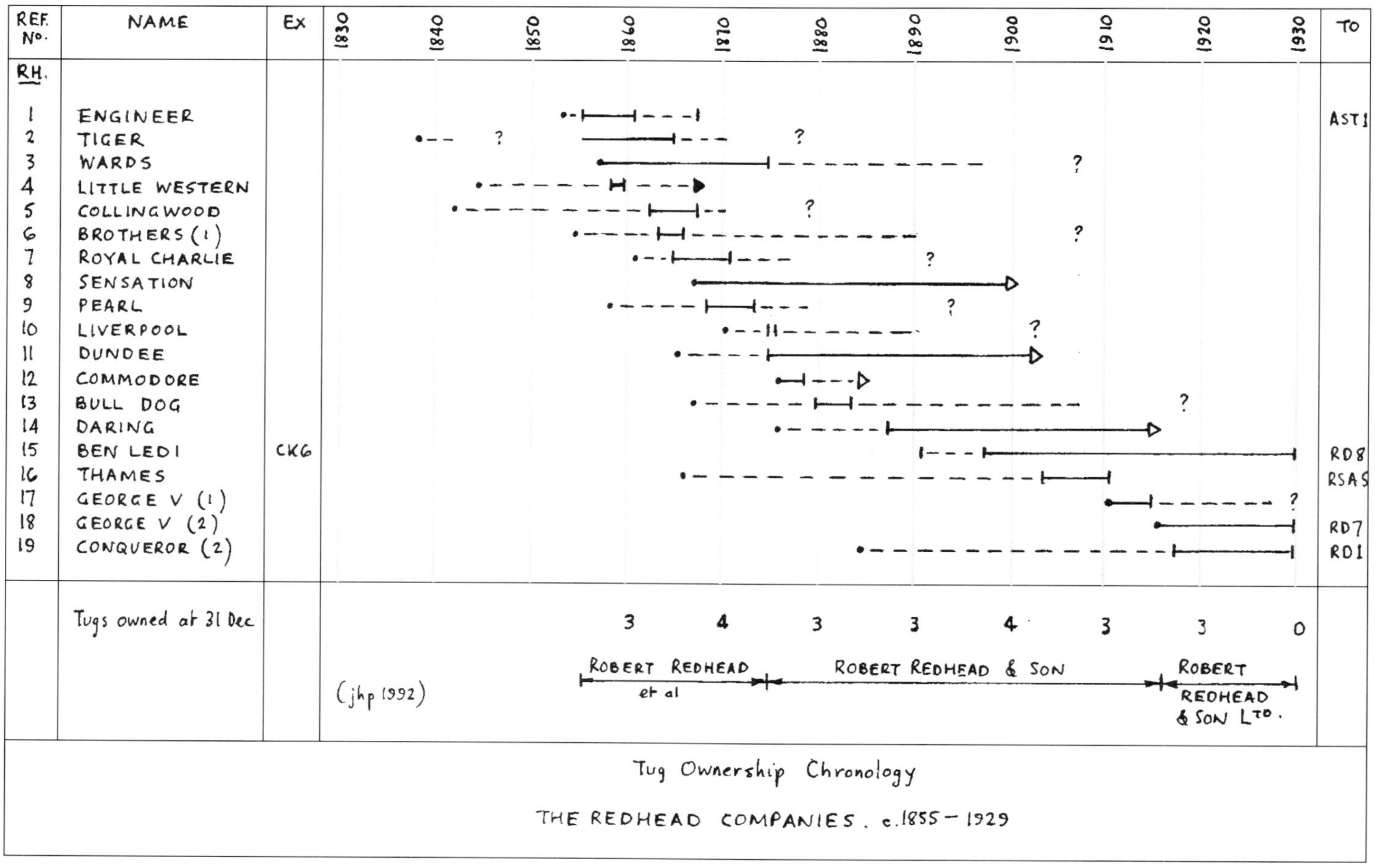

APPENDIX B TO CHAPTER FOURTEEN

The Redhead Companies Fleet List

RH 1 **ENGINEER** **(1855-1861)**
ON 2296 47 g. 72.3 x 14.5 x 7.9 ft.
B: On River Tyne.
E: 25 NHP.

PR: 1853 ?; By 1858 Newcastle; By 1864 Shields.

1853:	Completed. Owner not known
1855:	Purchased by Robert Redhead et al, Newcastle.
1861:	Sold to John W. Robson, North Shields.
1867:	Purchased by John Anderson and Robert Chater, North Shields.
1870:	Not in register.

RH 2 **TIGER** **(1855-1864)**
ON 15971
B: At North Shields.
E: 25 NHP.

PR: 1838 ?; By 1845 Newcastle; By 1864 Sunderland.

1838:	Completed. Owner not known.
By 1855:	Purchased by G. Redhead, Walker, and J. Redhead, North Shields.
By 1864:	Sold to John Taylor et al, Sunderland.
1871:	Not in register.

RH 3 **WARDS** **(1857-1874)**
ON 12908 54 g. 77.1 x 15.1 x 8.1 ft. Wood Paddle
B: John Robson, Gateshead.
E: L. 1 Cy. 27" x 39". 25 NHP. By Builder.

PR: 1857 Newcastle; 1869 Newcastle; 1874 Hull; 1875 Grimsby.

04.1857:	Completed for Robert Redhead and William Gray.
09.1874:	Sold to F. Peat, Hull.
1875:	Sold to W. Cooper, Grimsby.
c.1890:	Sold to E.A. Durrant, Yarmouth.
1898:	Not in register.

RH 4 **LITTLE WESTERN** **(1858-1859)**
ON 26957 36 g. 75.4 x 10.4 x 9.6 ft. Wood Paddle
B: At Gateshead.
E: L. 1 Cy. 14 NHP.

PR: 1844 ?; 1855 Newcastle; 1858 Newcastle; 1861 Middlesbrough.

1844:	Completed for E. Hutchinson.
1853:	Sold to B. Bootiman, Newcastle.
09.1858:	Purchased by Robert Redhead and William Gray.
03.1859:	Sold to R. Armstrong et al, Newcastle.
05.1860:	Sold to Jane Dixon, Middlesbrough.
09.1862:	Sold to R. Walker, Felling.
06.1863:	Sold to J. Wake et al, Newcastle.
01.1868:	Broken up at North Shields.

RH 5 **COLLINGWOOD** **(1862-1867)**
ON 2477 29 g. 54.5 x 12.7 x 6.7 ft. Wood Paddle
B: At North Shields.
E. 14 NHP.

PR: 1842 Newcastle; By 1855 Sunderland.; By 1864 Newcastle.

1842:	Completed. Owner not known.
1846:	Sold to F. Welch, Sunderland.
1852:	Sold to R. Lister, Sunderland.
1854:	Sold to L. Gale et al, Sunderland.
1857:	Sold to G. Dowell, Sunderland.
1862:	Purchased by Robert Redhead and William Gray, Newcastle.
1867:	Sold.
1871:	Not in register.

RH 6 **BROTHERS (1)** **(1863-1866)**
ON 8754 57 g. 71.5 x 14.9 x 8.0 ft. Wood Paddle
B: At North Shields.
E: 30 NHP.

PR: 1854 Newcastle; 1866 Port Glasgow; 1870 Greenock; 1882 Greenock.

1854:	Completed for J. Shield, Newcastle and D. Renwick, North Shields.
By 1858:	Owners now D. Renwick and J. Stoker, North Shields.
1863:	Purchased by Robert Redhead and William Gray, Newcastle.
1866:	Sold to J. Laird, Port Glasgow.
1870:	Sold to A. Robertson, Greenock; later D. Robertson and then J. Robertson.
1882:	Sold to W. Park, Greenock.
c.1890:	Not in register.

RH 7 **ROYAL CHARLIE** **(1864 - c.1871)**
ON 29722 48 g. 71.1 x 15.0 x 8.0 ft. Wood Paddle
B: At North Shields.
E: 20 NHP.

PR: 1861 Shields; 1864 Newcastle.

1861:	Completed for J. & W. Hall, Shields.
1864:	Purchased by Robert Redhead and William Gray, Newcastle.
c.1871:	Sold to G.P. Ward et al, Newcastle.
1878:	Not in register.

RH 8 **SENSATION** **(1867-1901)**
ON 56054 91 g. 90.2 x 18.6 x 9.7 ft. Wood Paddle
B: T. Hepple, Low Walker.
E: L. 1 Cy. 33¼" x 50". 40 NHP. By Builder.

PR: 1867 Newcastle; 1869 Newcastle.

1867:	Completed for Robert Redhead and William Gray.
c.1874:	Ownership now Robert and Mark H. Redhead, trading as Robert Redhead & Son.
11.1901:	Foundered off Hartlepool.

RH 9 **PEARL** **(1868-1873)**
ON 21571 72 g. 86.0 x 16.7 x 8.7 ft. Wood Paddle
B: At North Shields.
E: L. 1 Cy. 40 NHP.

PR: 1858 Greenock; 1868 Newcastle; 1873 Boulogne.

1858:	Completed, owner not known.
By 1865:	Owned by J. Hendry, Greenock.
11.1868:	Purchased by Robert Redhead and William Gray, Newcastle.
10.1873:	Sold to French owners, no further information.

RH 10 **LIVERPOOL** **(1874-1875)**
ON 58868 116 g. 105.9 x 19.3 x 9.6 ft. Wood Paddle
B: J.P. Rennoldson, South Shields.
E: L. 1 Cy. 37" x 50". 60 NHP. By Builder.

PR: 1870 Middlesbrough; 1873 Dublin; 1874 Newcastle; 1875 Harwich; 1877 Hull.

03.1870:	Completed, as VICTOR, for Halder & Co., Middlesbrough.
03.1873:	Sold to Dublin owners.
04.1874:	Purchased by Robert and Mark H. Redhead, trading as Robert Redhead & Son, and renamed LIVERPOOL.
08.1875:	Sold to W.J. Watts, Harwich.
1897:	Sold to H. Booker, Hull.
c.1886:	Sold to J.F. Ealand.
1891:	Not in Lloyds Register.

RH 11 **DUNDEE** **(1874-1903)**
ON 51379 89 g. 94.4 x 17.9 x 9.4 ft. Wood Paddle
B: J. Cooper & Co. North Shields.
E: L. 1 Cy. 50 NHP. J.P. Rennoldson, South Shields. (1869).

PR: 1865 Newcastle; 1869 Newcastle; 1875 Newcastle.

12.1865:	Completed, as HUSSAR, for Tyne Steam Shipping Co.Ltd.
05.1869:	Sold to J. Ormston Jnr., Newcastle, later J. Ormston Snr.
10.1874:	Ownership now Robert Redhead & J. Ormston Snr., renamed DUNDEE.
04.1883:	Ownership now Robert Redhead, J. Ormston Snr. & M.H. Redhead, trading as Robert Redhead & Son.
09.1896:	Owners now Robert Redhead and M.H. Redhead only.
05.1903:	Sank after collision off Trow Rocks, South Shields.

RH 12 **COMMODORE** **(1876-1878)**
ON 70252 158 g. 117.9 x 19.6 x 10.5 ft. Iron Paddle
B: J.T. Eltringham, South Shields. (YN 50).
E: L. 1 Cy. 40¼" x 56". 75 NHP. J.P. Rennoldson, South Shields.

PR: 1876 Newcastle; 1878 Limerick; 1882 Cork.

04.1876:	Completed for Robert Redhead & Son.
10.1878:	Sold to Citizen Steam Towing Co.Ltd., Limerick.
1882:	Sold to Queenstown Towing Co.Ltd.
09.1885:	Lost at Morris Head, near Roberts Cove, Cork.

RH 13 **BULL DOG** **(1879-1883)**
ON 56483 67 g. 78.1 x 16.9 x 8.9 ft. Wood Paddle
B: Richard Stobbs, North Shields.
E: L. 1 Cy. 28" x 48". 28 NHP. Readhead & Softley, South Shields.

PR: 1867 North Shields; 1879 Newcastle.

1867: Completed for Henry Jobes and J.H. Newton, North Shields.
1872: Sold to S.T. Freeman, Westminster.
06.1879: Purchased by Robert Redhead & Son.
05.1883: Sold to J.W. Taylor, F. Wilton et al, Sunderland.
1892: Sold to J.E. Dawson, Sunderland.
1900: Sold to T. Robson, Byker, and later transferred to Mrs. M. Robson.
1902: Sold to Fergus Graham, Durham.
1903: Sold to H. Farrier, Newcastle.
1907: Sold to M. Swinburne, Durham.

RH 14 **DARING** **(1887-1916)**
ON 73715 135 g. 106.9 x 19.1 x 10.0 ft. Iron Paddle
B: John Readhead & Co., South Shields. (YN 123).
E: L. 2Cy. 28" x 51". 70 NHP. By Builder.

PR: 1876 London.

1876: Completed for William Sandford, Gravesend.
1887: Purchased by Robert Redhead and M.H. Redhead, trading as Robert Redhead & Son.
05.1903: Transferred to M.H. Redhead and Susannah H. White.
06.1916: Transferred to Robert Redhead & Son Ltd.
12.1916: Foundered south of St. Mary's Island, Northumberland. Total loss but all five crew saved.

RH 15 **BEN LEDI** **(1897-1929)**
See Cook fleet list, CK 6.

Thames *(Local History Library, South Shields Central Library).*

RH 16 **THAMES** **(1903-1910)**
ON 54651 116 g. 96.0 x 17.6 x 8.8 ft. Iron Paddle
B: J. & W. Dudgeon, Cubitt Town.
E: L. 2 Cy. 25⅜" x 32¼". 50 NHP. J.A. Young, Blackwall, London.

PR: 1866 Swansea; 1888 Milford; 1903 Newcastle; 1922 Sfax.

1866: Completed for Alexander Brogden, M.P., Ulverston.
1876: Sold to Great Western Railway Co.
08.1903: Purchased by M.H. Redhead and Susannah H. White, trading as Robert Redhead & Son.
10.1910: Purchased by Northern Tug Co.Ltd. (Robert Stephenson Allen).
12.1912: Sold to Thomas H. Nelson, South Shields.
1922: Sold to E.L. Loisseaux, renamed NEPTUN.
c.1925: Owners restyled Soc. Francais d'Entrepéneurs Maritime.
1928: Not in Lloyds Register.

George V (1) *when new (Company archives).*

RH 17 **GEORGE V (1)** **(1910-1914)**
ON 129759 188 g. 100.00 x 23.2 x 11.9 ft. Steel Screw
B: Sir W.G. Armstrong, Whitworth & Co.Ltd., Low Walker (YN 831)
E: T. 3 Cy. 15½", 25" & 41" x 27". 750 IHP. 12 K. J.P. Rennoldson & Sons, South Shields.

PR: 1910 Newcastle; 1914 Reval.

09.1910: Completed, for £6,500, for Mark H. Redhead and Susannah H. White, trading as Robert Redhead & Son.
03.1914: Sold for £10,000 to Russian Government, renamed SALVAGE STEAMER No.2.
c.1916: Lengthened to 103.5 feet.
1928: Not in Lloyds Register.

RH 18 **GEORGE V (2)** **(1915-1929)**
ON 133567 224 g. 106.1 x 24.4 x 12.3 ft. Steel Screw
B: J.P. Rennoldson & Sons Ltd., South Shields. (YN 294).
E: T. 3 Cy. 16", 26" & 43" x 27". 850 IHP. 12½ K. By Builder.

PR: 1915 Newcastle; 1987: Gibraltar.

01.1915: Completed, for £9,500, for Mark H. Redhead and Susannah H. White, trading as Robert Redhead & Son.
05.1915: Hired by Royal Navy. 11.1915: Returned to owners.
06.1916: Transferred to Robert Redhead & Son Ltd.
1917: Hired by Royal Navy. 1919: Returned to owners.
08.1929: Transfered to Redhead & Dry's Tugs Ltd.
05.1944: Transferred to France, Fenwick Tyne & Wear Co.Ltd.
08.1954: Re-engined by P.K. Harris & Sons Ltd., Appledore. Converted to diesel propulsion. Oil. 2 SA. 12 Cy. 8½" x 10". 1086 BHP. Tonnage now 217g
01.1959: Chartered to Tyne Tugs Ltd.
05.1977: Transferred, for £7,000, to Lawson-Batey Tugs Ltd.
01.1978: Sold, for £11,500, to Coastal Marine Services, Plymouth.
07.1981: Sold to Bankside Shipping & Towage Co.Ltd., Queenborough.
08.1982: Sold to N.E. Murray, Queenborough.
02.1984: Sold to Estuary Tugs Ltd., Sark.
03.1987: Sold to Lavington International Ltd., Gibraltar.
1992: Still in existence.

RH 19 **CONQUEROR (2)** **(1917-1929)**
ON 89619 177 g. 123.5 x 20.1 x 10.1 ft. Iron Paddle
B: Hepple & Co., North Shields. (YN 386).
E: L. 2 Cy. 32" x 58". 140 NHP. 10 K. By Builder.

PR: 1884 London; c.1890 Boulogne; 1905 Middlesbrough; 1913 Dundee; 1917 Newcastle.

1884: Completed, as CONQUEROR, for William Sandford, Gravesend.
c.1890: Sold to Soc.Anon. du Remorquage Boulonnais, renamed CONQUÉRANT.
c.1897: Sold to Soc. Nouvelle du Remorquage, Boulonnais.
c.1904: Sold to Mrs. R. Lennard, Middlesbrough, renamed CONQUEROR.
1905: Sold to B.O. Davies, Middlesbrough.
c.1910: Sold to J. Nicholson, Leith.
1913: Sold to Victor A. Cappon, Dundee.
03.1917: Purchased by Robert Redhead & Son Ltd.
08.1929: Transferred to Redhead & Dry's Tugs Ltd.
05.1944: Transferred to France, Fenwick Tyne & Wear Co.Ltd.
05.1956: Sold, for £2,150, to British Iron & Steel Corporation. Allocated to J.J.King & Co.Ltd., Gateshead for breaking up.

• CHAPTER FIFTEEN •

The Dry Companies

Thomas Allan Dry. c.1857 - 1889

In 1857 it is known that Thomas Allan Dry was the part-owner of the tug *William* which had been built on the Tyne in 1842. It is believed that he kept his share in this vessel for only a very short time; probably from 1857 to 1858. Also in 1857, along with others, T.A. Dry had purchased the tug *Ocean Bride* (33/1857). Even though this was a new vessel she changed hands, going to a group of Sunderland owners, after just five years. In 1864 another new wooden paddle tug, *James Conley* (72/1864) was built for T.A. Dry, M.P. Martin and W.G. Dry. T.A. Dry was recorded as living at Willington Quay at this time, whereas earlier his address had been given as Newcastle. Again, after a very short time, this tug was sold, in 1866, to an owner living at Oswestry. She was re-registered at Aberystwyth, so presumably she was working from that port. Eighteen years later *James Conley* was back on the Tyne, with owners who are outside the scope of this history. Who James Conley was has so far not been discovered.

After having purchased two newly-built tugs T.A. Dry, with others, bought in 1869 the three-year-old *Great Emperor* (1) (82/1866). His partners in the ownership of this tug varied over the years and included J. Winlo, who was the owner of a wooden floating dock on the River Tyne. Again, along with others, T.A. Dry acquired the wooden paddle tug *Brothers* (2) (75/1869), certainly by 1873 and possibly from new. Initially he had a minority share in the vessel, but by 1876 he held the largest number of sixty-fourth shares. In 1874 the new paddle tug *Great Britain* (1) (100/1874) was built for T.A. Dry and his colleagues. At almost 100 feet long she was much larger than anything owned previously. She had been built at South Shields by J. Winlo, who, no doubt, held a number of shares in the vessel. In fact, in 1879, the ownership is recorded as being with J. Winlo & Co. This could simply mean that Winlo was by then the majority shareholder and not necessarily that T.A. Dry had relinquished his interest in *Great Britain* (1).

By 1871 T.A. Dry had moved from his earlier address to 17 Mitre Street, South Shields, and in 1875 had moved again to 25 Thrift Street in the same town. *Brothers* (2) had disappeared from the register by 1878.

The following year T.A. Dry and J. Tully purchased the wooden paddle tug *Test* (82/1853) which had been built by Andrew Woodhouse at South Shields. By that time T.A. Dry had moved premises once again to 16 Long Row, South Shields, only to move again to No.17 four years later. In 1883 the iron paddle tug *Washington* (122/1870) was purchased, following ownership on the Thames and at Sunderland. She was to become very well known on the Tyne for the next sixty-nine years, ending her days in the colours of France, Fenwick Tyne & Wear Co.Ltd. At the end of the same year another iron paddle tug *Gauntlet* (87/1869) was purchased. She too had started her working life on the Thames.

John Dry. 1889 - 1908

On the death of T.A. Dry in 1889, ownership of his towage business was transferred to his son John Dry. The fleet at that time consisted of four tugs:-

Name	Gross tonnage/Year	Hull construction	Means of propulsion
Great Emperor (1)	(82/1866)	Wood	Paddle
Test	(82/1853)	Wood	Paddle
Washington	(122/1870)	Iron	Paddle
Gauntlet	(87/1869)	Iron	Paddle

- with an average age of almost twenty-five years.

In 1891 *Test* was sold to Middlesbrough owners and four years later *Great Emperor* (1) was sold for further service on the Tyne. Both survived until early in the twentieth century. In 1893 the company's address had changed once more, to 3 Thrift Street, South Shields. All of these locations in South

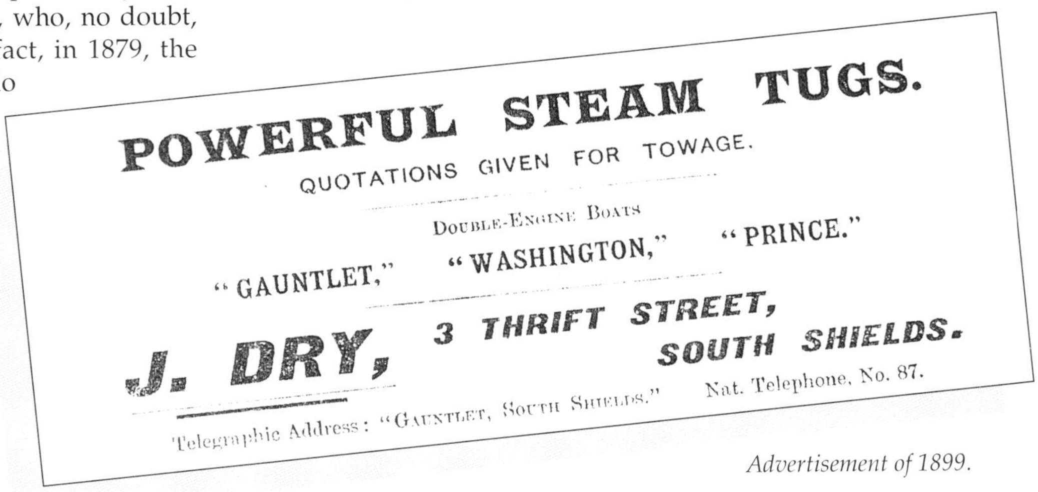

Advertisement of 1899.

Mauretania leaving the Tyne for trials in September 1907 with the Dutch tugs Cyclop and Poolzee at her head. A George Brown tug lies on her starboard quarter and a Dry tug is following astern. Two Lawson tugs are laid at the buoys. *(Author's Parry' collection)*

Shields since 1875 were quite close to each other, in the narrow streets near the riverside to the north-west of the Market Place, an area in more recent times occupied by Tyne Dock Engineering Co. Ltd.'s and Brigham & Cowan Ltd.'s drydocks.

In 1895 the London-based wooden paddle tug *Prince* (89/1859) was purchased by John Dry. She was a double-engined vessel and had been built at Rye on the south coast of England - an exception at this time, since so many of the tugs referred to in this volume had been built on the Tyne. Five years later the 200 IHP iron paddle tug *Snowdon* (106/1885) was purchased from the Portmadoc Steam Tug Co. - the name clearly reflecting her Welsh ownership. Even though built for owners on the west side of Great Britain, she, nevertheless, had been built in South Shields.

When, in September 1907, the world-famous Cunard liner *Mauretania* (31,938/1907) left her builders at Wallsend for trials her head tugs were Dutch-owned, much to the chagrin of the local tug owners and crews. However, when leaving the Tyne to enter service on 22nd October 1907 her towage was entrusted to John Dry who used his tugs *Washington, Snowdon, Gauntlet* and *Prince* together with Brown's *President* and Crosthwaite's *Champion.* Thus was honour restored.

The John Dry Steam Tugs Ltd. 1908 - 1929

In 1908 John Dry decided to convert his business into a limited liability company under the title of The John Dry Steam Tugs Ltd. with a base in offices in Dean Street, South Shields. This location was just off the Market Place, and near the South Shields/North Shields ferry landing. It has been suggested that the Dry company had connections, probably financial, with the Wallsend shipbuilders, Swan Hunter & Wigham Richardson Ltd. At the formation of the company the fleet comprised:-

Washington	(122/1870)	Iron	Paddle
Gauntlet	(87/1869)	Iron	Paddle
Prince	(89/1859)	Wood	Paddle
Snowdon	(106/1885)	Iron	Paddle

From this time, and probably from an earlier date, the funnels of the company's tugs were painted black with a broad red band, and the house flag carried a large blue 'D' on a red background.

In the year following incorporation the new company purchased their first screw tug. She was *Great Emperor* (2) (171/1909) and had been newly built for Dry by J.T. Eltringham & Co. at South Shields. Her compound engine of 650 IHP made her, by a very large margin, the most powerful tug the company had owned to date. John Dry was so proud of her that he would only allow her to be used for 'special jobs' and such was her low level of activity that she well deserved the nickname of the 'Mill Dam Lightship', given to her by the local tug crews. It is interesting to note that, although *Great Emperor* (2)'s name is so recorded in the Custom House Register, on the vessel itself the name is painted as *Gt. Emperor.* In this volume the full version of the name has been used to correspond with the official record.

The size of the Dry fleet had now reached five, a number which was not to be exceeded during the further twenty years of the company's independent existence.

In 1909 the shipbreakers Hughes Bolckow bought their first warship which was then lying in the Kyles of Bute. She was the ironclad barbette battleship *Collingwood* of 9,500 tons displacement and measuring 325 feet long x 68 feet beam. *Great Emperor* (2), soon after her completion and the paddle tug *Washington,* were entrusted with the job of getting this unwieldy bulk to the Tyne. The tow of 630 miles took eleven days, during which very heavy weather was encountered off Cape Wrath. On arrival off the Tyne on 15th June further assistance was given by *Snowdon* in nursing the old warship to her moorings at Dunston.

Again, *Great Emperor* (2) was involved in a 'special job' when the company obtained a contract to tow the French trawler *Rorqual,* the largest of that nationality afloat and which had been ashore in Iceland, from Reykjavik to the Tyne for repair. The tug, commanded by John Dix, left Shields on 23rd March 1910 for Iceland. Very bad weather was encountered on the way and, in the vicinity of the Faroes, with waves 28 feet high, she lost some 5 tons of her deck cargo of coal overboard. She made landfall on 28th March at Vestmann Island just off the south coast of Iceland and some 100 miles east of her destination. Together with about 100 trawlers, *Great Emperor* (2) took shelter under the north-east corner of the island the following day before proceeding further. When the weather eased she steamed round Cape Reykjanes and reached Reykjavik on 31st March. After about a week's delay the return journey with *Rorqual* in tow was commenced. Tempestuous seas were encountered again and as a consequence the tug's crew had to live in the fore cabin because of the seas continually breaking aboard. Nevertheless it was said that the crew of seventeen on *Rorqual* seemed to be oblivious to the weather and the movement of their ship. The tug and tow arrived safely in the Tyne after their 1,000-mile return trip on 14th April 1910, having been away for a total of 22 days.

Great Emperor (2), assisted by *Nestor* and also temporarily by *Washington,* left the Tyne for Harwich on 22nd February 1912 towing an Admiralty pontoon which had been built by Swan Hunter at Wallsend. The pontoon was 267 feet long x 91 feet beam and the voyage took about three days. Not all such business was, however, successful. On 1st May 1912 *Great Emperor* (2) towed the Tyne General Ferry Co.'s Newcastle landing stage, 160 feet long x 20 feet beam, away for whaling company owners at Sandefjord, Norway. Unfortunately, it never reached its destination since it had to be abandoned three days later, when 190 miles out, in high seas and gale force winds.

Prince was broken up at Hylton on the River Wear in 1912, temporarily reducing the fleet to four in number. Jimmy (Punch) Stewart, father of Hector Stewart referred to in the France Fenwick chapter, had been her skipper.

In April 1914 the firm of Robert Stephenson Allen of Newcastle was taken over by The John Dry Steam Tugs Ltd. At that time R.S. Allen was the owner of only one tug, the funnel colours of which were white with a deep black top. She was the iron paddle tug *Malta* (113/1875) and had worked on the Thames after being built by J.T. Eltringham at South Shields for the first eleven years of her life when she bore the name *Benachie.* Due to the change, R.S. Allen was appointed manager of the Dry company. *Snowdon* came to grief in September 1916 when she was cut through the bow and sunk by the steamer *Glencona.* She was, however, soon salvaged and returned to service. During the First World War two of the company's tugs were hired by the Royal Navy: *Great Emperor* (2) for seven months in 1917, and *Gauntlet* for over two years, to September 1919. The latter tug worked out of Hartlepool and was renamed *Gaunt* for the period of her hire. Whilst there, skippered by George Dry, she was visited by Their Royal Highnesses King George V and Queen Mary.

John Dry died on 4th October 1921 at the age of 69 years. He had started work on tugs when he was 13 years old and had been a skipper before he was 21. By 1925 the following were the recorded officials of the company:-

Chairman:	T.E. Thirlaway
Managing Director:	R.S. Allen
Secretary:	J.T. Alderson
Marine Superintendent:	W.H. Morrison

The company's office was at Mercantile Marine Buildings, Mill Dam, South Shields. In September 1929 an agreement was made with the tug company of Robert Redhead & Son Ltd. that the two companies should merge. Their combined operations on the River Tyne were to trade under the title of Redhead & Dry's Tugs Ltd. The Dry tugs contributed to the new organisation were:-

Washington	(122/1870)	Iron	Paddle
Gauntlet	(87/1869)	Iron	Paddle
Snowdon	(106/1885)	Iron	Paddle
Great Emperor (2)	(171/1909)	Steel	Screw
Malta	(113/1875)	Iron	Paddle

-with an average age of over forty-seven years!

Thus ended the seventy-two years existence of one of the Tyne's smaller tug companies.

APPENDIX A TO CHAPTER FIFTEEN

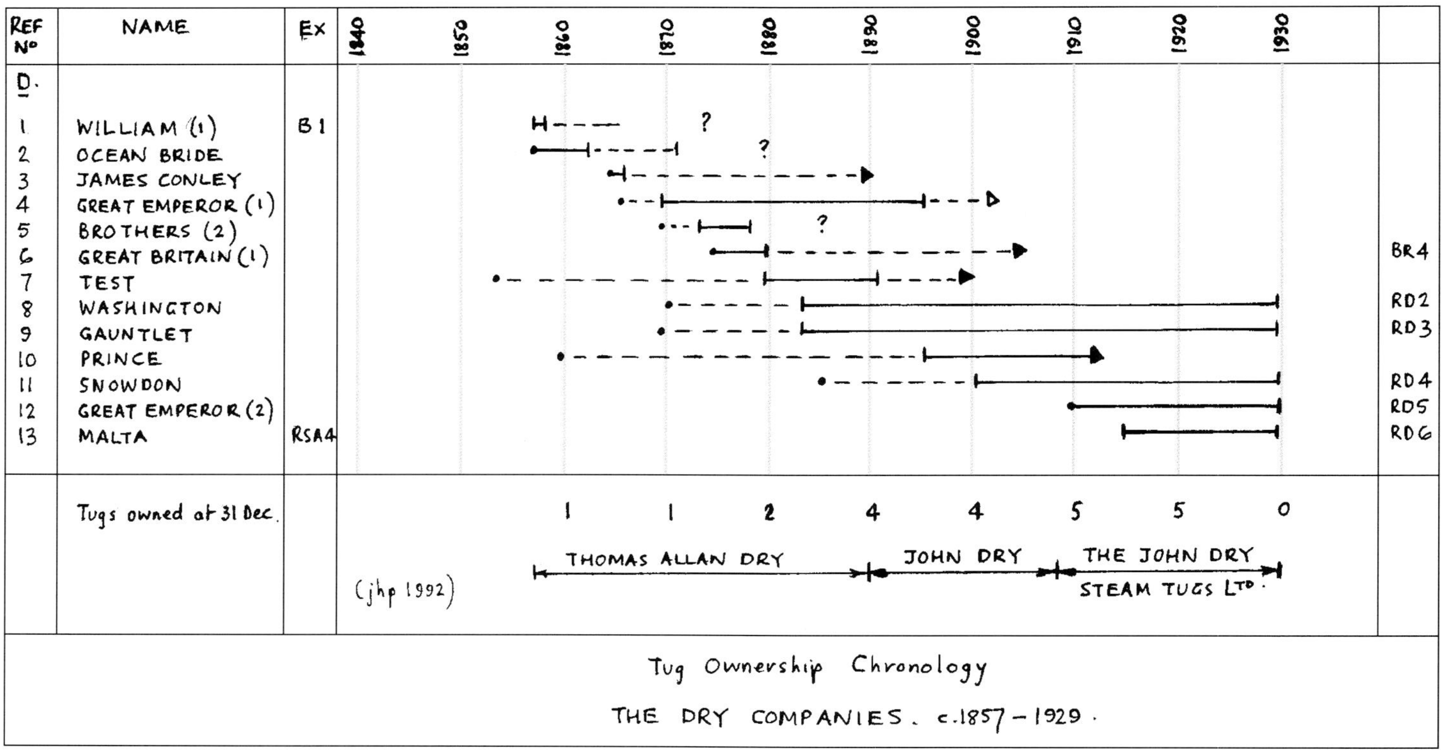

APPENDIX B TO CHAPTER FIFTEEN

The Dry Companies Fleet List

D 1 **WILLIAM (1)** **(1857-1858)**
See Batey fleet list, B 1.

D 2 **OCEAN BRIDE** **(1857-1862)**
ON 20258 33 g. 74.0 x 15.0 x 7.4 ft. Wood Paddle
B: At Gateshead, Co. Durham.
E: 20 NHP.

PR: 1857 Newcastle; 1863 Sunderland.

12.1857:	Completed for T.A. Dry et al, Newcastle.
12.1862:	Sold to G. Venus et al, Sunderland.
1871:	Not in Mercantile Navy List.

D 3 **JAMES CONLEY** **(1864-1866)**
ON 49743 72 g. 81.8 x 17.1 x 9.1 ft. Wood Paddle
B: At Willington Quay.
E: L. 1 Cy. 30" x ?. 30 NHP.

PR: 1864 Shields; 1866 Aberystwyth; 1867 Caernarvon; 1884 North Shields.

1864:	Completed for T.A. Dry & M.P. Martin, Willington Quay, and W.G. Dry, Gateshead.
1866:	Sold to T. Savin, Oswestry.
1871:	Sold to Portmadoc Steam Tug Co.Ltd.
1884:	Sold to S. Young, North Shields.
04.1886:	Sold to M.S.S. Dixon, South Shields.
05.1890:	Sold to J. Graham, Whickham.
12.1890:	Vessel broken up.

D 4 **GREAT EMPEROR (1)** **(c1869-1895)**
ON 56102 82 g. 89.6 x 17.6 x 9.3 ft. Wood Paddle
B: Readhead, Softley & Co., South Shields.
E: L. 1 Cy. 32¾" x 48". 35 NHP. By Builder.

PR: 1866 South Shields; 1871 South Shields; 1889 South Shields.

06.1866:	Completed for J. Dawson.
c.1869:	Purchased by T.A. Dry, J. Lawson et al.
c.1870:	Owners now T.A. Dry and R.W. Atkinson.
c.1877:	Owners now T.A. Dry, R.W. Atkinson and J. Winlo.
07.1889:	Transferred to John Dry.
10.1895:	Sold to R. Hedley, Jarrow.
09.1902:	Foundered 1/2 mile S.E. of Port Mulgrave on voyage Jarrow to Port Mulgrave, due to hull damage caused by engine failure.

D 5 **BROTHERS (2)** **(1873-1877)**
ON 58718 75 g. 81.4 x 17.3 x 9.1 ft. Wood Paddle
B: At South Shields.
E: 30 NHP.

PR: 1869 South Shields.

02.1869:	Completed, Owners not known.
By 1871:	Owned by J.R. Walker, South Shields.
1873:	Owners now J.R. Walker, T.A. Dry et al.
1878:	Not in register.

D 6 **GREAT BRITAIN (1)** **(1874-1879)**
ON 63030 100 g. 99.7 x 18.2 x 9.9 ft. Wood Paddle
B: J. Winlo & Co., South Shields.
E: L. 1 Cy. 35" x 50". 60 NHP. J. Readhead & Co., South Shields.

PR: 1874 South Shields; 1886 South Shields.

04.1874:	Completed for T.A. Dry et al.
1879:	Sold to J. Winlo & Co., South Shields.
1896:	Purchased by G. Brown.
1905:	Broken up at Shields.

D 7 **TEST** **(1879-1891)**
ON 7633 82 g. 79.5 x 15.5 x 9.2 ft. Wood Paddle
B: Andrew Wouldhouse, South Shields.
E: L. 1 Cy. 33" x 45". 105 IHP. 8 K. J.P. Almond, North Shields.

PR: 1853 Shields; 1895 Middlesbrough.

1853:	Completed for A.R. Grant, W. Gibson and T. Hunter, North Shields.
By 1858:	Sold to Joseph Cowen, Blaydon.
1879:	Purchased by T.A. Dry and J. Tully, South Shields.
1891:	Sold to C.C. Duncan, Middlesbrough.
02.1897:	Sold to G. Robinson, Middlesbrough.
05.1900:	Broken up.

Washington. *(Local History Library, South Shields Central Library).*

D 8 **WASHINGTON** **(1883-1929)**
ON 63655 122 g. 104.9 x 18.4 x 10.1 ft. Iron Paddle
B: Readhead, Softley & Co., South Shields. (YN 58)
E: L. 2 Cy. 27" x 46". 50 NHP. By Builder.

PR: 1870 London; 1879 Sunderland; 1883 North Shields.

1870:	Completed for W.G. Ives et al, Gravesend.
c.1874:	Sold to J. Martin, London.
c.1879:	Transferred to Miss S.C. Martin, London.
1879:	Sold to Mrs. M. Nicholson, Sunderland.
12.1883:	Purchased by T.A. Dry. 07.1889: Transferred to John Dry.
05.1908:	Transferred to The John Dry Steam Tugs Ltd.
08.1929:	Transferred to Redhead & Dry's Tugs Ltd.
05.1944:	Transferred to France, Fenwick Tyne & Wear Co.Ltd.
04 1952:	Sold to J.J. King & Co.Ltd., Gateshead, for breaking up.

D 9 **GAUNTLET** **(1883-1929)**
ON 63506 87 g. 93.3 x 17.0 x 8.4 ft. Iron Paddle
B: Readhead, Softley & Co., South Shields.
E: L. 2 Cy. 25¼" x 45". 60 NHP.

PR: 1869 London; 1882 North Shields.

08.1869:	Completed, as GAUNTLET, for T.W. Elliott, London.
05.1882:	Sold to J. Tully, Sunderland.
12.1883:	Purchased by Thomas A. Dry, South Shields.
07.1889:	Transferred to John Dry, South Shields.
05.1908:	Transferred to The John Dry Steam Tugs Ltd., South Shields.
05.1917:	Hired by Royal Navy. 09.1917: Renamed GAUNT.
09.1919:	Returned to owner, renamed GAUNTLET.
08.1929:	Transferred to Redhead & Dry's Tugs Ltd.
06.1935:	Broken up at Sunderland.

Prince *nearest camera with* Snowdon *behind* Malta *of R.S. Allen. (Local History Library, South Shields Central Library).*

D 10 **PRINCE** **(1895-1912)**
ON 28352 89 g. 90.3 x 18.5 x 9.0 ft. Wood Paddle
B: At Rye.
E: L. 2 Cy. 27½" x 42". 52 NHP. 10 K. J.A. Young, Blackwall.

PR: 1860 London; 1874 London; 1895 South Shields.

1859:	Completed for F. Spicer, Bermondsey.
c.1868:	Transferred to J.I. Spicer, Old Ford.
1874:	Sold to D.S.H. Oates et al for service on R. Thames.
1882:	Sold to E. Edwards, Poplar.
12.1895:	Purchased by John Dry, South Shields.
05.1908:	Transferred to The John Dry Steam Tugs Ltd.
1912:	Broken up at Hylton, River Wear.

D 11 **SNOWDON** **(1900-1929)**
ON 80242 106 g. 96.0 x 18.7 x 9.4 ft. Iron Paddle
B: Lawson & Eltringham, South Shields.
E: L. 2 Cy. 27" x 48". 200 IHP. 10 K. Baird & Barnsley, North Shields.

PR: 1885 Caernarvon; 1900 South Shields; 1938 Sunderland.

1885:	Completed for Portmadoc Steam Tug Co., Portmadoc.
11.1900:	Purchased by John Dry, South Shields (later owned jointly with T. Robinson).
05.1908:	Transferred to John Dry Steam Tugs Ltd.
08.1929:	Transferred to Redhead & Dry's Tugs Ltd.
09.1938:	Purchased by R.L. Cook Towage Co.Ltd.
11.1947:	Transferred to France, Fenwick Tyne & Wear Co.Ltd.
1950:	Broken up.

D 12 **GREAT EMPEROR (2)** **(1909-1929)**
ON 119907 171 g. 95.4 x 22.1 x 10.8 ft. Steel Screw
B: J.T. Eltringham & Co., South Shields. (YN 269).
E: C. 2 Cy. 21" & 43" x 27". 650 IHP. 11 K. Baird Bros., North Shields.

PR: 1909 South Shields.

01.1909:	Completed for The John Dry Steam Tugs Ltd.
05.1917:	Hired by Royal Navy. 12.1917: Returned to owners.
08.1929:	Transferred to Redhead & Dry's Tugs Ltd.
05.1944:	Transferred to France, Fenwick Tyne & Wear Co.Ltd.
02.1959:	Sold to British Iron & Steel Corporation and allocated to J.J. King & Co.Ltd., Gateshead, for breaking up.

D 13 **MALTA** **(1914-1929)**
See Allen fleet list, RSA 4

APPENDIX C TO CHAPTER FIFTEEN

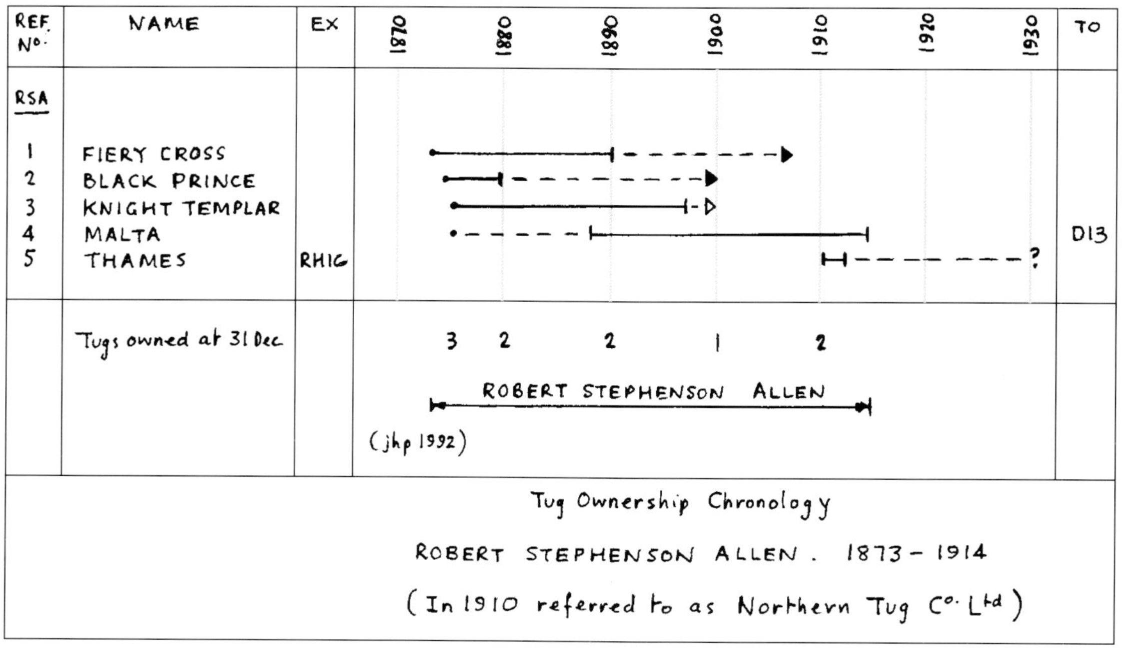

APPENDIX D TO CHAPTER FIFTEEN

R.S. Allen Fleet List

RSA 1 **FIERY CROSS** **(1873-1830)**
ON 63028 105 g. 93.7 x 19.1 x 9.8 ft. Wood Paddle
B: Richard Stobbs, North Shields.
E: L. 1 Cy. 35" x 50". 45 NHP. T.R. Scott & Co., North Shields.

PR: 1873 South Shields; 1890 South Shields; 1899 Middlesbrough.

06.1873: Completed for Robert Stephenson Allen et al, South Shields.
07.1890: Sold to Charles Dyble et al.
c.1895: Sold to Tees Tug & Lighter Co. Ltd., Middlesbrough.
11.1899: Sold to Eusebius Derwent, Middlesbrough.
02.1907: Sank on River Tees at Newport High Wharf, later refloated.
06.1907: Broken up at Hylton, River Wear.

RSA 2 **BLACK PRINCE** **(1874-1879)**
ON 69857 104 g. 97.8 x 18.9 x 9.8 ft. Wood Paddle
B: Richard Stobbs, North Shields.
E: L. 1 Cy. 35½" x 50". 42 NHP. T.R. Scott & Co., North Shields.

PR: 1874 South Shields; 1890 Bo'ness; 1895 Grangemouth.

07.1874: Completed for Robert Stephenson Allen, Westoe, South Shields.
1879: Sold to G. Chisholm, Shields.
1886: Sold to H. Stobbs.
1889: Sold to High & Shallcross, Dundee.
1890: Sold to W. Malcolm.
1891: Sold to J. Hogg, Bo'ness.
1892: Sold to Forth Towing Co.Ltd.
1895: Transferred to Grangemouth & Forth Towing Co.Ltd.
1900: Sold to M. McMillan, for breaking up.

RSA 3 **KNIGHT TEMPLAR** **(1875-1897)**
ON 69862 88 g. 88.8 x 17.8 x 9.4 ft. Wood Paddle
B: C.W. Dodgin & Co., North Shields.
E: L. 1 Cy. 32" x 48". 35 NHP. T.R. Scott & Co., North Shields.

PR: 1875 South Shields.

08.1875: Completed for Robert Stephenson Allen et al, South Shields.
1897: Sold to Thomas Gibson, South Shields.
04.1900: Abandoned 8 miles off Tyne when 'seeking'. Hot well gave way causing hole in hull.

RSA 4 **MALTA** **(1888-1914)**
ON 70712 113 g. 103.2 x 18.2 x 9.4 ft. Iron Paddle
B: J.T. Eltringham, South Shields. (YN 41)
E: L. 2 Cy. 27½" x 50". 90 NHP. J.P. Rennoldson, South Shields.

PR: 1875 London; 1888 South Shields.

1875: Completed, as BENACHIE, for J.F. Gibbs, London.
05.1886: Sold to William Watkins, London, renamed MALTA.
10.1888: Purchased, for £1,350, by Robert Stephenson Allen, Newcastle.
04.1914: Transferred to The John Dry Steam Tugs Ltd.
08.1929: Transferred to Redhead & Dry's Tugs Ltd.
05.1944: Transferred to France, Fenwick Tyne & Wear Co.Ltd.
03.1951: Sold to J.A. White & Co., North Queensferry, for breaking up.

RSA 5 **THAMES** **(1910-1912)**
See Redhead fleet list, RH 16

• CHAPTER SIXTEEN •

Redhead & Dry's Tugs Ltd.

The First Decade. 1929 - 1939

Redhead & Dry's Tugs Ltd. was formed in September 1929 when Robert Redhead & Son Ltd. took over the business and assets of The John Dry Steam Tugs Ltd. The eight tugs thus forming the combined fleet were:-

Name	Gross Tonnage/Year	Hull Construction	Means of propulsion	
Ben Ledi	(114/1878)	Iron	Paddle	Ex. Robert Redhead & Son Ltd
George V (2)	(224/1915)	Steel	Screw	
Conqueror (2)	(177/1884)	Iron	Paddle	
Washington	(122/1870)	Iron	Paddle	Ex. The John Dry Steam Tugs Ltd
Gauntlet	(87/1869)	Iron	Paddle	
Snowdon	(106/1885)	Iron	Paddle	
Great Emperor (2)	(171/1909)	Steel	Screw	
Malta	(113/1875)	Iron	Paddle	

- with an average age of over forty-three years. It will be noted that there was still a preponderance of paddle tugs but at least they all had double disconnecting engines.

In the new organisation James Patton, now sixty-eight years of age, continued as managing director, and his son, Alexander, was secretary and manager. The funnel colours adopted for the combined fleet were those of the Redhead company - yellow with a black top. Evidence as to the design of the house flag is lacking but it would be reasonable to assume that the Redhead flag, with suitably amended initial letters, would have been used. The company held the important contract for handling Royal Navy and Royal Fleet Auxiliary ships in and around the Tyne, and as a consequence *George V* (2) was kept on permanent standby for these customers. Most of the Tyne tug fleets had contracts with one or more shipping companies, and it is of interest to note one of the unwritten rules of the Tyne towage companies - if an inward bound ship had passed the North Shields ferry landing without its contracted tugs in attendance, then it was open to a tug of any company to offer its services. In November 1929 the company's tugs attended six ship launches in four days.

In 1930 the 600 IHP steel screw tug *Hannah Jolliffe* was purchased from Swansea Steam Tug Owners Ltd. and renamed *Robert Redhead* (178/1900). Although already thirty years old she was to give a further thirty-seven years' service on the Tyne. As has already been described, Redhead & Dry's bought, in 1933, a one-third share in the Joseph Crosthwaite (Tugs) Ltd. business along with Lawson-Batey and France Fenwick. The Crosthwaite company, nevertheless, continued to trade under its own title until 1949. James and Alexander Patton were appointed to the Crosthwaite board as representatives of Redhead & Dry's. Two of the old paddle tugs, *Ben Ledi* and *Gauntlet*, went to the breakers in 1935, and in 1938 *Snowdon* was sold to R.L. Cook Towage Co.Ltd., Sunderland for further service. The fleet was thus

Part of the Redhead & Dry fleet in the early 1930s. Right to left are Robert Redhead, Great Emperor *(2),* Gauntlet, Snowdon *and* Malta. *(Local History Library, South Shields Central Library).*

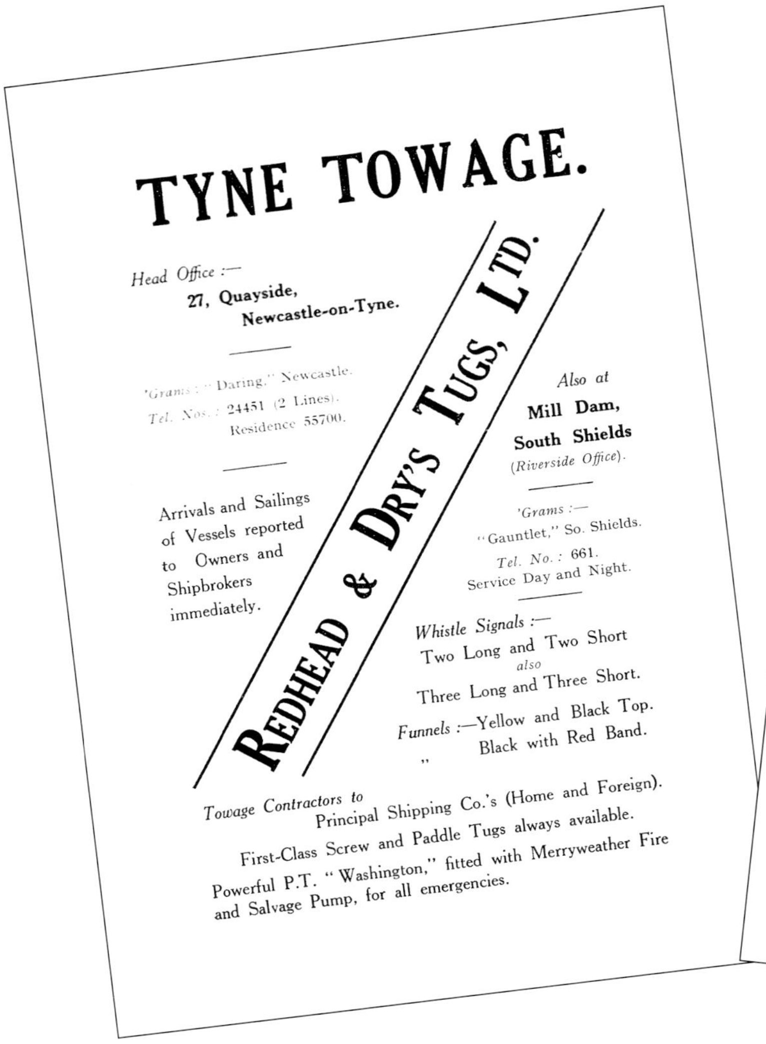

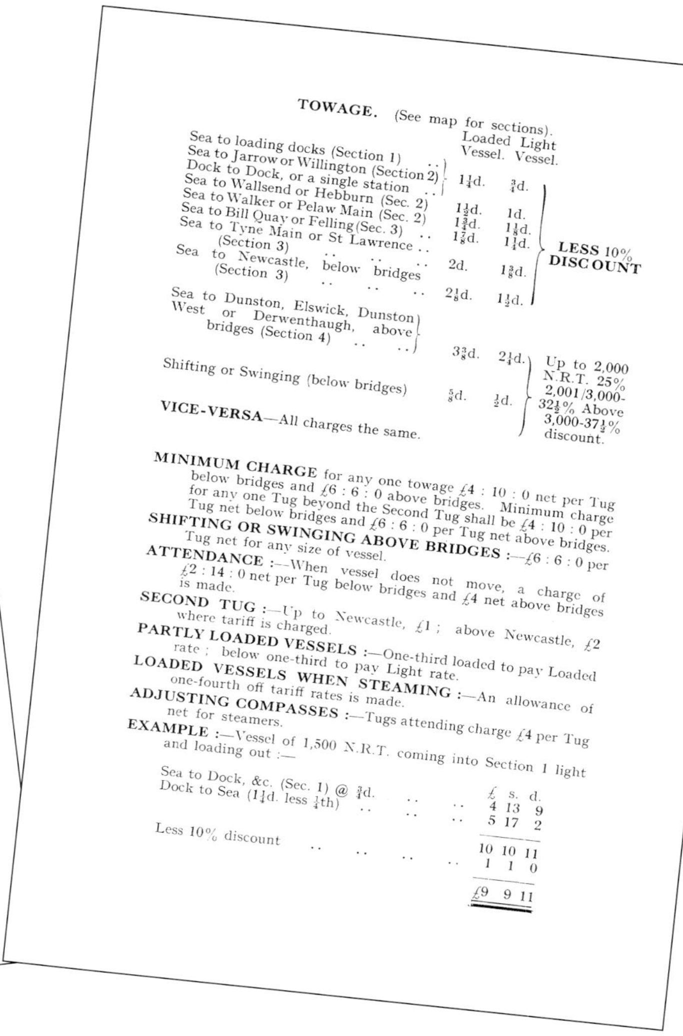

TOWAGE. (See map for sections).

	Loaded Vessel.	Light Vessel.	
Sea to loading docks (Section 1) .. / Sea to Jarrow or Willington (Section 2) / Dock to Dock, or a single station ..	$1\frac{1}{4}$d.	$\frac{3}{4}$d.	LESS 10% DISCOUNT
Sea to Wallsend or Hebburn (Sec. 2)	$1\frac{1}{2}$d.	1d.	
Sea to Walker or Pelaw Main (Sec. 2)	$1\frac{3}{4}$d.	$1\frac{1}{8}$d.	
Sea to Bill Quay or Felling (Sec. 3) ..	$1\frac{7}{8}$d.	$1\frac{1}{4}$d.	
Sea to Tyne Main or St Lawrence (Section 3)	2d.	$1\frac{3}{8}$d.	
Sea to Newcastle, below bridges (Section 3)	$2\frac{1}{8}$d.	$1\frac{1}{2}$d.	
Sea to Dunston, Elswick, Dunston West or Derwenthaugh, above bridges (Section 4)	$3\frac{3}{8}$d.	$2\frac{1}{4}$d.	Up to 2,000 N.R.T. 25% 2,001/3,000-$32\frac{1}{2}$% Above 3,000-$37\frac{1}{2}$% discount.
Shifting or Swinging (below bridges)	$\frac{5}{8}$d.	$\frac{1}{2}$d.	

VICE-VERSA—All charges the same.

MINIMUM CHARGE for any one towage £4 : 10 : 0 net per Tug below bridges and £6 : 6 : 0 above bridges. Minimum charge for any one Tug beyond the Second Tug shall be £4 : 10 : 0 per Tug net below bridges and £6 : 6 : 0 per Tug net above bridges.
SHIFTING OR SWINGING ABOVE BRIDGES :—£6 : 6 : 0 per Tug net for any size of vessel.
ATTENDANCE :—When vessel does not move, a charge of £2 : 14 : 0 net per Tug below bridges and £4 net above bridges is made.
SECOND TUG :—Up to Newcastle, £1 ; above Newcastle, £2 where tariff is charged.
PARTLY LOADED VESSELS :—One-third loaded to pay Loaded rate ; below one-third to pay Light rate.
LOADED VESSELS WHEN STEAMING :—An allowance of one-fourth off tariff rates is made.
ADJUSTING COMPASSES :—Tugs attending charge £4 per Tug net for steamers.
EXAMPLE :—Vessel of 1,500 N.R.T. coming into Section 1 light and loading out :—

	£	s.	d.
Sea to Dock, &c. (Sec. 1) @ $\frac{3}{4}$d.	4	13	9
Dock to Sea ($1\frac{1}{4}$d. less $\frac{1}{4}$th)	5	17	2
	10	10	11
Less 10% discount	1	1	0
	£9	9	11

Company advertisement and towage rates on the Tyne in 1931. As Lawson-Batey had done, Redhead & Dry's Tugs retained their original funnel colours for some time after the merger of 1929.

reduced to six tugs - three paddle and three screw.

In 1938 William Rutherford began work with the company as an office boy in the offices over the Custom House at the Mill Dam, South Shields. Two years later he was promoted to 'Water clerk' - a misleading job description if ever there was one! An alternative job title was 'running fitter' which did not help matters for the layman one little bit. His job entailed running around the river, either by bus or on foot, to give orders to the tugs. He was very well equipped - with a pea whistle - to call the attention of the tugs by 'blowing the knocks'. This was the name for calling a tug by long and short blasts on a whistle. The code used by Redhead & Dry's Tugs Ltd. was two long and two short blasts. This system was also used by ships when they needed to call a tug of a particular company and the whistle codes were often given in tug company advertisements. Bill Rutherford, like many more, went into the Army in 1942, and, by the time he returned home in 1947, his company had been taken over and so he found himself employed by France, Fenwick Tyne & Wear Co.Ltd., and working shifts in the Mill Dam office. His career continued with steady advancement until he retired in 1988 as Towage Manager of Tyne & Wear Tugs Ltd., after forty-nine years' service.

Into the Second World War. 1939 - 1944

Like all other north-east coast tug companies, Redhead & Dry's were involved in the rescue and salvage of vessels damaged by enemy attack during the Second World War. One such event, in December 1939, with the war only three months old, was the salvage of *Eskdene* (3,829/1934). The struggle by *George V* (2) and *Hendon* (2) to get this damaged vessel into harbour has been described in the France Fenwick chapter. *Robert Redhead* and *Great Emperor* (2) were also involved in the tricky operation to beach *Eskdene* on the Herd Sands at South Shields, just inside the Tyne entrance. As has been related in the Lawson-Batey chapter *Great Emperor* (2), along with two Lawson-Batey tugs, set out on the afternoon of 14th December 1939 to the rescue of the tanker *Atheltemplar* (8,939/1930). The destroyer HMS *Kelly*, whose commander was Captain Lord Louis Mountbatten, and HMS *Mohawk* were in attendance as protection from further enemy attack. On reaching the scene of the casualty, about eleven miles from the Tyne entrance, HMS *Kelly* struck a mine and *Great Emperor* (2) was detailed to go to her assistance. After the tug was made fast it was possible to maintain a speed of about three knots to the Tyne. When near the harbour entrance *Robert Redhead* doubled-up with *Great Emperor* (2) on the bow and *Washington* took hold aft. Together they were able to bring HMS *Kelly* into the river shortly before midnight and to take her back to Hawthorn-Leslie's shipyard at Hebburn where she had been built and where she had just been repaired following an earlier misfortune. When *Atheltemplar* reached the coast on 15th December in tow of the Lawson-Batey tugs it was found necessary to beach her south of the Tyne. *George V* (2) and *Great Emperor* (2) were sent to assist in this operation.

The salvage of *Nutfield* (1,561/1919), in extremely bad weather, nearly caused the loss of *Robert Redhead* which had gone to the steamer's assistance after she had been damaged by enemy action. At 10pm. on 4th December 1942 the tug was ordered out to give what help she could. Visibility was almost nil. *Nutfield,* when located, was seen to have a large hole in her side, a heavy list to starboard, and she was very much down by the head. *Robert Redhead,* alone, started to tow the ship towards the shore without being sure of their position, due to the poor weather conditions. A radio request was made to put some lights on the beach so that the tug skipper could see where he was but this was misinterpreted and searchlights were directed at the two vessels, thus blinding their crews! What the lights did reveal was that the tug was in broken water and close in to a rocky shore and liable to go aground at any moment. However, a second attempt, now with lights played onto the beach, was successful in getting *Nutfield* aground when the examination vessel then took off her crew. *Robert Redhead* and *George V* (2) returned on 23rd December 1942 and succeeded in refloating the damaged vessel and towing her into the Tyne, where she was repaired.

In August 1943 the newly-built Ministry of War Transport tug *Empire Mustang* (242/1943) was placed under the management of Redhead & Dry's Tugs Ltd. She was one of many tugs of several sizes and designs that were built during the war for the Ministry and which were managed by tug companies throughout the United Kingdom.

Redhead & Dry's were taken over, in May 1944, by France, Fenwick Tyne & Wear Co.Ltd., and with this merger the latter company also acquired Redhead & Dry's one-third holding in Joseph Crosthwaite (Tugs) Ltd. The tugs transferred in this deal were:-

George V (2)	(224/1915)	Steel	Screw
Conqueror (2)	(177/1884)	Iron	Paddle
Washington	(122/1870)	Iron	Paddle
Great Emperor (2)	(171/1909)	Steel	Screw
Malta	(113/1875)	Iron	Paddle
Robert Redhead	(178/1900)	Steel	Screw

- and also the management of *Empire Mustang* (242/1943).

James Patton, the managing director of Redhead & Dry's, at the age of eighty-two years, retired after fifty-eight years' service with the company and its forebears. Retire he might from direct involvement in this towage company, but he was to remain involved in the towage business, for he was still chairman of the Tyne Mutual Marine Insurance Association. He did not retire from that Board until 1949 after thirty-four years' service, nineteen as chairman. He still drove a car when he was ninety-six years old, and didn't stop working until he died, aged ninety-seven, on 18th February 1959. His son, Alexander Patton, was towage manager of the France, Fenwick Tyne & Wear Co.Ltd. fleet on the Tyne until 1958.

A fine overhead view of George V (2) *and* Robert Redhead *passing through the swing bridge at Newcastle in the 1930s or early 1940s.*
(Local History Libary, South Shields Central Libary).

APPENDIX A TO CHAPTER SIXTEEN

REF. No.	NAME	EX	1900 1910 1920 1930 1940 1950	To
RD.				
1	CONQUEROR (2)	RH19		F28
2	WASHINGTON	D8		F30
3	GAUNTLET	D9		
4	SNOWDON	D11		CK17
5	GREAT EMPEROR (2)	D12		F32
6	MALTA	D13		F31
7	GEORGE V (2)	RH18		F29
8	BEN LEDI	RH15		
9	ROBERT REDHEAD			F33
	Tugs owned at 31 Dec.		9 7 6 0	
	Managed tug.			
M1	EMPIRE MUSTANG			F/MS
			(jhp 1992)	

Tug Ownership Chronology.

REDHEAD & DRY'S TUGS LTD. 1929-1944

APPENDIX B TO CHAPTER SIXTEEN

Redhead & Dry's Tugs Ltd. Fleet List

RD 1 **CONQUEROR (2)** **(1929-1944)**
See Redhead fleet list, RH 19

Washington *(Local History Library, South Shields)*

RD 2 **WASHINGTON** **(1929-1944)**
See Dry fleet list, D 8

RD 3 **GAUNTLET** **(1929-1935)**
See Dry fleet list, D 9

RD 4 **SNOWDON** **(1929-1938)**
See Dry fleet list, D 11.

RD 5 **GREAT EMPEROR (2)** **(1929-1944)**
See Dry fleet list, D 12.

RD 6 **MALTA** **(1929-1944)**
See Allen fleet list, RSA 4.

RD 7 **GEORGE V (2)** **(1929-1944)**
See Redhead fleet list, RH 18.

RD 8 **BEN LEDI** **(1929-1935)**
See Cook fleet list, CK 6.

Robert Redhead *(Local History Library, South Shields)*

RD 9 **ROBERT REDHEAD** **(1930-1944)**
ON 110646 178 g. 103.0 x 23.6 x 12.1 ft. Steel Screw
B: J.P. Rennoldson & Sons, South Shields. (YN 202).
E: T. 3 Cy. 15½", 25" & 41" x 27". 600 IHP. 11½ K. By Builder.

PR: 1900 Liverpool; 1923 London; 1925 Swansea; 1930 Newcastle.

1900:	Completed, as HANNAH JOLLIFFE, for W. & T. Jolliffe Ltd., Liverpool.
1908:	Sold to The Alexandra Towing Co.Ltd., Liverpool.
08.1914:	Hired by Royal Navy. 10.1919: Returned to owners.
1923:	Sold to James Dredging, Towage & Transport Co.Ltd., Southampton.
1924:	Sold to Swansea Steam Tug Owners Ltd., Swansea.
03.1930:	Purchased by Redhead & Dry's Tugs Ltd., renamed ROBERT REDHEAD.
05.1944:	Transferred to France, Fenwick Tyne & Wear Co.Ltd.
01.1959:	Chartered to Tyne Tugs Ltd. until 05.1962 when transferred to Sunderland.
11.1967:	Sold to Clayton & Davie Ltd., Dunston, for breaking up.

RD/M 1 **EMPIRE MUSTANG** **(1943-1944)**
ON 169308 242 g. 106.7 x 26.7 x 11.6 ft. Steel Screw
B: J.S. Watson Ltd., Gainsborough. (YN 1543).
E: T. 3 Cy. 16", 26" & 43" x 30". 1000 IHP. C.D. Holmes & Co.Ltd., Hull.

PR: 1943 Hull; 1947 Middlesbrough; 1958 Newport, Mon.

08.1943:	Completed for Ministry of War Transport. Redhead & Dry's Tugs Ltd. appointed managers.
05.1944:	Management transferred to France, Fenwick Tyne & Wear Co.Ltd.
1946:	Owners now Ministry of Transport.
1947:	Sold, for £16,250, to Tees Towing Co.Ltd., renamed DUNDAS CROSS.
01.1958:	Sold to Newport Screw Towing Co.Ltd., renamed DUNEAGLE.
1965:	Sold to Tsavliris (Salvage & Towage) Ltd., renamed NISOS SYROS.
c.1975:	Sold to Maritime Commercial Enterprises Ltd., Piraeus.
c.1976:	Broken up.

The Lambton Companies

The Earl of Durham. 1858 - 1896.

The Earls of Durham, who lived at Lambton Castle some seven miles inland from the port of Sunderland, owned a number of collieries in that area. The coal from these mines was transported by rail to staiths on the south side of the River Wear just upstream from the Wearmouth Bridge. The Lambton Railway and the Lambton Staiths took their name from the Earl's residence. To assist the colliers calling at these staiths to load coal for shipment to the south of England, the Earl of Durham owned a number of tugs. The first of these which has been identified was, not surprisingly, *Lambton* (48/1852). She was a small wooden paddle tug of only 62 feet length and is known to have been owned by George Lambton the second Earl by at least 1858, though because of her name it is probable that she was involved in the working of the Lambton staiths from the time of her building.

In 1872 a second tug, *Snowdrop* (77/1866), was purchased from a Sunderland owner. Not until 1890 was a third tug obtained by John George Lambton, the third Earl, this time from R. Lee of Cardiff. She was the iron-hulled paddler *Her Majesty* (86/1877) and, although built on the Tyne, had worked at Cardiff since her completion sometimes as a trawler with the registration CF70. At about the same time *Lambton* was sold to C. O'Connor of St. Anthony's for further service on the Tyne. In 1894 the iron paddle tug *Balmoral* (90/1871) was acquired from North Shields owners for the Earl of Durham to become her sixth owner. This tug was one of those single-engined paddle tugs which had a heavy bogie on deck which could be moved by chains to port or starboard to assist in steering by creating a deeper immersion of the paddle wheel on one side or the other. In 1873, in heavy seas, the bogie crashed into the bulwarks trapping the leg of the skipper, David Boag, and causing him to be severely injured - another example of the dangers inherent in the operation of the early tugs.

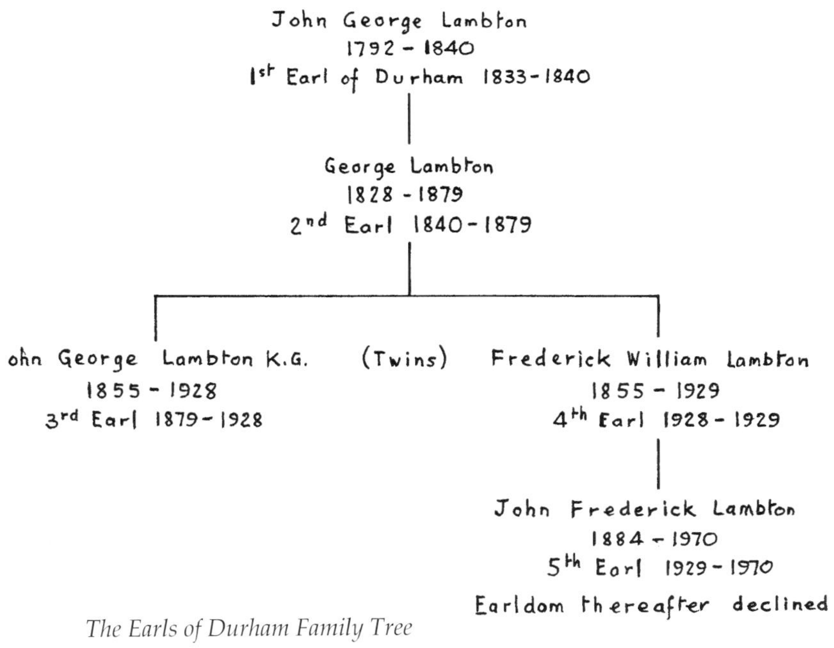

The Earls of Durham Family Tree

The Lambton Collieries Ltd. 1896 - 1911

In 1896 the Earl of Durham sold out to Sir James Joicey who also had collieries adjacent to the River Wear inland from Sunderland. Joicey renamed the business The Lambton Collieries Ltd., but this, and his original company, continued for the time being to operate separately. The three tugs transferred to the new company were:-

Vessel name	Gross tonnage/year	Hull construction	Means of propulsion
Snowdrop	(77/1866)	Wood	Paddle
Her Majesty	(86/1877)	Iron	Paddle
Balmoral	(90/1871)	Iron	Paddle

The iron paddle tug *Eddystone* (82/1877) was purchased from H.H. Doeg of North Shields in June 1896. *Snowdrop* disappeared from the register in 1898, and in 1900 *Her Majesty* was sold to T. Cook of Sunderland. This latter vessel was immediately replaced by the wooden paddle tug *Norman Prince,* purchased from W.H. Storey of North Shields and soon renamed *Hetton* (91/1885).

The company's first newly-built tug *Thomas Nicholson* (132/1903), named after the company's coal fitter, was delivered in November 1903 from Hepple & Co., South Shields. She was considerably larger than the tugs previously owned by the company and she was fitted with a double disconnecting engine producing 425 IHP. Her iron hull was over 96 feet long. In the following month *Balmoral* was sold to North Shields owners. January 1904 saw a sister vessel to *Thomas Nicholson* delivered from the same builders. She was named *Houghton* (133/1904) and was to have a life of sixty years. *Eddystone* was sold to R.B. Stewart of South Shields in 1907 and was replaced by a second sister to *Thomas Nicholson* in October of that year. She did differ slightly in that her hull

LH 7 **THOMAS NICHOLSON** **(1903-1936)**
ON 118625 132 g 96.5 x 20.1 x 9.9 ft. Iron Paddle
B: Hepple & Co.Ltd., South Shields.
E: L. 2 Cy. 28" x 48". 425 IHP. 10 K. By Builder.

PR: 1903 Newcastle

11.1903:	Completed for The Lambton Collieries Ltd.
08.1911:	Transferred to The Lambton & Hetton Collieries Ltd.
12.1924:	Transferred to The Lambton, Hetton & Joicey Collieries Ltd.
03.1936:	Sold, for £575, to T.W. Ward, Inverkeithing, for breaking up.

LH 8 **HOUGHTON** **(1904-1945)**
ON 118629 133 g. 96.0 x 20.0 x 9.7 ft. Iron Paddle
B: Hepple & Co.Ltd., South Shields. (YN 538).
E: L. 2 Cy., 28" x 48". 425 IHP. 10 K. By Builder.

PR: 1904 Newcastle.

01.1904:	Completed for The Lambton Collieries Ltd.
1911:	Owners restyled The Lambton & Hetton Collieries Ltd.
1924:	Owners restyled The Lambton, Hetton & Joicey Collieries Ltd.
11.1945:	Transferred to France, Fenwick Tyne & Wear Co.Ltd.
10.1964:	Sold to Clayton & Davie Ltd., Dunston, for breaking up.

Lumley *on trials,1907. (P.N.Thomas).*

LH 9 **LUMLEY** **(1907-1945)**
ON 125432 135 g. 96.3 x 20.2 x 9.9 ft. Steel Paddle
B: Hepple & Co.Ltd., South Shields. (YN 571)
E: L. 2 Cy. 28" x 50". 425 IHP. 10 K. By Builder.

PR: 1907 Newcastle.

10.1907:	Completed for The Lambton Collieries Ltd.
1911:	Owners restyled The Lambton & Hetton Collieries Ltd.
1924:	Owners restyled The Lambton, Hetton & Joicey Collieries Ltd.
11.1945:	Transferred to France Fenwick Tyne & Wear Co.Ltd.
11.1955:	Sold, for £1,500, to British Iron & Steel Corporation. Allocated to J.J. King & Co.Ltd., Gateshead, for breaking up.

Eppleton Hall. *(Company archives).*

LH 10 **EPPLETON HALL** **(1914-1945)**
ON 133548 166 g. 100.5 x 21.1 x 10.8 ft. Steel Paddle
B: Hepple & Co.Ltd., South Shields. (YN 632).
E: L. 2 Cy. 30" x 51". 500 IHP. 10 K. By Builder.

PR: 1914 Newcastle; 1969 San Francisco.

02.1914:	Completed for The Lambton & Hetton Collieries Ltd.
1924:	Owners restyled Lambton, Hetton & Joicey Collieries Ltd.
11.1945:	Transferred to France, Fenwick Tyne & Wear Co.Ltd.
11.1964:	Sold to Seaham Harbour Dock Co.
11.1967:	Sold to Clayton & Davie Ltd., Dunston, for breaking up, but this did not proceed very far.
05.1969:	Sold, for £2,500, to Scott Newhall, San Francisco. Restored by R.B. Harrison & Son, Bill Quay, when converted to oil firing and had sails fitted.
09.1969:	Left River Tyne for U.S.A. via Lisbon, Madeira and Georgetown.
03.1970:	Arrived San Francisco. Transferred to Maritime Museum at that port.

• CHAPTER EIGHTEEN •

The Cook Companies

Robert Lumley Cook. c.1865 - 1908

The earliest connection of Robert Lumley Cook with the towage business which has been discovered was his shared ownership with W. and J. Brown of the tug *Belmont* (48/1861) in 1865. This single-engined wooden paddle tug had been built on the Tyne by Hepple & Landells at Low Walker and was 73 feet long. For the next fifteen years or so R.L. Cook and his partners carried out general towage operations at Sunderland as a joint enterprise. Sometimes he would have the majority shareholding in the tugs and on other occasions only a minority holding. Other members of the Cook family were also part owners of the tugs from time to time. In 1866 the small wooden paddle tug *Wansbeck* (29/1835) was purchased from Elizabeth Hawks of Whickham. Members of the Brown family again shared the ownership with R.L. Cook. Four years later *Blue Bell* (55/1862) was added to the fleet and in October 1872, along with W. and J. Brown and J. Hazard, R.L. Cook purchased *North Star* (58/1861) from Henry Carr of Newcastle.

Blue Bell was sold to Tyne owners in 1873 and *Belmont* left the partnership in 1875 also transferring to the Tyne for South Shields owners. Two tugs were thus left to carry on business on and about the River Wear and in the docks at Sunderland. In 1878 a somewhat larger wooden paddle tug named *Rescue* (1) (82/1869) was brought to Sunderland from Guernsey where she had worked for the past nine years after having been built on the Tyne. *Wansbeck* and *North Star* were sold later that year to William Almony of Monkwearmouth and Fairlie Downs respectively, leaving *Rescue* (1) to soldier-on alone for the next eight years. The iron paddle tug *Ben Ledi* (114/1878) was purchased by R.L. Cook in 1886 from J.N. Newton of Aberdeen. She had originally been a well-known tug on the River Thames when owned by J.F. Gibb. She stayed with the Cook establishment for five years before being sold to Ayr Harbour Trustees for £1,700. In 1891 she was replaced by *Cruiser* (2) (84/1880), a wooden paddle tug which had spent her early years at Yarmouth.

Earl of Glamorgan (82/1869), a single-engined wooden paddle tug was purchased for £700 in 1892 from Palmers Shipbuilding & Iron Co.Ltd. As her name suggests she had spent the first years of her life in Cardiff even though she had been built on the Tyne. *Rescue* (1) foundered off Seaham Harbour in 1897, but was replaced by the wooden paddle tug *Brothers* (3), which had been owned since new by John Anderson and Robert Chater of North Shields. She was immediately renamed *Rescue* (2) (96/1883). Three years later another tug, *Stag* (3) (123/1885) was purchased from the same owners. Her skipper was Robert Cook Baister. *Earl of Glamorgan* was broken up in 1904 at the age of thirty-five years.

Elizabeth Sarah Cook. 1908 - 1912.

Robert Lumley Cook died on 18th September 1908 from which date his widow, Elizabeth Sarah Cook, became the owner of the three tugs then in the fleet.

Vessel name	Gross tonnage/year	Hull construction	Means of propulsion
Cruiser (2)	(84/1880)	Wood	Paddle
Rescue (2)	(96/1883)	Wood	Paddle
Stag (3)	(123/1885)	Wood	Paddle

It is likely that Robert Cook Baister was in charge of the operation of the tugs on her behalf.

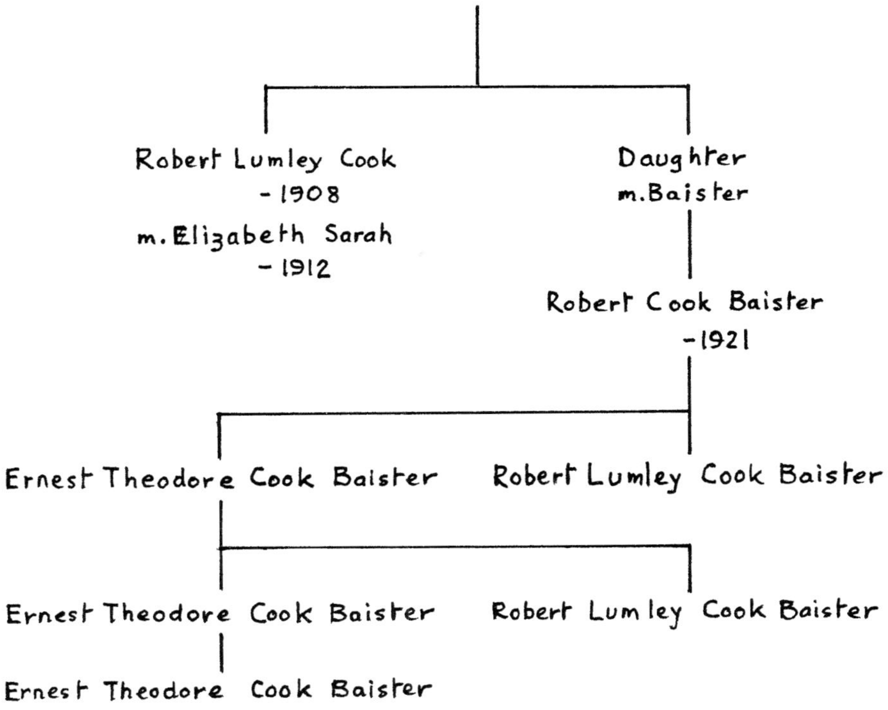

The Cook and Baister Family Tree

Robert Cook Baister. 1912 - 1921

Mrs. Elizabeth Cook died on 14th October 1912 and the ownership of the three tugs *Cruiser* (2), *Rescue* (2) and *Stag* (3) passed to Robert Cook Baister, who, it is believed, was her nephew. The business was then being run from 12 James Williams Street, Sunderland. At this time, and probably from an earlier date the tugs carried a white funnel with a black top and with a red five-pointed star on the white part. It is not known whether a house flag was used. *Cruiser* (2) was broken up in 1913.

In 1915 the iron paddle tug *Betefdoe* was purchased from J.S. Wilson of Bo'ness and renamed *Robert L. Cook* (131/1875). She had previously been owned on the Thames, in Belgium, on the Thames again, and in France, before going to Scotland. Being 107 feet in length she was obviously a large and powerful tug for her time. Not surprisingly she had been built by Readhead at South Shields. At about the same time *Rescue* (2) was broken up. *Pero Gomez* (96/1869) was acquired in 1919 also from J.S. Wilson after being under the control of the Royal Navy for five years during and after the First World War. She was registered in the name of Robert Lumley Cook Baister, a son of Robert Cook Baister, as were *Cinema Star* (69/1884) ex *Goole No.5*, and *Hutton* (85/1893) which were purchased in 1920.

Robert Lumley Cook Baister. 1922 - 1924.

Robert Cook Baister died on 31st December 1921 and the two tugs registered in his name were bequeathed to his son Robert Lumley Cook Baister. With the three tugs which the latter had acquired recently this brought the fleet up to its maximum size of five vessels.

Stag (3)	123/1885)	Wood	Paddle
Robert L. Cook	(131/1875)	Iron	Paddle
Pero Gomez	(96/1869)	Iron	Paddle
Cinema Star	(69/1884)	Iron	Screw
Hutton	(85/1893)	Wood	Paddle

It is surprising that *Hutton*, when considering her date of build, had a wooden hull. She had been built for the North Eastern Railway Co. for work in Middlesbrough Dock and more recently had been owned at Hull.

In June 1922 *Pero Gomez* was converted to a lighter, and *Stag* (3) was broken up at Sunderland in October of the same year.

R.L. Cook Towage Co.Ltd. 1924 - 1947.

In February 1924 the Baister business was converted into a limited liability company under the title of R.L. Cook Towage Co.Ltd. At the time of the change the tugs owned by the new company were:

Robert L. Cook	(131/1875)	Iron	Paddle
Cinema Star	(69/1884)	Iron	Screw
Hutton	(85/1893)	Wood	Paddle

- with an average age of forty years.

The following month a second screw tug, *Wellington*, was purchased from The Alexandra Towing Co.Ltd. who had operated her at Liverpool. On take-over she was renamed *Stag* (5) (148/1897). Shortly afterwards *Hutton* was broken up. In April 1927 the company purchased another paddle tug which had begun her life with the North Eastern Railway Co. This vessel, which had recently operated at West Hartlepool in the ownership of the London & North Eastern Railway Co., had an iron hull even though built in the same year as *Hutton*. She was *Clarence* (131/1893), built on the Tyne by J.P. Rennoldson & Sons, with a double engine developing 350 IHP. At fifty-three years of age *Robert L. Cook* was broken up in 1928 and in August 1932 *Cinema Star* was sold to the River Wear Commissioners with whom she operated

A company advertisement of c.1927 showing the ill-fated Stag *(5) (Les Turner).*

for the next thirty-two years, only being scrapped when she was eighty years old! *Clarence* went to T. Young & Sons Ltd., Sunderland, for breaking up in 1938, and in September of the same year the iron paddle tug *Snowdon* (106/1885) was purchased from Redhead & Dry's Tugs Ltd.

When the destroyers HMS *Fame* and *Ashanti* ran aground at Whitburn in October 1940 (see also Chapter Eight), *Snowdon* was used to carry the rocket apparatus and seven members of the Sunderland (South Side) Volunteer Life Brigade to the scene so that life-saving attempts could be made both from the land and the sea.

During the last months of the Second World War, *Stag* (5) was on charter to France, Fenwick Tyne & Wear Co.Ltd., and is known to have operated on the River Tyne during this period. Whilst doing so she retained her white funnel with red star colour scheme. In November 1947 the R.L. Cook Towage Co.Ltd. was taken over by France, Fenwick Tyne & Wear Co.Ltd. and the two tugs transferred with the business were:-

Stag (5)	(148/1897)	Steel	Screw
Snowdon	(106/1885)	Iron	Paddle

Both survived for only another three years under their new ownership, one of them being lost in tragic circumstances.

APPENDIX A TO CHAPTER EIGHTEEN

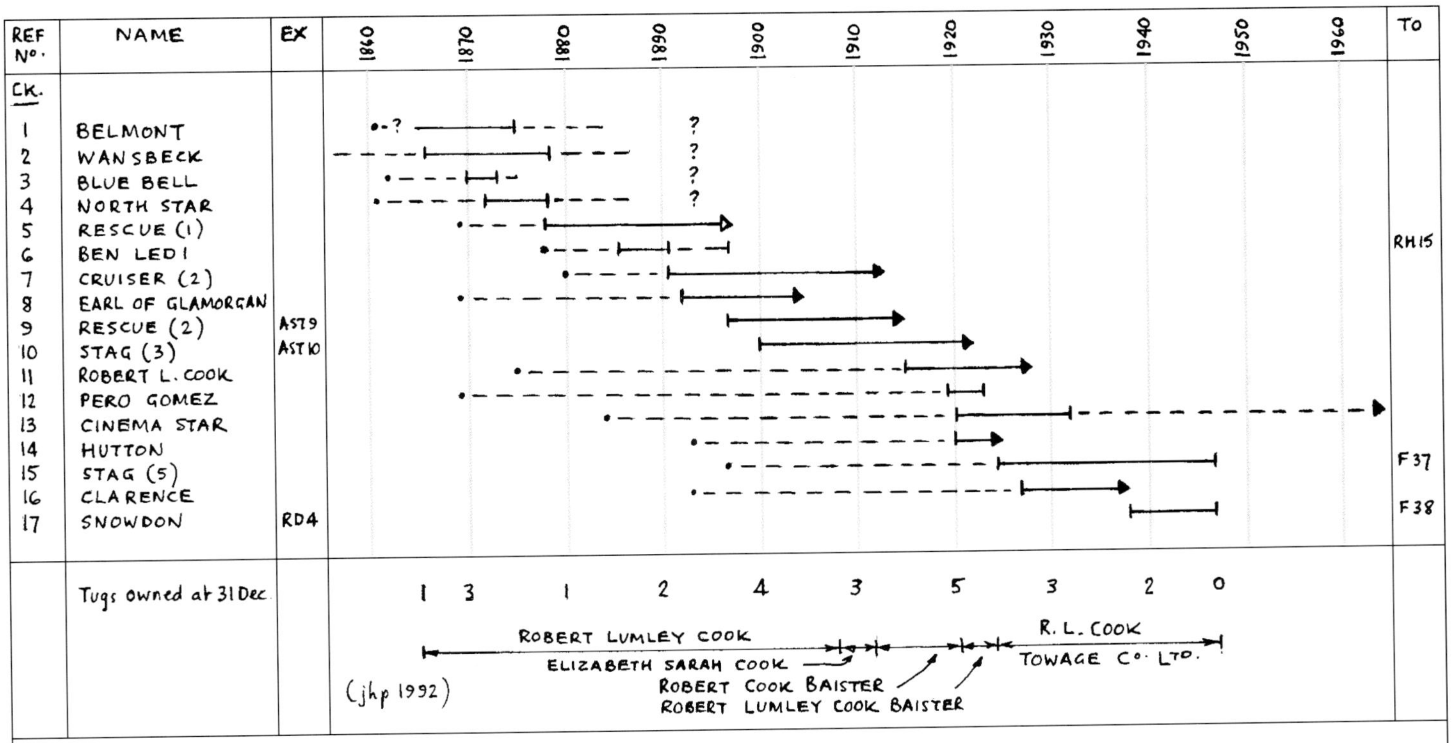

Tug Ownership Chronology

THE COOK COMPANIES. c. 1865–1947.

APPENDIX B TO CHAPTER EIGHTEEN

The Cook Companies Fleet List

CK 1 **BELMONT** **(By 1865-1875)**
ON 43754 48 g. 73.0 x 15.1 x 7.8 ft. Wood Paddle
B: Hepple and Landells, Low Walker.
E: L. 1 Cy. 25" x 46". 20 NHP. By Builder.

PR: 1861 Sunderland; 1872 Sunderland; 1875 South Shields.

1861: Owners not known.
By 1865: Purchased by W. & J. Brown and Robert Lumley Cook, Sunderland.
09.1875: Sold to Jonathan Hall et al, South Shields.
c.1879: Sold to Ralph Robson, St. Anthony's.
1885: Not in register.

CK 2 **WANSBECK** **(1866-1878)**
ON 20784 29 g. 55.7 x 11.9 x 6.7 ft. Wood Paddle
B: At North Shields.
E. 18 NHP.

PR: 1835 Newcastle; By 1855 Sunderland.

1835: Completed, owner not known.
1848: Sold to T. & W. Winter, Sunderland.
c.1858: Owners now T. & J. Winter and Elizabeth Hawks.
1866: Purchased by E. Brown, Robert Lumley Cook et al, Sunderland.
1874: Owners now M.W. Cook, R.L. Cook, M. Thurlbeck and M. Wilson.
1878: Sold to W. Almony, Monkwearmouth.
1887: Not in register.

CK 3 **BLUE BELL** **(1870-1873)**
ON 44496 55 g. 73.7 x 15.8 x 8.5 ft. Wood Paddle
B: At North Shields.
E: L. 1 Cy. 27" x 45". 24 NHP. Richard Landells, North Shields.

PR: 1862 Sunderland; 1873 Newcastle.

09.1862: Completed for M. Thurlbeck et al, Sunderland.
1870: Owners now M. Thurlbeck and Robert Lumley Cook, Sunderland.
10.1873: Sold to John Todd, Gateshead, and E. Ord, Felling.
03.1875: Sold foreign. No further information.

CK 4 **NORTH STAR** **(1872-1878)**
ON 29108 58 g. 73.2 x 15.5 x 8.0 ft. Wood Paddle
B: Thomas Hepple, Low Walker.
E: L. 1 Cy. 26" x 45" 22 NHP. Hepple & Landells, Low Walker.

PR: 1861 Newcastle; 1870 Newcastle; 1872 Sunderland.

1861: Completed, owner not known.
By 1864: Owned by H. Carr, Newcastle.
10.1872: Purchased by W. & J. Brown, J. Hazard and R.L. Cook, Sunderland.
07.1878: Sold to F. Downs et al.
1887: Not in register.

CK 5 **RESCUE (1)** **(1878-1897)**
ON 63476 82 g. 84.3 x 18.0 x 9.4 ft. Wood Paddle
B: W. Hepple & Son, Newcastle.
E: 33 NHP. By Builder.

PR: 1869 Guernsey; 1878 Sunderland.

1869: Completed for A. Collinette, Guernsey.
1872: Sold to W.W. Bird Snr., Guernsey.
1878: Purchased by Robert Lumley Cook et al, Sunderland.
1897: Foundered off Seaham Harbour.

CK 6 **BEN LEDI** **(1886-1891)**
ON 79641 114 g. 90.0 x 18.6 x 10.2 ft. Iron Paddle
B: Hall Russell & Co., Aberdeen. (YN 213).
E: L. 2 Cy. 25" x 48". 50 NHP. By Builder.

PR: 1878 London; 1883 Aberdeen; 1891 Ayr; 1897 Newcastle.

1878: Completed for J.F. Gibb, London.
1882: Sold to J.N. Newton, Aberdeen.
1886: Purchased by Robert Lumley Cook, Sunderland.
04.1891: Sold, for £1,700, to Ayr Harbour Trustees.
05.1897: Purchased by R. Redhead and M.H. Redhead, trading as Robert Redhead & Son.
05.1903: Transferred to M.H. Redhead and Susannah H. White.
08.1914: Hired by Royal Navy. 09.1915: Returned to owner.
06.1916: Transferred to Robert Redhead & Son Ltd.
08.1929: Transferred to Redhead & Dry's Tugs Ltd.
1935: Vessel broken up.

CK 7 **CRUISER (2)** **(1891-1913)**
ON 79230 84 g. 88.2 x 18.4 x 9.5 ft. Wood Paddle
B: Wouldhave & Johnston, North Shields.
E: L. 1 Cy. 32" x 48". 34 NHP. Baird & Barnsley, North Shields.

PR: 1880 Yarmouth; 1880 North Shields; 1891 Sunderland.

1880: Completed for George Baker, Yarmouth.
05.1891: Purchased by Robert Lumley Cook et al, Sunderland.
09.1908: Transferred to Mrs. Elizabeth S. Cook, Sunderland.
10.1912: Transferred to Robert Cook Baister, Robert Lumley Cook Baister et al.
06.1913: Broken up.

CK 8 **EARL OF GLAMORGAN** **(1892-1904)**
ON 63391. 82 g. 87.6 x 17.8 x 9.4 ft. Wood Paddle
B: At North Shields.
E: 45 NHP.

PR: 1869 Cardiff; 1879 Sunderland; 1884 Newcastle.

06.1869: Completed for H. Marquand & W.H. Martin, Cardiff.
03.1879: Sold to Palmers Shipbuilding & Iron Co.Ltd.
09.1892: Purchased, for £700, by Robert Lumley Cook, Sunderland.
1904: Broken up.

CK 9 **RESCUE (2)** **(1897-1915)**
See Anchor Steam Tug fleet list, AST 9 as BROTHERS (3).

Stag *(3)*. *(World Ship Photo Library)*.

CK 10 **STAG (3)** **(1900-1922)**
See Anchor Steam Tug fleet list, AST 10.

CK 11 **ROBERT L. COOK** **(1915-1928)**
ON 73609 131 g. 107.3 x 19.0 x 9.7 ft. Iron Paddle
B: John Readhead & Co., South Shields. (YN 115).
E: L. 2 Cy. 29" x 50⅝". 56 NHP. 11 K. By Builder.

PR: 1876 London; c.1877 Belgium; 1877 London; c.1878 France; 1913 Bo'ness; 1915 Sunderland.

1875: Completed, as CHAMPION, for T.W. Elliott, London.
c.1877: Sold to Belgian owners, renamed AMERICA.
1877: Sold to J. Watkins, London, renamed CHAMPION.
c.1878: Sold to French owners.
12.1913: Sold to J.S. Wilson, Bo'ness, renamed BETEFDOE.
05.1915: Purchased by Robert Cook Baister, Sunderland, renamed ROBERT L. COOK.
12.1921: Transferred to Robert Lumley Cook Baister.
02.1924: Transferred to R.L. Cook Towage Co.Ltd.
1928: Broken up.

Pero Gomez *at Dundee c.1908. (Graham Langmuir collection, courtesy P.N. Thomas).*

CK 12 **PERO GOMEZ** **(1919-1922)**
ON 60534 96g. 88.0 x 17.6 x 9.1 ft. Iron Paddle
B: T. Hepple & Co., Low Walker.
E: L. 1 Cy. 32⅜" x 50". 35 NHP. By Builder.

PR: 1869 Swansea; 1890 North Shields; 1902 Bo'ness.

1869:	Completed for R. Rosser, Swansea.
04.1890:	Sold to A. Robertson, North Shields.
1899:	Sold to T. Devlin, Leith.
1902:	Sold to J.S. Wilson, Bo'ness.
12.1914:	Hired by Royal Navy. 03.1919: Returned to owners.
1919:	Purchased by Robert Lumley Cook Baister, Sunderland.
06.1922:	Converted to a lighter.

Cinema Star *when owned by River Wear Commissioners. (John Clarkson, Longton).*

CK 13 **CINEMA STAR** **(1920-1932)**
ON 67846 69 g. 75.2 x 16.2 x 8.9 ft. Iron Screw
B: Thomas Scott, Goole.
E: C. 2 Cy. 17" & 34" x 21". 50 NHP. By Builders.

PR: 1884 Goole; 1920 Sunderland.

1884:	Completed, as GOOLE No.5, for The Goole & Hull Steam Towing Co. Ltd.
04.1918:	Requisitioned by Royal Navy. 12.1919: Returned to owners.
10.1920:	Purchased by Robert Lumley Cook Baister, renamed CINEMA STAR.
02.1924:	Transferred to R.L. Cook Towage Co.Ltd.
08.1932:	Sold, for £900, to River Wear Commission.
1964:	Broken up.

Hutton *in Middlesbrough Dock when owned by NER Co. (P.N. Thomas).*

CK 14 **HUTTON** **(1920-1925)**
ON 98779 85 g. 76.5 x 18.9 x 9.5 ft. Wood Paddle
B: J.P. Rennoldson & Sons, South Shields.
E: L. 2 Cy. 27" x 45". 250 IHP. 10 K. By Builder.

PR: 1893 Middlesbrough; 1911 Hull.

07.1893:	Completed for North Eastern Railway Co.
01.1911:	Sold to R.W. Wheeldon and T. Commander, Hull.
1920:	Purchased by Robert Lumley Cook Baister, Sunderland.
1924:	Transferred to R.L. Cook Towage Co.Ltd.
1925:	Broken up.

Stag *(5), (Local History Library, South Shields Central Library).*

CK 15 **STAG (5)** **(1924-1947)**
ON 106863 148 g. 92.0 x 20.1 x 10.7 ft. Steel Screw
B: Cumming & Ellis, Inverkeithing. (YN 25).
E: C. 2 Cy. 21" & 42" x 27". 490 IHP. 11 K. J. Cran & Co., Leith.

PR: 1897 Liverpool; 1924 Sunderland.

1897:	Completed, as WELLINGTON, for The Alexandra Towing Co. Ltd.
03.1924:	Purchased by R.L. Cook Towage Co.Ltd., Sunderland, renamed STAG.
c.1944:	Chartered to France, Fenwick Tyne & Wear Co.Ltd.
11.1947:	Transferred to France, Fenwick Tyne & Wear Co.Ltd.
02.1950:	Sank on passage River Tyne to River Wear whilst towing CITY OF MANCHESTER. Only two survivors.

CK 16 **CLARENCE** **(1927-1938)**
ON 102702 131 g. 95.5 x 19.2 x 9.8 ft. Iron Paddle
B: J.P. Rennoldson & Sons, South Shields. (YN 148).
E: L. 2 Cy. 28" x 50". 350 IHP. 10¼ K. By Builder.

PR: 1893 West Hartlepool.

07.1893:	Completed, for £5,450, for North Eastern Railway Co.
01.1923:	Transferred to London & North Eastern Railway Co.
03.1927:	Sold to T. Young, Sunderland.
04.1927:	Purchased by R.L. Cook Towage Co.Ltd., Sunderland.
1938:	Broken up by T. Young & Sons Ltd., Sunderland.

CK 17 **SNOWDON** **(1938-1947)**
See Dry fleet list, D 11.

• CHAPTER NINETEEN •

Clyde Shipping Co. Ltd.

The company seal illustrating Industry *of 1815.* *(Clyde Shipping Co. Ltd.).*

The origins of Clyde Shipping Co.Ltd. go back to 1815 when the early partners of the Clyde Shipping Co. purchased the rudimentary steam paddle boat *Industry* which had been built at Fairlie, Ayrshire in the previous year. In 1856 the partnership was dissolved and the assets and goodwill of the company were offered for sale. The buyers were George Jardine Kidston (1835-1909), his brothers John Pearson and Richard, and three others, only the former being active in the running of the business. In 1893 when the partners were the same three Kidston brothers and James Cuthbert (1843-1901) it was decided that the status of the business be changed to that of a limited liability company under the title of Clyde Shipping Co.Ltd. The company seal adopted then, and still in use today, incorporated a drawing of *Industry* and the date 1815. At this time the company possessed some twenty-four tugs and a fleet of coasters. The chairman of the new company was G.J. Kidston and two of his sons were to follow him in that office until in 1920 William Cuthbert, son of James, took over the position until 1946.

In the latter part of 1982 Clyde Shipping Co.Ltd. made an offer to purchase Lawson-Batey Tugs Ltd. The assets included the thirteen tugs then based in the Tyne, Blyth and Wear areas of north-east England. About a year later agreement was reached and the takeover went ahead. Clyde Shipping Co.Ltd. had a wholly-owned subsidiary named Clyde Shipping Tugs Ltd. and it was this latter firm, which operated tugs on the River Clyde, which actually took over Lawson-Batey in April 1983. The management of Lawson-Batey remained in local hands although two Clyde directors, Dr. L.J. Paterson and W.M. Cuthbert, were appointed to the board, the former becoming chairman on the death of T.A. Greenwell in 1987. In 1986, D.F. Hamilton, who was the Lawson-Batey managing director, joined the board of Clyde Shipping Tugs Ltd. Outwardly little changed as a result of the takeover but in March 1985 the new tug *Seasider* (2) (164/1985) was purchased by Clyde Shipping Co.Ltd. and was chartered to the Lawson-Batey subsidiary company Tyne Tugs Ltd., which in the following year was restyled Tyne & Wear Tugs Ltd.

So Clyde Shipping Co.Ltd., 175 years old in 1990, is now strictly the sole surviving tug-owning company of all those companies from the seventeen lines of descent which have been described in this 'family' of tug-owning companies, and whose stories have been covered in this volume.

Seasider *on trials before going on charter to Tyne Tugs Ltd. in 1985.* *(Walter Fussey & Son).*

• CHAPTER TWENTY •

The Charter Companies

Tyne Tugs Ltd. 1959 - 1986.

The Lawson-Batey, Ridley and France Fenwick Alliance.

The two major towage companies operating on the River Tyne since 1920 had been Lawson-Batey Tugs Ltd. and France, Fenwick Tyne & Wear Co.Ltd. Whereas the former company's operations were based mainly on the Tyne, the latter company, as its title suggests, worked also at Sunderland on the River Wear some eight miles further south. Late in 1957 conversations took place regarding the possible merger of operations on the Tyne to improve efficiency and benefit from the economies of scale. At that time it was not known what form such a merger would take or even if it would come about at all. Ridley Tugs Ltd., from 1958 a wholly-owned subsidiary company of Lawson-Batey, also operated on the Tyne and was expected to take part in the scheme. Through 1958 the discussions became more formal and the idea emerged that a new company should be formed which would be responsible for operating the combined fleet. The ownership of the vessels would remain with Lawson-Batey, France Fenwick and Ridley. A formal agreement was signed on 31st October 1958.

The new company was officially incorporated on 28th November 1958 with the intention of commencing operations not later than 1st January 1959. In the event, it was on this latter date that co-operation between the tug-owning companies became effective. The new private charter company, not surprisingly, was entitled Tyne Tugs Ltd. The original capital of the company was £2,400 divided into 1200 'A' shares and 1200 'B' shares of £1 each. These shares were issued thus: 1000 'A' shares to Lawson-Batey, 200 'A' shares to Ridley and 1200 'B' shares to France Fenwick. As Lawson-Batey owned the Ridley company the financial arrangement was effectively 50/50 between Lawson-Batey and France Fenwick. The directors were to number not less than four, nor more than six, and the first to be appointed were:-

Frank Sewell Batey	Chairman of Lawson-Batey
John Francis Batey	Managing director of Lawson-Batey
Francis Christian Batey	Managing director of Ridley Tugs
Kenneth Raymond Pelly, MC	Chairman of France Fenwick
Andrew Desmond Pelly, DFC	Director of France Fenwick
Tom McEwan	Director of France Fenwick

K.R. Pelly was appointed chairman of directors, William Richard Nicholson, managing director of France Fenwick, was appointed secretary and J.F. Batey became the managing director of Tyne Tugs Ltd. The registered office of the company was Mercantile Chambers, Quayside, Newcastle upon Tyne. William Rutherford, ex France Fenwick, and George Hedley, ex Lawson-Batey, were appointed joint managers of the South Shields office. F.G. Hutton was towage manager and E. Heal was marine superintendent.

The agreement between the participating companies was for Lawson-Batey to contribute ten tugs, Ridley two and France Fenwick twelve. They were:-

Name	Gross Tonnage/Year	Means of propulsion	
Joffre	(260/1916)	Steam	Lawson-Batey Tugs Ltd.
Francis Batey	(151/1914)	Steam	
Tynesider (1)	(262/1942)	Steam	
Waysider	(166/1919)	Steam	
Southsider	(239/1940)	Steam	
Eastsider	(162/1924)	Steam	
Quaysider	(137/1955)	Diesel	
Westsider (1)	(153/1913)	Steam	
Roughsider	(143/1958)	Diesel	
Hillsider (1)	(163/1911)	Steam	
Impetus	(141/1954)	Diesel	Ridley-Tugs Ltd.
Maximus	(141/1956)	Diesel	
Hendon (2)	(241/1924)	Steam	France, Fenwick Tyne & Wear Co. Ltd.
Cullercoats	(181/1898)	Steam	
Wearmouth	(182/1929)	Steam	
Fulwell	(149/1899)	Steam	
George V (2)	(224/1915)	Diesel	
Robert Redhead	(178/1900)	Steam	
Earsdon	(234/1912)	Steam	
Beamish	(242/1944)	Steam	
Ashbrooke	(119/1955)	Diesel	
Alnwick	(119/1955)	Diesel	
Bamburgh	(122/1956)	Diesel	
Marsden (4)	(122/1956)	Diesel	

This was the plan, but, as with most plans, there was a hiccup. *Earsdon* sank in the Tyne after a collision in October 1958 and was so badly damaged that she had to be scrapped. In order to maintain the important balance of contribution to the new scheme, France Fenwick had to find a replacement tug. It so happened that the Ridley company had a third tug,

Westsider *(2) towing the stern half of the Norwegian bulk-carrier* Happy Dragon *into the Tyne from Sunderland in October 1966.*
(Company archives).

France Fenwick's Prestwick *towing the bow half of* Happy Dragon *out of Sunderland bound for the Tyne in February 1967. (Company archives).*

up at this time, leaving only sixteen tugs in operation and soon, when two other tugs had to go for special survey, only fourteen were available. In December further adjustments to the fleet took place. *Southsider* was brought back into service, whilst *Cullercoats* was laid-up at Stanhope Buoys, South Shields. *Wearmouth,* which had spent the last 30 years working on the Tyne, was taken off charter and returned to Sunderland, where she had started her working life in 1929. The fleet, therefore, at the end of 1966 comprised:-

Tynesider (1)	Steam	*Beamish*	Diesel
Southsider	Steam	*Ashbrooke*	Diesel
Eastsider	Steam	*Alnwick*	Diesel
Quaysider	Diesel	*Bamburgh*	Diesel
Roughsider	Diesel	*Marsden* (4)	Diesel
Impetus	Diesel	*Appelsider*	Diesel
Maximus	Diesel	*Alnmouth*	Diesel
Cullercoats	Steam	*Westsider* (2)	Diesel
George V (2)	Diesel		

- seventeen in all (two of them laid up) with an average age of nineteen years.

Whilst these revisions to the tug fleet were taking place, changes were also occurring amongst the company personnel. The chairman, Sir K.R. Pelly, (he had been knighted in 1961) retired on 30th June 1966, and was succeeded by his son, A.D. Pelly, who had been a director since the establishment of Tyne Tugs in 1958. The secretary, W.R. Nicholson, was appointed a director at the same time. In November, because of the reduction in work available, there was a surplus of sixteen crewmen. Ten of these were made redundant and six retained for relief duties. Five others were given one month's notice.

In September 1966 a rather unusual charter was arranged for *George V* (2) captained by Hector Stewart. A German submarine had foundered on the Dogger Bank in the North Sea and the British Colombia Television Broadcasting Company wished to film the search for, and the hoped-for subsequent raising of, the vessel. The location was 225 miles east of the Tyne and *George V* (2) spent six days at sea carrying the TV personnel and their equipment whilst the submarine was located and raised. The submarine was lifted and carried back to Germany beneath a pontoon. It is presumed that all the crew had perished. No doubt this was useful business for the company but it is difficult to understand the reason for the morbid interest of a Canadian TV company in events in the North Sea.

A large bulk carrier, *Happy Dragon* (30,815/1967), building at Austin & Pickersgill, Sunderland, had to be launched in halves and each section towed to the Tyne for joining together in the large Palmer's drydock at Hebburn. The stern section was towed to the Tyne in October 1966 by four of the Tyne Tugs fleet. The bow portion took the same journey in February 1967, but this time the tow was carried out by four of France Fenwick's Wear fleet. Obviously, 'fair shares for all'.

In 1967 the Tyne Tugs' management was looking very strongly for sea-towage work to offset the continuing decline in river towage. This effort was only partly successful because there was great competition for such work. In April the new Lawson-Batey tug *Northsider* (2) (156/1967) was placed on charter and in September her sister *Ironsider* (156/1967) also arrived. The skipper of the latter tug was Brian Walker who was well known not only for his excellence as a tugmaster but also for his fine display of potted plants in the wheelhouse! *Southsider* went back to lay-up in May 1967 to join *Eastsider* and *Cullercoats.* In September *Westsider* (2) was sent to Dundee to tow the paddle ferry *B.L. Nairn* (395/1929) to Blyth for demolition and at the end of 1967 *Northsider* (2) towed the Fleetwood trawler *Matanuska* 60 miles into the Tyne.

In April 1968 it was learnt that William France, Fenwick & Co.Ltd. had been taken over by Jessel Securities Ltd., a London finance company. Since France, Fenwick Tyne & Wear Co.Ltd. was a subsidiary of William France Fenwick then control of the former company also passed to Jessel. Naturally, the directors of Tyne Tugs were concerned as to what effect this change would have on the joint ownership of their company. However, Jessel soon made it clear that they had no intention of interfering in the Tyne Tugs arrangement and that France Fenwick would remain in being as a company. The 10-year charter agreement between Lawson-Batey, Ridley and France Fenwick expired on 31st December 1968. All parties were highly satisfied with the arrangement and wished to continue with the scheme. A new agreement was signed in May 1968 for a further period of ten years to commence on 1st January 1969.

In January 1968, the large tanker *Regent Westminster* (54,908/1968), fitting out at Swan Hunter & Tyne Shipbuilders Ltd. at Wallsend, broke loose and drifted downstream, eventually grounding on the opposite side of the river at Hebburn. It took nine of the Tyne Tugs' fleet to recover her from this predicament.

Several changes to the fleet occurred during the year. *Cullercoats,* laid up on the Tyne was, in January, towed by *Quaysider* to South Dock, Sunderland, from where in May she was sold for breaking up. February saw the other two laid-up tugs, *Southsider* and *Eastsider* taken away for demolition at Blyth, and, finally, *Tynesider* (1) was taken off charter by Lawson-Batey when she was transferred to Blyth Tug Co. These changes meant that Tyne Tugs had a wholly diesel-powered fleet for the first time. The fleet numbered fifteen tugs but the management felt that the present level of business could be handled by fourteen. The surplus tug would be either *Impetus* or *Roughsider,* but in the meantime *Marsden* (4) and *Maximus* were laid-up, in November 1968. An application for an increase in towage rates of 7.5% had been reduced by agreement to 4%.

On 31st December 1968 five of the senior tug skippers retired. They were Alex Brown (50 years' service); Harry Jobling (51 years); Dave McRitchie (51 years) whose father, brother and then he himself had been master of *Corsair;* Hector Stewart (54 years - 47 as master) then 69 years old, and James Elsdon (50 years). What a tale these five could have told about their long experience with steam tugs! Unfortunately, the author has been able to talk to only one of them, Hector Stewart, still alert when aged ninety-one.

By the time the new charter agreement came into force at the beginning of 1969 the fleet consisted of:-

Tug	Status	Owner
Quaysider	Laid-up	Lawson-Batey Tugs Ltd.
Roughsider		Lawson-Batey Tugs Ltd.
Appelsider		Lawson-Batey Tugs Ltd.
Westsider (2)		Lawson-Batey Tugs Ltd.
Northsider (2)		Lawson-Batey Tugs Ltd.
Ironsider		Lawson-Batey Tugs Ltd.
Impetus		Ridley-Tugs Ltd.
Maximus	Laid-up	Ridley-Tugs Ltd.
Alnmouth		France, Fenwick Tyne & Wear Co. Ltd.
Alnwick		France, Fenwick Tyne & Wear Co. Ltd.
Ashbrooke		France, Fenwick Tyne & Wear Co. Ltd.
Bamburgh		France, Fenwick Tyne & Wear Co. Ltd.
Beamish		France, Fenwick Tyne & Wear Co. Ltd.
George V (2)		France, Fenwick Tyne & Wear Co. Ltd.
Marsden (4)	Laid-up	France, Fenwick Tyne & Wear Co. Ltd.

The five senior skippers who retired on 31st December 1968. From left to right: back row, Alex Brown, Harry Jobling; front row, Dave McRitchie, Hector Stewart, Jimmy Elsdon. *(The Sheilds Gazette).*

Not many vessels navigating the Tyne require ten tugs to keep them under control. The builders of the first VLCC's built on the Tyne were taking no chances when Esso Northumbria *left the river in February 1970.* *(R.L.Palmer).*

In the ten years since Tyne Tugs Ltd. was formed in 1959 the size of the fleet had reduced from twenty-four to fifteen tugs. At the same time the number of jobs carried out had reduced from about 15,000 to 11,000 per year and the downward trend was to continue still further. In February *Quaysider* was brought back into service and *Beamish* joined *Maximus* and *Marsden* (4) in lay-up. In May *Impetus* was sold for further service in Lubeck.

George Hedley, joint manager of the South Shields office, died in January 1969 aged 63. He had held that position with Lawson-Batey since 1955 and had started forty-nine years earlier as a junior with the Batey company. Bill Rutherford became the sole manager of Tyne Tugs Ltd.

Jessel Securities Ltd. were, in February 1969, trying to dispose of France, Fenwick Tyne & Wear Co.Ltd., as has been described in Chapter Eight.

A.D. Pelly, chairman of Tyne Tugs since 1966, resigned from the board in April 1969 on his leaving Jessel Securities to take up another appointment. Tom McEwan, a director since 1958, was appointed chairman in his place, and John Keville joined the board to represent France Fenwick. In July, using an agreed formula, the company made an application to the Port of Tyne Authority to raise the towage rates by 46%. The Port Authority did not take kindly to this request and eventually approved a 20% increase. Because of the continuously reducing volume of business the tug crews were very concerned about their future and registered their protest by going on strike for 24 hours on 21st October 1969. This was followed by a further short stoppage ten days later.

At the beginning of 1970 the Tyne Tugs fleet comprised fourteen tugs. Two, *Maximus* and *Beamish*, were laid-up with attempts to sell them meeting with little response; another, *Marsden* (4), was undergoing survey prior to transfer to France Fenwick's Sunderland fleet in May. However, *Beamish* was soon sold and eventually found her way abroad.

In February 1970 the first of the Tyne-built supertankers *Esso Northumbria* (126,543/1970) left her builders' yard at Wallsend in the charge of ten of the company's tugs. This much-publicised event created such interest all along the banks of the river that enormous crowds gathered to see her go. She was not a pretty vessel but much film must have been exposed that day. The author still remembers very clearly the crowds gathered at South Shields and also the fact that it took over three hours to drive out of the town after the tanker had departed. This type of interest and excitement was typical of what had occurred many times before and would occur again later, because of the affinity which Tynesiders had with their river. Although such scenes are a thing of the past other events like the Tall Ships Race have provided a similar surge of interest in marine matters.

In 1970 towage revenue had improved somewhat, no doubt due to the large increase in the 1969 tariff rates. A further increase of 10% took place in October 1970. On 19th January 1971 F.S. Batey and F.C. Batey resigned their directorships and their places were taken by S.S.W. Dalgliesh and G.H.R. Towers, who had been directors of Lawson-Batey since 1946 and 1951 respectively. Towage charges were no longer controlled by the Department of the Environment through the Port of Tyne Authority. A further rise of 6% took place from June 1971 and in the same year arrangements had been made under which Tyne Tugs Ltd. would carry out the towage of ships calling at the new Alcan establishment at Blyth.

In May *George V* (2) and *Quaysider* were laid-up and, as a result, thirteen men had to be made redundant. *Appelsider* was surplus to requirements and was time-chartered at £100 per day to Cementation Ltd. and John Howard Ltd. The latter company had an option to purchase her for £120,000 if required, but this was not taken up. Whilst working away she took the funnel colours of the Howard company. *Maximus* was withdrawn from charter and Ridley Tugs transferred her to Blyth Tug Co. This was the last tug to be owned by the Ridley company.

At the beginning of September 1971 the tugmen applied for a wage increase of from 42 to 48%, claims which the management described as insupportable. Basic weekly rates at that time were: Master £20.50; Engineers £19.00; Mates £17.90; and Deckhands £17.60. One would have thought that these did not show adequate differentials for responsiblity. By the end of the month the company had offered 7.5%, increased a month later to 10%. By 29th October, the crews, who were members of the North East Coast Tugboatmen's Association, decided to strike, having turned down an offer of arbitration. The Association secretary, Harry Malcolm, conducted negotiations on the men's behalf. Six days later, the Department of Trade became involved. On 11th October the offer was raised to 15% which resulted in the tugboatmen returning to work four days later, but not before they had rejected the latest offer. Finally an offer of a 21% increase was made which was agreed. Moral -

Alnwick *having a slight collision with* Derwent Fisher *(1,100/1966).* *(Company archives).*

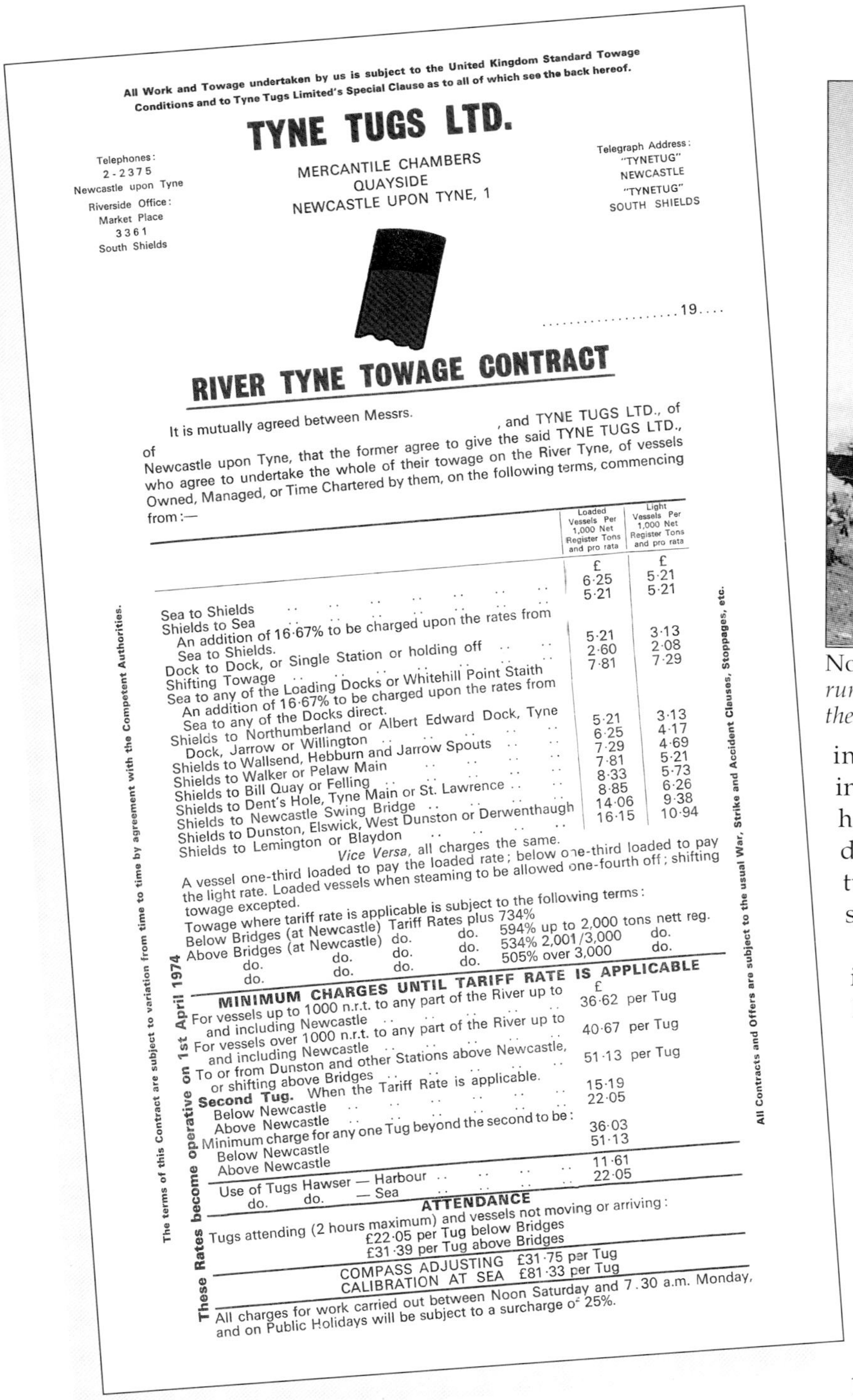

All Work and Towage undertaken by us is subject to the United Kingdom Standard Towage Conditions and to Tyne Tugs Limited's Special Clause as to all of which see the back hereof.

TYNE TUGS LTD.

Telephones: 2-2375 Newcastle upon Tyne
Riverside Office: Market Place 3361 South Shields

MERCANTILE CHAMBERS
QUAYSIDE
NEWCASTLE UPON TYNE, 1

Telegraph Address: "TYNETUG" NEWCASTLE "TYNETUG" SOUTH SHIELDS

....................19....

RIVER TYNE TOWAGE CONTRACT

It is mutually agreed between Messrs. , and TYNE TUGS LTD., of
of
Newcastle upon Tyne, that the former agree to give the said TYNE TUGS LTD., who agree to undertake the whole of their towage on the River Tyne, of vessels Owned, Managed, or Time Chartered by them, on the following terms, commencing from:—

	Loaded Vessels Per 1,000 Net Register Tons and pro rata	Light Vessels Per 1,000 Net Register Tons and pro rata
	£	£
Sea to Shields	6·25	5·21
Shields to Sea	5·21	5·21
An addition of 16·67% to be charged upon the rates from Sea to Shields.		
Dock to Dock, or Single Station or holding off	5·21	3·13
Shifting Towage	2·60	2·08
Sea to any of the Loading Docks or Whitehill Point Staith	7·81	7·29
An addition of 16·67% to be charged upon the rates from Sea to any of the Docks direct.		
Shields to Northumberland or Albert Edward Dock, Tyne Dock, Jarrow or Willington	5·21	3·13
Shields to Wallsend, Hebburn and Jarrow Spouts	6·25	4·17
Shields to Walker or Pelaw Main	7·29	4·69
Shields to Bill Quay or Felling	7·81	5·21
Shields to Dent's Hole, Tyne Main or St. Lawrence	8·33	5·73
Shields to Newcastle Swing Bridge	8·85	6·26
Shields to Dunston, Elswick, West Dunston or Derwenthaugh	14·06	9·38
Shields to Lemington or Blaydon	16·15	10·94

Vice Versa, all charges the same.

A vessel one-third loaded to pay the loaded rate; below one-third loaded to pay the light rate. Loaded vessels when steaming to be allowed one-fourth off; shifting towage excepted.

Towage where tariff rate is applicable is subject to the following terms:
Below Bridges (at Newcastle) Tariff Rates plus 734%
Above Bridges (at Newcastle) do. do. 594% up to 2,000 tons nett reg.
do. do. do. 534% 2,001/3,000 do.
do. do. do. do. 505% over 3,000 do.

MINIMUM CHARGES UNTIL TARIFF RATE IS APPLICABLE

	£
For vessels up to 1000 n.r.t. to any part of the River up to and including Newcastle	36·62 per Tug
For vessels over 1000 n.r.t. to any part of the River up to and including Newcastle	40·67 per Tug
To or from Dunston and other Stations above Newcastle, or shifting above Bridges	51·13 per Tug
Second Tug. When the Tariff Rate is applicable.	
Below Newcastle	15·19
Above Newcastle	22·05
Minimum charge for any one Tug beyond the second to be:	
Below Newcastle	36·03
Above Newcastle	51·13
Use of Tugs Hawser — Harbour	11·61
do. do. — Sea	22·05

ATTENDANCE

Tugs attending (2 hours maximum) and vessels not moving or arriving:
£22·05 per Tug below Bridges
£31·39 per Tug above Bridges

COMPASS ADJUSTING £31·75 per Tug
CALIBRATION AT SEA £81·33 per Tug

All charges for work carried out between Noon Saturday and 7.30 a.m. Monday, and on Public Holidays will be subject to a surcharge of 25%.

The terms of this Contract are subject to variation from time to time by agreement with the Competent Authorities.

These Rates become operative on 1st April 1974

All Contracts and Offers are subject to the usual War, Strike and Accident Clauses, Stoppages, etc.

Tyne Tugs Ltd. towage contract, 1974.

always ask for twice as much as you expect to accept, or offer half as much as you are willing to concede!

Whilst all this was going on, the newly-built supertanker *Texaco Great Britain* (125,942/1971) was waiting to sail on trials from Wallsend. It so happened that three Norwegian tugs had brought a tow into the Tyne and thus the shipbuilders engaged *Boss* (181/1960), *Beni* (249/1967) and *Baus* (479/1970) to sail her on 8th November 1971, much to the surprise of the Tyne Tugs crews.

Having completed her charters in the south-east of England, *Appelsider* was sold to Dublin Port & Docks Board and renamed *Coliemore*. There had been a proposal to charter *Alnmouth* to John Howard Ltd. at about this time, but this did not materialise. Due to the falling level of business a further increase in towage rates of 33.3% was imposed from 17th May 1972, but a further rise of 5%, needed in December, was blocked by the government-imposed prices and wages freeze. One must sympathise with river and port towage companies who were badly hit by the drop in trade to and from the United Kingdom. They had little alternative to increasing their rates, since they really had no means of influencing the number of ships requiring towage in their home port area. Advertising would not help; all they could do was to try to obtain coastal work, but, of course, every tug company in the UK. with suitable tugs, was doing the same thing.

Northsider *(2) aground and damaged after attempting to prevent* Oregis *running ashore on the Black Middens just inside the Tynemouth pier on the 10th March 1974.* *(Author).*

Because of a disagreement over wages, the tug crews imposed a restriction on overtime working from 7th November 1973 to the end of the month. They stipulated that they would only work from 6am. to 8pm. Monday to Friday, and 6am. to 11am. on Saturday with no work on Sundays. Incredible though it may seem the Pay Board refused to recognise that tugmen worked unusual and unsocial hours! Nevertheless, the dispute was resolved with improved overtime rates and payment for unsocial hours.

Sir K.R. Pelly, MC, who had been chairman of Tyne Tugs from its formation in 1958 to 1966, died on 30th September 1973 in his eightieth year. Tom McEwan, the company chairman, retired from the board in March 1974. Sir Rupert Speir, who was chairman of the shipowners Common Bros. Ltd., joined the Tyne Tugs board and was elected chairman from 1st April. Serious concern was expressed by the management at the delays being experienced in the delivery of the two new high-powered tugs currently on order. These tugs were meant to handle the supertankers and large bulk-carriers being built on the Tyne, and to carry out coastal and North Sea towage.

On 10th March 1974 *Oregis* (6,448/1955), newly converted from an ore-carrier to an offshore research vessel, went aground on the Black Middens rocks on the north side of the Tyne entrance due to engine failure, suffering serious hull damage. *Northsider* (2), one of the tugs trying to refloat her, went aground on the same rocks, losing a propeller blade and receiving rudder and hull damage in the process. The skipper of the tug at the time, John Farthing, and his crew had to be rescued by lifeboat, two of whose crew were awarded the Bronze Medal for their bravery. *Northsider* (2) had to have eighteen shell plates renewed and the whole repair was to cost £33,000. She was not refloated until 24th March and was out of action until 16th April. Later, in September 1974, seven tugs salvaged the Greek *Aliki* (8,453/1953), aground on the same Black Middens rocks.

Towage rates were still increasing. A rise of 18% was applied from 1st April 1974 and a further 12% was added in November. Another strike of tugmen took place from noon on 1st October, this time as a token gesture since they returned to work at 8am. the following day. This involved seventy men from the ten tugs then in the fleet. The strike was over pay and in particular over the 'strings which were attached' thereto. Average weekly earnings quoted at the time were £64 for a master and £56 for a deckhand. When salvage awards were made it was often a long wait for the company and the tug crews before they received their financial benefits. A case in point was when *Northsider* (2), *Alnmouth*, *Westsider* (2) and *Quaysider* were involved in the salvage of *Aldora* (10,538/1964) off Blyth on 8th February 1972, when they did not get their reward of £16,000 in total until early in 1975. Increase of towage rates seemed now to be on a regular basis of twice a year. 1975 was no exception - 33.3% in March and 3% in November.

As has been recounted in the Lawson-Batey chapter, J.F. Batey, the company's managing director, was tragically killed by a terrorist bomb in London on 13th November 1975. His place on the board of directors was taken by F.S. Batey, who had previously served in that office from 1958 to 1971, but the position of managing director was not filled until 16th March 1976, when T.A. Greenwell took over.

In March 1975 *Northsider* (2) was contracted for North Sea work at £550 per day for three months. This was just the sort of medium-term charter that the company was wanting, to keep their tugs occupied whilst jobs on the Tyne continued their steady decline, albeit at a slightly slower rate. At about this time coastal towage constituted 14% of total revenue.

Jessel Securities Ltd., the owners of France Fenwick since 1968, had again been trying to sell that company throughout 1975 without success. Late in that year they were in financial difficulties and went into voluntary liquidation. Lawson-Batey, though naturally very interested in what was to happen to France Fenwick, did not immediately make an offer, though before long they did so in a joint bid with a European company. This was rejected by the liquidators of Jessel. When T.A. Greenwell had become managing director, F.S. Batey stood down from the board and W.R. Nicholson was appointed deputy managing director. The twice-yearly towage rate increases that year were 25% and 4% thus making the 1976 charges over four times what they had been at the beginning of 1969!

In July 1976 the tug crews banned all overtime work over a long-simmering grievance prompted by the new pay code - the so-called 'Social Contract'. They complained that they received lower rates of pay for overtime than for normal hours. This ban lasted for about six weeks. There was also a ban on Tyne men working at Blyth and Sunderland, and subsequently on sea-towage also.

In December, the new 3,180 BHP Lawson-Batey tug *Cragsider* (266/1976) was delivered and placed on charter to Tyne Tugs Ltd. She was to be a very valuable addition because of her much greater size and power than any of the tugs which had preceded her, though she was to have some teething troubles early in 1977. An early charter arranged for her with Shell Exploration Ltd. involved towing a barge from Peterhead to Rotterdam and from Rotterdam to Lerwick. This lasted for twelve days and brought in £1,100 per day plus the cost of bunkers, which compared favourably with the £750 per day received by the smaller tugs for sea towage. The delivery of *Cragsider's* sister, *Hendon* (3), was expected in May 1977, but, as will be seen, this never materialised. For seagoing work, seven days per week, the rates of pay at that time were: Masters £176 per week, Engineers £168 and Deckhands £146.

Lawson-Batey Alone

Jessel Securities Ltd. had, in 1977, been taken over by the London & Yorkshire Trust Ltd. who were now the owners of France, Fenwick Tyne & Wear Co.Ltd. Lawson-Batey negotiated with London & Yorkshire to purchase the towage business of France Fenwick which included the five tugs they had on charter to Tyne Tugs Ltd. and the five tugs on the River Wear at Sunderland. The deal included France Fenwick's shareholding in Tyne Tugs but excluded the purchase of the new tug *Hendon* (3) (266/1977) which was later to be sold separately, to The Alexandra Towing Co.Ltd. Lawson-Batey thus became the sole owners of Tyne Tugs and all the tugs chartered to the latter company, namely:

Quaysider	*Alnmouth*
Roughsider	*Alnwick*
Westsider (2)	*Ashbrooke*
Northsider (2)	*Bamburgh*
Ironsider	*George V* (2)
Cragsider	

Cragsider *when new in 1976.* *(Turners Photography Ltd.).*

When London & Yorkshire Trust took over Jessel, a shuffling of the Tyne Tugs board took place with the France Fenwick representatives, Sir Rupert Speir the chairman and J.E. Keville, resigning, and their places being taken temporarily by Michael E.A. Keeling and Henry E. King on 4th May 1977. Within a fortnight they too resigned and F.S. Batey rejoined, and D.F. Hamilton joined the board of directors. The latter also became company secretary, and F.S. Batey was elected chairman. The other directors continuing in their existing posts were: T.A. Greenwell (managing), W.R. Nicholson (deputy managing), S.S.W. Dalgliesh and G.H.R. Towers.

In 1977, Her Majesty Queen Elizabeth II toured Great Britain to celebrate the Jubilee of her accession to the throne. When visiting Tyneside, *Northsider* (2) and *Ironsider* had the honour of berthing the Royal Yacht *Britannia* on 15th and assisting with the sailing on 18th July. The company's managing director, T.A. Greenwell, was High Sheriff of Tyne & Wear at the time and he had escorted the Royal Party and dined aboard the Royal Yacht during the celebrations, and had, of course, paid a special interest in the successful berthing of the vessel.

George V (2), now 63 years old, was sold by her owners in January 1978. Although very old she had, of course, been converted from steam to diesel propulsion and completely refitted in 1954. In February, the office at Mercantile Chambers, Quayside, Newcastle was closed and all administrative activities were then concentrated at 27 Market Place, South Shields, which became the registered office of the company. In May, *Cragsider* spent ten days working for the Department of Trade on oil dispersal duties off the Norfolk coast, due to damage to the tanker *Eleni V* (12,680/1958). W.R. Nicholson retired from his post of deputy managing director on 31st December 1978.

The charter agreement between the tug-owning companies (now in fact only Lawson-Batey) and Tyne Tugs had now completed a second ten-year period. Therefore, from 1st January 1979, a new demise charter needed to be drawn up. This was similar to the previous ones, except that the minimum hire rates were to be £600 per month for *Cragsider,* £500 per month for *Northsider* (2) and *Ironsider,* and £400 per month for the other tugs of which there were six in operation - *Ashbrooke* having been laid-up since November 1978. Edgar Heal, the superintendent engineer, who had been with the company and Lawson-Batey for more than twenty-five years, died on 4th February 1979. He was succeeded by G.D. Dawson, his deputy, and J. Nutman was employed as an additional superintendent engineer. During 1979 D.F. Hamilton was appointed deputy managing director. The relentless increase in towage charges continued, rises of 10% and 15% being applied on 1st January 1979 and 1980. These rises were brought about by inflation and by the ever-reducing number of jobs available - there had been 742 fewer jobs in 1980 than in 1979.

In June 1980 *Roughsider's* charter to Tyne Tugs was terminated when she was transferred to Sunderland to work with the Wear fleet, where she replaced *Cornhill.* Michael Francis Pyman was appointed to the board of directors in September. The new tug *Wearsider* (150/1980) arrived on the Tyne in November 1980 and was soon operating there on charter to Tyne Tugs Ltd. In December *Cragsider* had the misfortune to lose the barge *Aanleg* which sank in appalling weather one mile off Whitby whilst under tow from Rotterdam to the River Tees. Ten days later, however, and not to be deterred, she towed a dock gate, in the vertical position, from Dundee to the River Clyde. This unwieldy item had a draught of 21 feet and the voyage was completed in twelve days. *Cragsider* may have spent Christmas at sea, but at the end of her tow, she was in the right place to celebrate the New Year!

In March 1981 a further new tug arrived on the scene, She was *Tynesider* (2) (150/1981) and was a sister to *Wearsider* delivered four months previously from Richard Dunston of Hessle. The new arrival replaced the twenty-five year old *Alnwick* which went to lay-up at Sunderland pending sale. Two months later *Alnwick* unfortunately sank at her berth in South Dock. A small hole, caused by corrosion in a shell plate under the main engine, was the cause. Eurosalve Ltd. of Folkestone were engaged to salvage her and agreed to buy the tug where she lay. They eventually repaired *Alnwick* and sold her on to other owners. She is still in service at the time of writing! Lawson-Batey received £40,000 insurance over this loss, but as a result, the insurers, who would pay 50% of the cost, insisted that *Ashbrooke* and *Bamburgh* which were for sale, should have their hull plating inspected in dry dock. In the event this was deferred when, in the meantime, a buyer was found for the tugs. *Quaysider* was laid-up at Stanhope Buoys on the Tyne.

The company was facing a difficult time, particularly due to the shiprepair yards being very short of work. It was therefore necessary to reduce the number of tugs, and in consequence the number of employees.

Four years previously, the company had had warning that from 1981 new regulations would come into force regarding the certification of tug crews. This meant that there would be a limit to the area in which tugs could get dispensation to operate without certificated personnel. That area was specified as 'just north of Newbiggin, east for 10 miles and from there down to Redcliffe, just north of Whitby'. The existing sea crews would be granted certificates of service taking into account their skills and experience but in future new certificates of competency would only be obtained by examination and after serving the relevant required sea-time.

During 1981, *Cragsider, Northsider* (2) and *Ironsider* had been fitted with modern VHF radio sets and the same three tugs were fitted, as operations permitted, with permanent oil-dispersal equipment at Tyne Dock Engineering Ltd.

On 1st January 1982 the towage tariffs went up by 8% having earlier been raised by 16% and 5% on 1st January and 1st July 1981 respectively. *Northsider* (2) suffered a serious breakdown at Methil in February 1982 with crankshaft damage and had to be towed back to the Tyne by *Tynesider* (2). She was not to return to service until May. A claim of £80,000 was made against her insurers. *Bamburgh* and *Ashbrooke,* which had been laid-up since March 1980 and November 1978 respectively, were sold in May 1982, to Greek interests, thus reducing the Tyne Tugs fleet to eight, namely:-

Quaysider	*Westsider* (2)
Northsider (2)	*Ironsider*
Cragsider	*Alnmouth*
Wearsider	*Tynesider* (2)

In June *Quaysider* was drydocked, cleaned up and taken to South Dock, Sunderland to continue her lay-up.

During 1982 S.S.W. Dalgliesh retired from the board after twelve years' service and on 1st January 1983 T.A.

Greenwell, having retired as managing director, was appointed vice-chairman. D.F. Hamilton, a director since 1977, became the new managing director and, at the same time, relinquished his position as company secretary, a post which was then filled by H.R. (Harry) Gilbert.

Sea towage work was becoming increasingly difficult to find, but one short job was obtained when the Royal Fleet Auxiliary *Sir Tristram* (4,473/1967) which had been severely damaged in the Falklands War arrived in the River Tees on board a semi-submersible deck-cargo vessel *Dan Lifter* (10,282/1982). After floating off, *Sir Tristram* was towed to the Tyne by *Cragsider* on 15th June 1983 for re-building at the Swan Hunter yard at Wallsend. In November, *Westsider* (2) was taken off charter and transferred to Blyth Tug Co.Ltd., leaving Tyne Tugs with a fleet of seven tugs. The company had an office and store at Bell Street, North Shields, which had been used by the marine superintendents. The lease of those premises was surrendered in December 1983 and George Dawson and John Nutman, the superintendent engineers, moved to the South Shields office. Tyne Dock Engineering Ltd. made available a storage hut for engine spares and other consumables. At the end of the year, F.S. Batey, aged 84, retired as chairman having held the post since 1977. He was succeeded by the vice-chairman, T.A. Greenwell, who thus became chairman of all the companies in the Lawson-Batey Group.

The latest rises in towage charges had then settled down to one increase per year having been advanced by 6.5% on 1st January 1983 and 7.5% on 1st January 1984. In July 1984, *Quaysider* was sold for further service and became *Towing Wizard*, thus leaving only six tugs available for operations on the Tyne.

In March 1985, Clyde Shipping Co.Ltd., who, since 1983, had been the owners of Lawson-Batey Tugs Ltd., placed on charter to Tyne Tugs Ltd. their new tug *Seasider* (2) (164/1985). This tug was a sister of *Holmsider* which had been chartered by Lawson-Batey to Wear Tugs Ltd. They differed from the other tugs in the fleet at that time in having controllable-pitch propellers. *Wearsider* was then transferred to the Sunderland fleet. Late in November, *Tynesider* (2) suffered an engine failure whilst at sea and had to be towed to the Tyne by *Seasider* (2), her repairs taking about three weeks. *Westsider* (2), which had been transferred to Blyth in 1983 and had since been laid-up in Sunderland, was brought back into service to stand-in for *Tynesider* (2).

It was at this point, in 1986, that Tyne Tugs Ltd. was to change its company name to Tyne & Wear Tugs Ltd., to combine the management of the tugs on both Tyne and Wear under one organisation. To understand this change we must leave the story of Tyne Tugs Ltd. for a while to look at what had been happening on the River Wear since the Lawson-Batey take-over of France, Fenwick Tyne & Wear Co.Ltd. in

Retirement party in December 1985 for Andy Leask who had completed fifty years' service. In the picture are, left to right, Alan Fenwick, Gordon Bell, Shauna Britton, Kay Atkinson, Andy Leask, John Hall, Jean Bell, Eddie Whitelaw (behind), Bill Rutherford, Douglas Hamilton, Tony Greenwell, Gordon Jackson and George Allen.

1977. Before we do so it may be useful to review the Tyne Tugs fleet as at 31st March 1986:

Westsider (2)	(151/1964)	Average age 14 years
Northsider (2)	(156/1967)	
Ironsider	(156/1967)	
Cragsider	(266/1976)	
Alnmouth	(173/1962)	
Tynesider (2)	(150/1981)	
Seasider (2)	(164/1985)	

All of these tugs carried the red and black funnel colours which had been Tyne Tugs' insignia since 1959.

Wear Tugs Ltd. 1978 - 1989.

It has already been related how The 'Ridley' Steam Tug Co.Ltd. came into being in 1920 and that, in 1945, Lawson-Batey Tugs Ltd. obtained a controlling interest in the company. In 1955 the Ridley company was re-titled Ridley Tugs Ltd. and in 1958 Lawson-Batey owned all the Ridley shares. Thus, Ridley Tugs Ltd. became a wholly-owned subsidiary of Lawson-Batey and was, at that time, the owner of one steam and two diesel tugs. By mid-1971 Ridley Tugs Ltd. had disposed of their last tug *Maximus* by transfer to Blyth Tug Co.Ltd. The future of the company was discussed and it was decided, for the time being, to keep the company in existence although not owning or operating any tugs. In 1977, as has been described, Lawson-Batey took over the tugs and certain other assets of France, Fenwick Tyne & Wear Co.Ltd. whose fleet consisted of ten tugs, five of which were chartered to Tyne Tugs Ltd. and the other five were working at Sunderland on the River Wear. The five tugs with Tyne Tugs remained so managed, but the five at Sunderland needed a separate arrangement by which they could operate.

This requirement was met by changing the title of Ridley Tugs Ltd. to Wear Tugs Ltd. on 24th January 1978, and by Lawson-Batey chartering their four newly-acquired Sunderland tugs to Wear Tugs Ltd. One tug, *Whitburn* (2), had been sold in the interim between the take-over of France Fenwick and the establishment of Wear Tugs Ltd. The four tugs initially chartered to Wear Tugs Ltd. were:-

Cornhill	(176/1943)
Prestwick	(119/1955)
Marsden (4)	(122/1956)
Dunelm (2)	(132/1964)

- as will be seen, a somewhat aging fleet. One pleasing feature of the new arrangements was that, although the tugs chartered to Tyne Tugs Ltd. had lost their original companies' funnel colours, those now chartered to Wear Tugs Ltd. retained the very attractive blue and white France Fenwick markings.

At the time of the formation of Wear Tugs Ltd. the board of directors consisted of F.S. Batey - chairman, T.A. Greenwell - managing director, and D.F. Hamilton, the latter also being company secretary. At the Annual General Meeting on 17th February 1978 S.S.W. Dalgliesh and G.H.R. Towers were also appointed directors. The registered office of Wear Tugs was now at 27 Market Place, South Shields, alongside Tyne Tugs and Lawson-Batey. However, the Sunderland office was kept open, with Gordon Bell as towage manager there. When the changeover took place, the personnel at Sunderland were all given notice of termination of employment by Lawson-Batey, but at the same time, as previously arranged, were notified of their re-employment by Wear Tugs - a rather unnerving experience, one would have thought.

Because of her age, *Cornhill* was taken out of class and repairs to her were kept to a minimum. The number of jobs handled at Sunderland in 1978 was 993, and appeared to be reducing in much the same way as they were on the Tyne. Increase in towage rates on the Wear had been 33.3% on 1st November 1976, 15% on 1st January 1978 and 12.5% on 1st January 1980. In February 1980 the tugs on the River Wear were carrying out about 73 jobs per month, and, because of the reducing work load, the Sunderland men were reluctant to allow employees of Tyne Tugs to work on the Wear.

Cornhill, which had been for sale since early in 1979, suddenly attracted a buyer in July 1980 when she was sold to W.G.S. Crouch & Sons Ltd., who renamed her *Hannah Spearing.* To replace her, *Roughsider* was transferred from Tyne Tugs Ltd. to Wear Tugs Ltd. In September 1980 Michael Francis Pyman was appointed a director of the company. Also, D.F. Hamilton became deputy managing director. During December the tug crews began an overtime ban after rejecting a pay offer, one result of which was the postponement of a launch.

Apart from towage work, Wear Tugs continued to carry out a small amount of agency work, in particular for *Cymbeline* (5,513/1966), an ex-William France Fenwick collier, and some small B.P. tankers.

In November 1981 a fatal accident occurred on *Roughsider* when the mate, Gordon Nixon, was struck on the side of his head when a polypropylene rope parted because of a jagged fairlead on the ship she was towing. This sad incident reminds one of the ever-present risk inherent in the operation of tugs. Whilst all possible precautions are taken there is always the mischance that a rope or wire can break. At the inquiry into this fatality it was suggested by the Health & Safety Executive that a steel shelter should be available to protect the men who had need to be on the after-deck of a tug during towing operations. Not surprisingly, after two years of trials on *Roughsider* it was jointly agreed by the company and the Executive that such a scheme was unpractical.

On 22nd February 1982 the mate of *Prestwick,* William Swinhoe, dived into the River Wear to save a man who was floating downstream. With the help of Ivor Swinhoe, the mate of *Marsden* (4), they managed to pull the man aboard *Marsden* (4) and resuscitated him. The Royal Humane Society awarded William Swinhoe a certificate on parchment in recognition of his bravery. This was presented to him by the Chief Constable of Northumbria later in the year. In addition, Wear Tugs Ltd. made a small monetary award to each of the men.

During 1982, S.S.W. Dalgliesh resigned from the board of directors, and, at the end of the year, T.A. Greenwell retired as managing director and became vice-chairman. D.F. Hamilton was appointed managing director from 1st January 1983 at the same time relinquishing his position as company secretary to H.R. Gilbert. At the end of 1983 F.S. Batey retired as chairman of Wear Tugs Ltd. and on 1st January 1984 he was succeeded by T.A. Greenwell.

The new Lawson-Batey tug *Holmsider* (164/1984) was delivered from Richard Dunston in June 1984 and was placed on charter to Wear Tugs to work at Sunderland. She had been appropriately named in recognition of one of the old

Telephone: South Shields 553361 (5 lines)
556911 (4 lines)

Reg. No 166371 (England)

Telegrams: "Tynetug" South Shields

WEAR TUGS LIMITED

Registered Office:
27, Market Place
South Shields
Tyne and Wear
NE33 1JF

Directors:
T. A. GREENWELL (Chairman)
D. F. HAMILTON (Managing Director)
M. F. PYMAN
G. H. R. TOWERS

Wear Tugs Ltd. letterhead c. 1983

streets of the town - 'Holmside'. By December 1984 it was apparent that the likelihood of requiring more than two tugs at any one time was very low, so two tug crews were laid off. When more than two tugs would be required for launches the two laid-up tugs would be re-activated. The number of jobs carried out in 1984 was approximately 900. In March 1985 *Wearsider* (150/1980) had her charter transferred from Tyne Tugs Ltd. to Wear Tugs Ltd. and moved to Sunderland.

Roughsider was sold in February 1986 for breaking up in Scotland. Due to the reducing level of business at Sunderland and the impending cessation of coal shipments from the port, it was decided that all of the Lawson-Batey tugs on charter to Wear Tugs Ltd. would be rechartered to Tyne Tugs Ltd. on 1st April 1986, and that Wear Tugs Ltd. would cease trading on that date. All of the business and assets of Wear Tugs Ltd., therefore, were transferred to Tyne Tugs Ltd., which then changed its name to Tyne & Wear Tugs Ltd. The tugs transferred were:-

Prestwick	(119/1955)	Laid up for sale
Marsden (4)	(122/1956)	
Dunelm (2)	(132/1964)	Laid up
Holmsider	(164/1984)	To service Sunderland
Wearsider	(150/1980)	

Very soon, the attractive funnel colours on these tugs were to disappear, possibly for ever. They had recently worn the funnel colours of the France, Fenwick Tyne & Wear Co.Ltd., which had originated in 1920, though with the Lawson-Batey house-flag. The new colours were to be those of the owning company, Lawson-Batey. The Tyne Tugs fleet was also similarly changed so that all tugs chartered to Tyne & Wear Tugs Ltd. now had the same colour scheme, no matter where they were working. The office staff - Gordon Bell, Alan Fenwick and Gordon Allan - and the seventeen tugboatmen were transferred to the employ of Tyne & Wear Tugs, and moved from Gladstone House, South Dock, Sunderland, to South Shields. Gordon Bell, the towage manager, retired six months later. The directors, M.F. Pyman and G.H.R. Towers, retired in April 1986 and also, in July, H.R. Gilbert, the secretary. J.G. Mucklow was appointed as a director and also took over as the company secretary.

In June 1987, the chairman, T.A. Greenwell, died, and he was succeeded by J.G. Mucklow. The company was still not trading, and, in 1988, it was declared 'dormant' until in February 1989 Wear Tugs Ltd. was struck off the Register of Companies. The company had been formally in existence since 1920 when it was named The 'Ridley' Steam Tug Co.Ltd. It had changed its name in 1955 to Ridley Tugs Ltd., had become a wholly-owned subsidiary of Lawson-Batey Tugs Ltd. in 1958, and continued to own tugs until 1971. In 1978 the company changed its name again to Wear Tugs Ltd. to become a tug-operating company on the River Wear until its demise in 1989.

Tyne & Wear Tugs Ltd. 1986 - 1990.

We left the story of Tyne Tugs Ltd. in 1986 when they were operating seven tugs on the River Tyne; six owned by Lawson-Batey and one by Clyde Shipping Co.Ltd. From 1st April 1986 it had been decided that the chartered fleets of Tyne Tugs and Wear Tugs should be combined under one company title of Tyne & Wear Tugs Ltd. and that Wear Tugs should be wound up. The combined fleet operated by Tyne & Wear Tugs was therefore:-

Ex Tyne Tugs Ltd.		**Ex Wear Tugs Ltd.**	
Westsider (2)	(151/1964)	*Prestwick*	(119/1955)
Northsider (2)	(156/1967)	*Marsden* (4)	(122/1956)
Ironsider	(156/1967)	*Dunelm* (2)	(132/1964)
Cragsider	(266/1976)	*Holmsider*	(164/1984)
Alnmouth	(173/1962)	*Wearsider*	(150/1980)
Tynesider (2)	(150/1981)		
Seasider (2)	(164/1985)		

- twelve tugs with an average age of almost sixteen years. Very soon, after this amalgamation, the fleet was re-painted with Lawson-Batey's funnel colours of a blue Maltese cross on a buff background. Thus, for the first time, all the tugs of Lawson-Batey and its associated companies working on the Blyth, Tyne and Wear rivers had the same livery.

In July 1986 H.R. Gilbert, the company secretary, retired. He was succeeded by J.G. Mucklow who was also appointed a director of the company. About this time G.H.R. Towers resigned from the board. The Tall Ships Race used

Newcastle as its starting base in July 1986. Tyne & Wear Tugs towed four of the largest sailing vessels and two escorting Royal Navy ships whilst they were in the river. A supporting event was a parade of Tyne working boats, in which *Tynesider* (2) took part.

Early in 1987 *Alnmouth* was sold to Greece where she was renamed *Thiela*. One rather interesting fact regarding efficiency of operation came to light when it was reported that the fuel consumption on *Holmsider* and *Seasider* (2), which both had controllable-pitch propellers, was between 35 and 45% higher than for the fixed propeller tugs *Tynesider* (2) and *Wearsider*. Nevertheless, this was thought to be worthwhile when the operating advantages were taken into account. In July 1987 *Holmsider* was transferred from Sunderland to the Tyne to replace *Westsider* (2), which was then laid-up in South Dock, Sunderland. This move left only one operating tug stationed at Sunderland - appropriately *Wearsider*. Five of the eleven crewmen working there were made redundant.

The company chairman, T.A. Greenwell, died on 23rd June 1987. He had served Tyne Tugs and then Tyne & Wear Tugs since 1976, first as managing director and then briefly as vice-chairman before becoming chairman in 1984. He was succeeded as chairman by M.F. Pyman who had been a director since 1980. In December 1987 Lawson-Batey sold their two oldest tugs *Prestwick* and *Marsden* (4) both for further service leaving the company with a fleet of nine tugs. At the end of the year W. Rutherford retired as towage manager after 49 years' service with Redhead & Dry, France Fenwick, Tyne Tugs and Tyne & Wear Tugs. He was succeeded, on 1st January 1988, by Gordon Jackson, who had until that time held a similar position in the Blyth office. Some work was being undertaken at Peterhead, but the forecast for future work on and around the Tyne was 'gloomy'. In March 1988, *Dunelm* (2) was sold to Fowey for further service.

Peter Henry P. Southern was appointed a director of Tyne & wear Tugs Ltd. in February 1989, thus bringing the strength of the board up to four. *Westsider* (2) was sold abroad in May 1989, and in June 1990 *Wearsider* was laid-up, making the one remaining tug crew at Sunderland redundant. The company were, however, searching for means of employing this latter tug and were able to charter her to North Sea Tugs Ltd. in December 1990 for a contract on the River Medway. As this story closes at the end of 1990 the fleet based on the Tyne numbers six tugs carrying out approximately 1,900 jobs per year on the Tyne, some 200 at Blyth and about 300 at Sunderland. A huge change indeed from the time in 1959 when Tyne Tugs Ltd. was formed and twenty-four tugs were carrying out over 15,000 jobs per year on the Tyne alone. Comments on some of the reasons for the decline, and on the future possibilities, will have been read at the end of the Lawson-Batey chapter and need not be repeated here.

APPENDIX A TO CHAPTER TWENTY

Tyne Tugs Limited 1958-1986
Tyne & Wear Tugs Limited 1986-

Board of Directors

Chairmen	(Sir)K.R. Pelly	1958-1966	Retired
	A.D. Pelly	1966-1969	Resigned
	T. McEwan	1969-1974	Retired
	Sir Rupert Speir	1974-1977	Resigned
	F.S. Batey	1977-1983	Retired
	T.A. Greenwell	1983	Vice-Chairman
	T.A. Greenwell	1984-1987	Died
	M.F. Pyman	1987-	
Managing Directors	J.F. Batey	1958-1975	Died
	T.A. Greenwell	1976-1982	Became Vice-Chairman
	W.R. Nicholson	1976-1978	Deputy M.D. Retired
	D.F. Hamilton	1979-1982	Deputy M.D.
	D.F. Hamilton	1983-	
Directors	F.C. Batey	1958-1971	Resigned
	F.S. Batey	1958-1971	Resigned
	F.S. Batey	1975-1976	Resigned. Later Chairman
	A.D. Pelly	1958-1966	Became Chairman
	T. McEwan	1958-1969	Became Chairman
	W.R. Nicholson	1966-1976	Became Deputy M.D.
	J.E. Keville	1969-1977	Resigned
	S.S.W. Dalgliesh	1971-1982	Retired
	G.H.R. Towers	1971-1986	Retired
	D.F. Hamilton	1977-1978	Became Deputy M.D.
	M.E.A. Keeling	May 1977 only	Resigned
	H.E. King	May 1977 only	Resigned
	M.F. Pyman	1980-1987	Became Chairman
	J.G. Mucklow	1987-	
	P.H.P. Southern	1989-	
Secretaries	W.R. Nicholson	1958-1977	
	D.F. Hamilton	1977-1982	
	H.R. Gilbert	1983-1986	
	J.G. Mucklow	1986-	

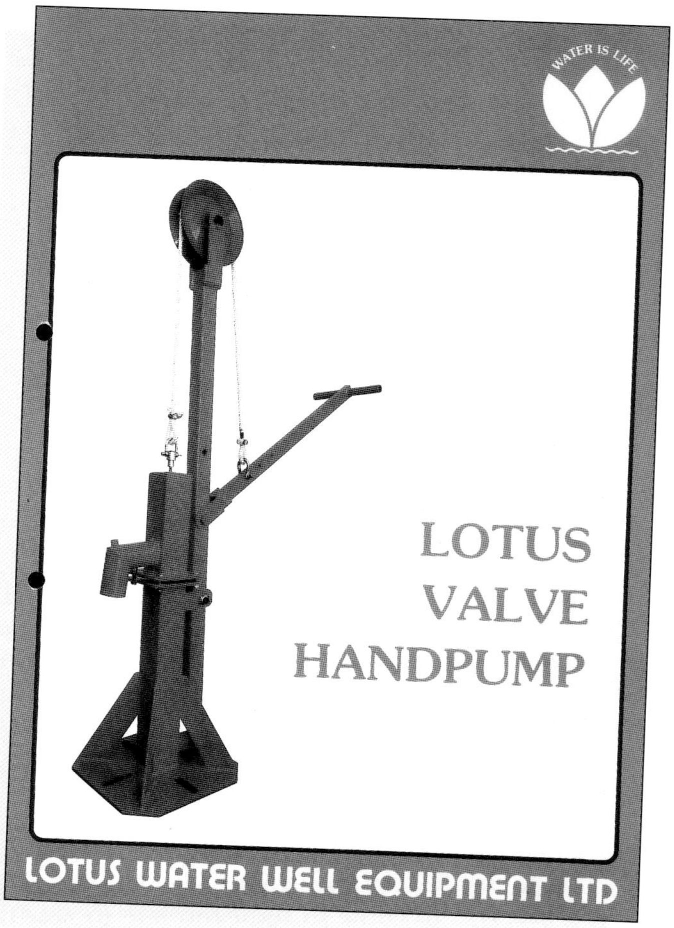

The Lotus Valve Handpump brochure.

Robert Redhead & Son already have a tug *Liverpool* (116/1870) included in their history (reference no. RH 10) which, as *Victor,* they had purchased from Dublin in 1874. But in fact they had owned another *Liverpool* (112/1863) ex *Flying Squall* (ON 45257) from 1871 to 1873 when they sold her to Dublin. So it would appear that an exchange was arranged of two very similar tugs between the Tyne and Dublin, with the reason not being at all obvious. It would explain, however, why Redhead renamed *Victor* as *Liverpool,* when otherwise it would have seemed a strange choice for a Tyne-based company. To the customers it must have appeared as though no change had been made. Even so the second *Liverpool* was sold to Harwich after only sixteen months with Redhead. This episode is a good example of the pitfalls waiting for the unwary small-ship researcher.

Iron Duke (85/1877) ex *Goole No.3* was very briefly owned by Joseph and Robert Dain Lawson and others in 1882-83. Another *Expert* (65/1866) ex *George Jackson* was owned for a while by John Robert and Joseph Lawson in 1875. When with later owners she was run down and sunk between Souter Point and Sunderland in 1886.

J. Anderson and R. Chater owned *Triumph* for approximately two years from 1874.

France Fenwick, in 1951, purchased an additional US-army tug of the same design as *Grangetown, Ryhope* and *Cornhill.* She was *ST.773* (146/1943), ex *DPC 75,* and was bought whilst lying at Piraeus. She was, however, never to reach the north-east of England for, whilst under tow by the tug *Zealandia* (479/1944), ex *Empire Winnie,* she sank near Malta on 15th September 1951. It does not appear that she was ever given a France Fenwick name.

In October 1992, *Marsden* (4) a one-time France Fenwick tug and owned elsewhere since December 1987, reappeared at Blyth under her original name. She was involved in handling vessels carrying the propeller sections for the power-generation windmills on Blyth north pier.

Douglas Hamilton, managing director of Tyne & Wear Tugs Ltd. and Lawson-Batey Tugs Ltd., was appointed chairman of the British Tugowners' Association in November 1992.

Late in 1992 an announcement was made that the car manufacturers Nissan were, at the end of 1993, to transfer their import and export shipping activities from the River Tees to a new purpose-built terminal located on the western part of Jarrow Slake. This was good news for Tyne & Wear Tugs and would mean that higher-powered tugs than those currently available would be needed to handle the large car-carriers. Conversely, the news that the last remaining shipbuilder on the Tyne, Swan Hunter, had no further orders beyond those now in hand was not welcome.

Since 1989 Tyne & Wear Tugs Ltd. have been involved with Newcastle University in the development and marketing of a simple hand-operated water well pump, for use particularly in third-world countries, where there is a serious shortage of easily-accessible water. A unique feature of this pump is its lotus-shaped valve so it is not surprising to find that the subsidiary company has been named "Lotus Water Well Equipment Ltd." Diversification into water pumping equipment may seem strange for a tug company but if this sort of enterprise is necessary to survive in the lean times, then, if such a pun can be allowed, the company are to be congratulated on 'taking the plunge'.

As this book goes to press there are encouraging signs that, at least on the River Tyne, business will improve and begin to reverse the steady decline which has taken place over the last thirty-five years. It is hoped that this will be so for the sake of the present day staff afloat and ashore.

Funnel and Flag Colours

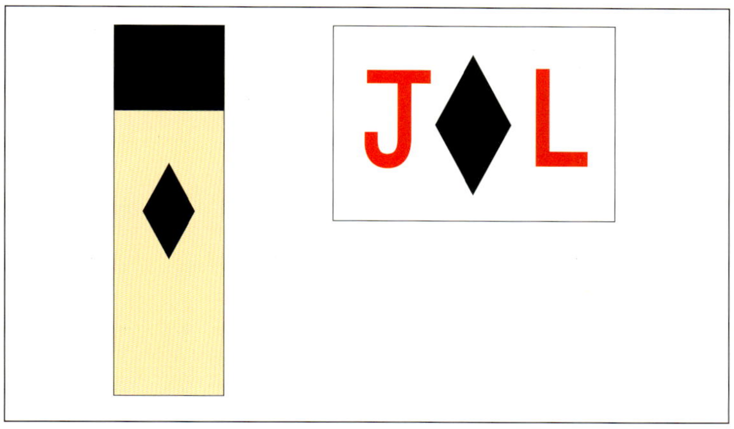

1. Lawson Tug Co. — pre 1873 - 1895
   Lawson Steam Tug-boat Co.Ltd. — 1895 - 1920

The funnel colours of the Lawson tugs were retained after the Lawson-Batey merger of 1920 until after the Second World War. The black diamond is known to have been painted on the paddle boxes c.1900.

2. John Batey
   J. Batey & Son — pre 1899 - 1906
   J. Batey & Son Ltd. — 1906 - 1920

The funnel colours of the Batey tugs were retained after the Lawson-Batey merger of 1920. A black 'B' is known to have been painted on the paddle boxes c.1905.

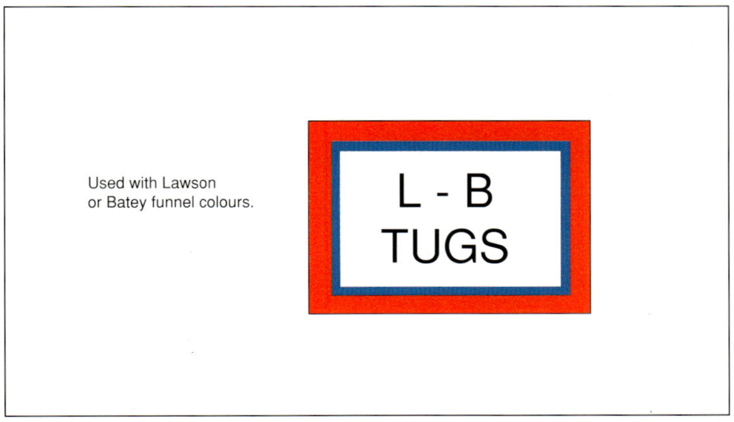

3. Lawson-Batey Tugs Ltd. — 1920 - post WW2.

This flag was used in conjunction with the Lawson and the Batey funnel colours. (See 1 and 2 above).

4. Lawson-Batey Tugs Ltd. — post WW2 - 1958
   Tyne & Wear Tugs Ltd. — 1986 -

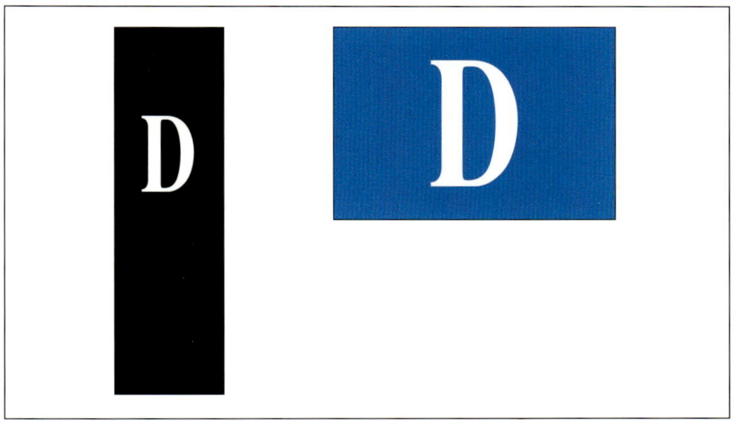

5. Blyth Steam Tug Co. — 1887 - 1923
   Blyth Tug Co.Ltd. — 1923 - 1958

The 'D' on the funnel and flag refers to George Dinsdale.

6. Blyth Tug Co.Ltd. — circa 1958

An interim plain funnel without embellishment.

Joseph Crosthwaite — circa 1854 - 1924
Joseph Crosthwaite (Tugs) Ltd. — 1924 - 1949

Details of house flag not known.

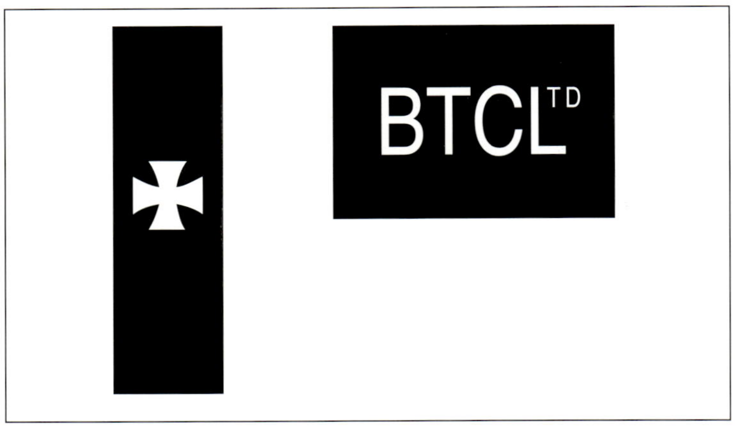

7. Blyth Tug Co.Ltd. 1958 - ?

8. Blyth Tug Co.Ltd. 1962 - 1984

Probably in existence for a while along with Item 7.

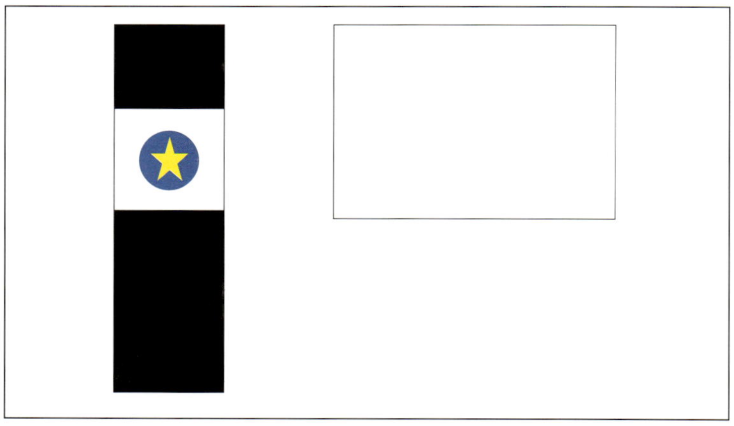

9. Joseph Ridley ? - 1920
The 'Ridley' Steam Tug Co.Ltd. 1920 - c.1939

The yellow star on a blue circle are the funnel colours of the Svenska Lloyd company with whom the Ridley company had contracts.

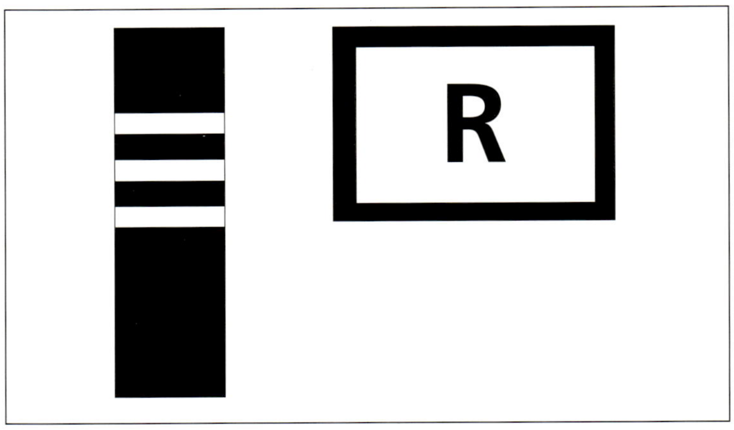

10. The 'Ridley' Steam Tug Co.Ltd. circa 1923 - 1955
Ridley Tugs Ltd. 1955 - 1958

The funnel markings are those of the Bergen Steamship Company with whom the Ridley company had contracts. It would appear that the funnel markings Items 9 and 10 were in existence at the same time in the 1920s and 1930s.

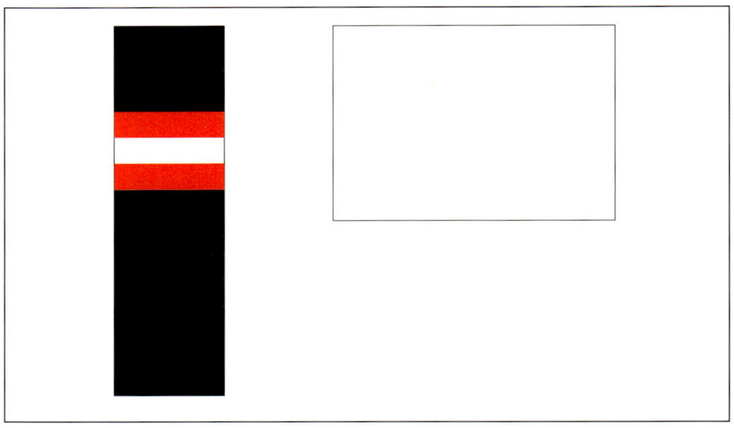

11. Axel H. Andersen 1899 - 1936

Andersen, being Danish, has used the red and white colours of the Danish flag.

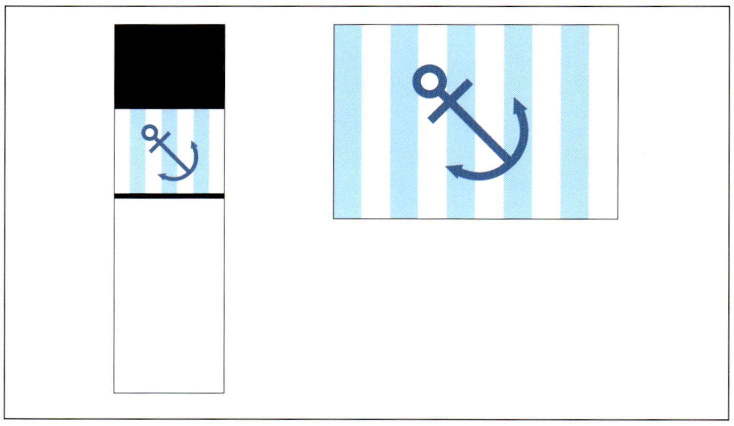

12. France, Fenwick Tyne & Wear Co.Ltd. 1920 - 1977

The combination of the Sunderland Towage Co.Ltd. and Anchor Steam Tug Co.Ltd. colours.

13. Sunderland Towage Co.Ltd. 1900 - 1918
France, Fenwick Tyne & Wear Co.Ltd. 1918 - 1920

The blue and white stripes are representative of a policeman's armband.

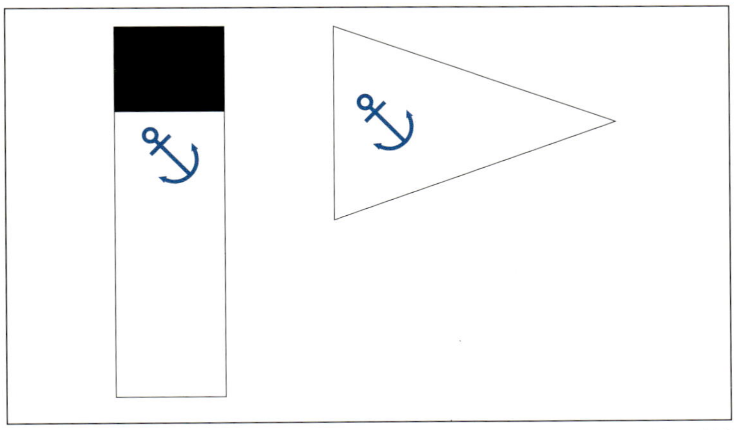

14. Anchor Steam Tug Co.Ltd. 1903 - 1920

15. John White ? - 1918
Coulson Tug Co.Ltd. 1918 - 1925

It is not known whether these companies had a house flag.

16. George Brown ? - 1923
Brown Tug Co.Ltd. 1923 - 1930

17. Anthony Irwing ? - 1911
Anthony Irwing & Son 1911 - 1926
Irwing Steam Tug Co.Ltd. 1926 - 1934

The back-to-back interlinked letters 'C' and 'C' represent the initial letters of Culliford and Clark. In the centre is a horizontal figure-of-eight. The shape of the surrounding shield varies from tug to tug.

It is interesting to note that Chanel at the present time also use the back-to-back 'C' and 'C' as their logo. Perhaps an early case of diversification by Culliford and Clark! If so, it must be an extreme example from colliers and tugs to cosmetics and haute couture!

18. Robert Redhead & Son Ltd. 1916 - 1929

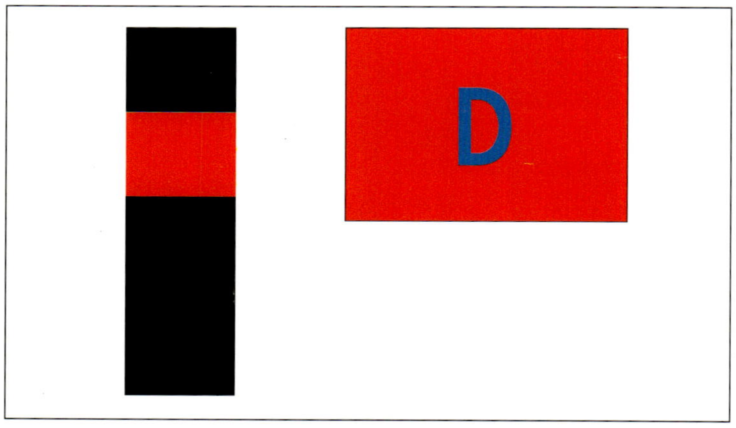

19. John Dry ? - 1908
The John Dry Steam Tugs Ltd. 1908 - 1929

Some tugs carried 'Prince of Wales' feathers on bow or paddle box.

20. Robert Stephenson Allen pre 1905 - 1914

21. Redhead & Dry's Tugs Ltd. 1929 - 1944

The funnel colours were those of the Redhead company pre-1929. It is not known what the flag colours were, but if one existed then it is likely that the design of the Redhead flag would have been used with suitably modified lettering.

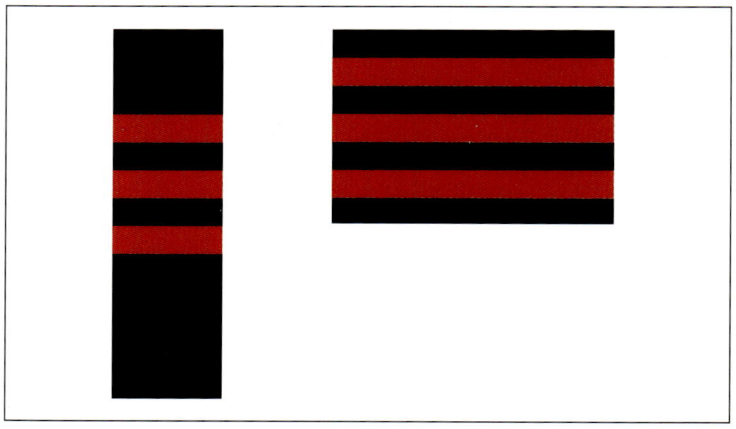

22. The Lambton Collieries Ltd. ? - 1911
The Lambton & Hetton Collieries Ltd. 1911 - 1924
The Lambton, Hetton & Joicey Collieries Ltd. 1924 - 1945

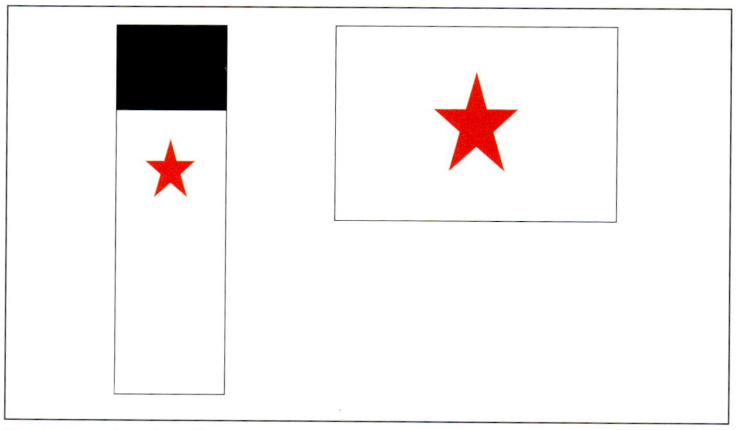

23. Robert Lumley Cook ? - 1908
Elizabeth Sarah Cook 1908 - 1912
Robert Cook Baister 1912 - 1921
Robert Lumley Cook Baister 1921 - 1924
R.L. Cook Towage Co.Ltd. 1924 - 1947

24. Tyne Tugs Ltd. 1959 - 1986

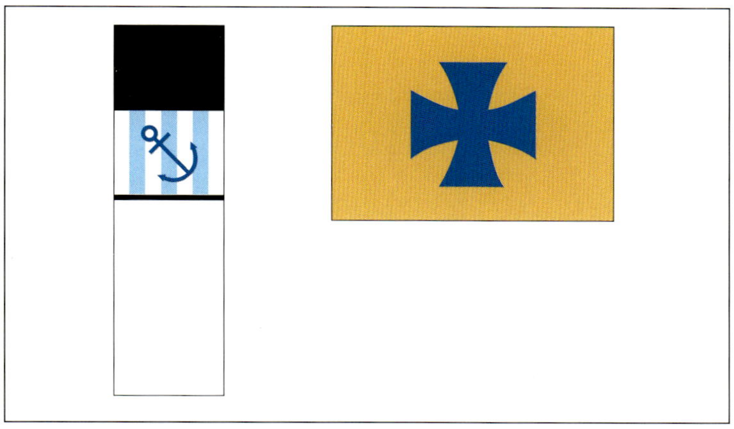

25. Wear Tugs Ltd. 1978 - 1989

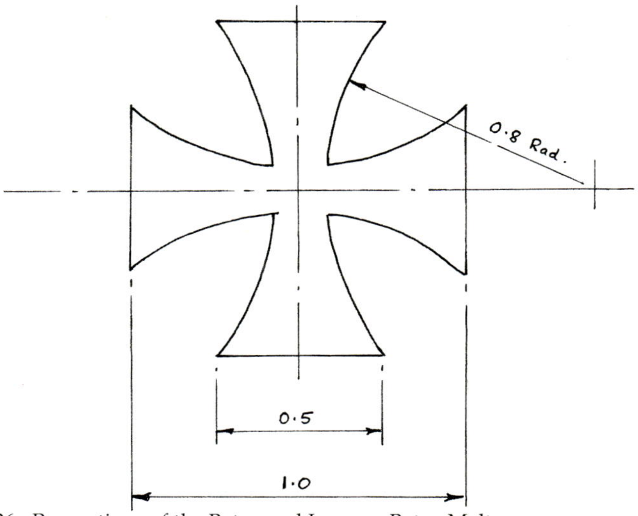

26. Proportions of the Batey and Lawson-Batey Maltese cross.

GENERAL APPENDIX 2

Tug Whistle Signals

When it was necessary for a merchant ship entering or in port to call for the assistance of tugs in the days before VHF radio was in common use a system of coded signals sounded on their steam whistles was used. The water clerks used the same codes to call attention to any of their tugs anchored in the river. The tug companies also referred to the codes on their contract documents and advertisements.

Lawson Steam Tug-boat Co.Ltd.	2 long, 4 short, 1 long
J. Batey & Son Ltd. Lawson-Batey Tugs Ltd.	} 3 long, 2 short
Blyth Tug Co.Ltd.	1 long, 4 short
The 'Ridley' Steam Tug Co.Ltd.	1 long, 2 short
France, Fenwick Tyne & Wear Co.Ltd.	1 long, 1 short
The John Dry Steam Tugs Ltd.	3 long, 3 short
Robert Redhead & Son Ltd. Redhead & Dry's Tugs Ltd.	} 2 long, 2 short

These signals were known colloquially as "Blowing the Knocks".

GENERAL APPENDIX 3

Origin of Tug Names

Agamemnon	In Greek legend, King of Mycenae.
Albert Edward	Prince of Wales, later King Edward VII Opened Albert Edward Dock in 1884.
Alnmouth	Town on Northumberland coast 27 miles north of the Tyne.
Alnwick	Town in Northumberland 4 miles inland from Alnmouth.
Amazon	South American river.
Annandale	Valley of the River Annan in S.W. Scotland.
Anthony Irwing	Sunderland tug owner.
Appelsider	See Lawson-Batey, Chapter 4.
Ashbrooke	District of Sunderland.
Australia	Country.
Bamburgh	Town on Northumberland coast 44 miles north of Tyne.
Beamish	Mining village in County Durham, 11 miles inland from Sunderland.
Ben Ledi	Scottish mountain.
Black Prince	Eldest son of King Edward III (1330-1376).
Black Watch	Scottish regiment.
Bob Chambers	Notable oarsman.
Bothal	Northumberland village 6 miles N.W. of Blyth.
Charles Dickens	English novelist (1812-1870).
Clarence	Clarence Railway Co.
Cleadon	Village 3 miles N. of Sunderland.
Collingwood	Admiral Lord Cuthbert Collingwood (1748-1810) (Presumed)
Cornhill	District of Sunderland.
Cragsider	Relates to Cragside House, Rothbury, Northumberland.
Criccieth	Welsh town on the Lleyn peninsular.
Cullercoats	Coastal village 2 miles north of Tyne.
Cumbrian	Of the county of Cumberland.
Dundee	Scottish coastal town on Firth of Tay.
Dunelm	Ancient form of modern Durham.
Earl of Beaconsfield	Benjamin Disraeli, Prime Minister.
Earl of Glamorgan	Edward Somerset (1601-1667).
Earsdon	Northumberland village 3.5 miles N.W. of Tynemouth.
Eddystone	Lighthouse in English Channel.
Eppleton Hall	Hall near Hetton-le-Hole, 7 miles SSW of Sunderland.
Francis Batey	Tyne tug owner (1842-1915).
Fulwell	District of Sunderland.
Ganges	Indian river.
General Gordon	Officer killed at Khartoum. (1833-1885).
George Dinsdale	Blyth tug owner.
George Peabody	Merchant, financier and philanthropist. (1795-1869).
George V	King of Great Britain 1910-1936. (1865-1936).
Gosforth	District of Newcastle.
Grangetown	District of Sunderland.
Greatham	Village 3 miles south of West Hartlepool.
Harwich	East coast port in Essex.
Hendon	District of Sunderland.
Her Majesty	Queen Victoria at 40th year of her reign.
Hercules	Greco-roman legendary hero.
Hetton	Colliery town 6.5 miles SSW of Sunderland.
Holmsider	Holmeside is one of the main streets in Sunderland.
Homer	Greek author.
Houghton	Colliery town 6 miles SSW of Sunderland.
Hutton	Hutton Hall, home of Pease family, near Guisborough, North Yorkshire.
Isaac Wilson	A Tees Conservency Commissioner, Middlesbrough.
Joffre	French World War I general (1852-1931).
John Batey	Tyne tug owner (1809-1897).
John Coulson	Tyne tug owner.
Joseph Crosthwaite	Tyne tug owner.
Jupiter	Planet.
King Edward VII	King of Great Britain 1901-1910. (1841-1910).
Lambton	Lambton Castle or Lambton village 6.5 miles SW of Sunderland.
Langton	Langton Dock, Liverpool, which had been named after William Langton, chairman of Mersey Docks & Harbour Board.
Lewis Crosthwaite	Tyne tug owner.
Liverpool	Major west coast seaport.
Lizard	Lighthouse at Lizard Point, Cornwall
Lord Derby	Edward Henry Stanley, 15th Earl, Foreign Secretary (Presumed).
Lord Warden	Highest office of Cinque Ports.
Lumley	Lumley Castle or Lumley village, 6.5 miles SW of Sunderland.
Malta	Mediterranean island.
Marsden	Coastal village between South Shields and Sunderland.
Mentor	Greek general.
Montrose	Scottish east coast port.
Monty	Familiar name of Field Marshall the Viscount Montgomery of Alamein GCB, DSO. (Presumed)
Nestor	King of Pylos in Greek legend.
Prestwick	Village 6 miles NW of Newcastle.
Prince of Wales	Albert Edward, later King Edward VII.
Robert L. Cook	Sunderland tug owner.
Robert Redhead	Tyne tug owner.
Roker	District of Sunderland.
Roughsider	Relates to the village and hall of Ruffside (referred to on earlier maps as Roughside) which was near to the home of the managing director, J.F. Batey at Edmondbyers.
Ryhope	Town 3 miles S of Sunderland.
Scotland	Country.
Seaburn	District of Sunderland.
Silksworth	Mining town 3 miles S of Sunderland.
Sir Colin Campbell	C-in-C of British Forces at relief of Lucknow, 1857. Later Lord Clyde (1792-1863)
Sir Walter Raleigh	Renowned Elizabethan seafarer. (1554-1618)
Snowdon	Mountain in North Wales.
Souter	Lighthouse near Marsden between South Shields and Sunderland.
Sunderland	Town at mouth of River Wear.
Taliesin	Welsh village.
Thames	River Thames.
Thomas Nicholson	Coal fitter for the Lambton Collieries.
Triton	In Greek mythology, a merman, demigod of the sea, son of Poseidon and Amphitrite
Ulysses	Roman name of Greek hero Odysseus.
W.D.S.	William Davies Stephens co-founder of Tyne Steam Shipping Co. in 1864.
Wansbeck	Northumberland river.
Wearmouth	Mouth of the River Wear.
Wexford	Irish port.
Whitburn	Coastal village 2.5 miles north of Sunderland.
William Fallows	A Tees Conservancy Commissioner, Middlesbrough.

Any suggestions as to why the following were so named would be welcome:

Cinema Star	Sunderland, 1920.
James Conley	Tyne, 1864.
Little Western	Probably Tyne, 1844.
Pero Gomez	Swansea, 1869.

GENERAL APPENDIX 4

END OF YEAR → / COMPANY	1880	1885	1890	1895	1900	1905	1910	1915	1920	1925	1930	1935	1940	1945	1950	1955	1960	1965	1970	1975	1980	1985	1990
LAWSON	5	9	7	6	7	8	9	10															
BATEY	6	7	8	6	6	6	6	6															
LAWSON - BATEY									17	15	13	11	7	7	9	10	8	8	6	5	15	12	6
BLYTH			3	4	4	4	3	2	3	3	4	3	3	3	4	5	4	4	1	1			
CROSTHWAITE	1	2	1	2	3	3	3	3	3	3	3	2	2	2									
RIDLEY	2	3	2	1	0	2	3	2	2	6	6	5	7	6	6	5	2	2	1				
FRANCE FENWICK									12	14	14	13	13	22	26	25	20	18	10	10			
SUNDERLAND					3	4	6	6															
ANCHOR	3	3	3	4	3	4	4	4															
COULSON	2	1	1	2	1	2	2	1	3														
BROWN			1	3	4	3	3	3	4	4													
IRWING	2	5	5	6	4	5	3	3	4	4	3												
REDHEAD	3	2	3	3	4	3	3	3	3	3													
DRY	2	4	4	3	4	4	5	5	5	5													
REDHEAD & DRY'S											9	7	6										
LAMBTON	2	2	2	3	3	4	4	5	4	4	4	4	3										
COOK	1	1	2	3	4	3	3	2	5	3	3	2	2	2									
CLYDE																						1	1
ANDERSON								1	1	1	1	1											
WHITE	1	2	2	2	2	2	2	2															
ALLEN	2	2	2	2	1	1	2																
TOTAL	32	43	46	50	53	58	61	58	66	65	60	48	43	42	45	45	34	32	18	16	15	13	7

(jhp 1992)

↑ Peak number of tugs was 68 in 1923.

NUMBERS OF TUGS OWNED BY THE COMPANIES IN THIS HISTORY.
1880 - 1990

GENERAL APPENDIX 5

The North East Coast Tugboatmen's Association

The North East Coast Tugboatmen's Association was formerly, from about 1900, The North East Coast Tugboatmen's and Fishermen's Association, and before that The Tyne Steam Packet Provident Society, and had its origins about the year 1838. Although beginning as a provident society, the organisation was officially registered as a Trade Union under the acts of 1853 and 1873 and was one of the first to do so. Initially, as one would expect of a provident society, its self-help aims were to support its members in times of need or distress, such as paying maternity benefit so that a doctor could be employed and death benefit so avoiding a pauper's funeral being necessary. The society was, therefore, as much a family association as it was a trade union looking after the tugboatmen's working rights.

In 1992 the Association claims to be the smallest non-affiliated trade union in the United Kingdom. There can be few trade unions where the whole of the working membership can meet together at regular intervals. In 1968 an approach was made to the Trade Union Council requesting that the Association become an affiliated society of that body. The then general secretary, Vic Feather, told the representatives that they should go away and join one of the big unions as they were too small to be considered for affiliation to the TUC. Not having travelled to London to be given advice, they returned home more convinced than ever that they should remain independent and make the most of the undoubted benefit of the close contact enjoyed between officers and men in a small group. This direct contact meant that problems did not get magnified by having to be passed through the many layers of a large organisation.

In earlier times the Association had two groups; one based in North Shields and the other in South Shields. As numbers reduced then the two sections combined. The headquarters were at Jubilee Buildings, Sibthorpe Street, North Shields, where, as well as an office, there was a meeting hall which was available for parties, wedding receptions and dances. These premises were sold some years ago.

On formal occasions regalia, which included a medallion with the appropriate motto "United to Assist", was worn. There was also a colourful banner which was paraded on suitable occasions. Unfortunately, the whereabouts of this banner is today not known.

In 1981 the Association made a 30-minute BBC TV programme about themselves and the activities of the Tyne tugs. As well as referring to their origins and to the aims of the society a number of interesting anecdotes were given by the members. Some of these are worth recording here:-

(a) Relationships between management and men have been generally very good. Disputes and strikes have been few and far between.

(b) The Association's best achievement was when they obtained one clear day off per week and payment by the hour. Prior to this the men worked such long hours that they only rarely saw their children awake and when paid 'by the ship' could spend hours, or even days, waiting for an inbound vessel with very little to show for it financially.

(c) Children, as young as 5 or 6 years old, would go on board their father's boat accompanied by their elder brothers and sisters. Thus the tugs almost became part of the family. When tugs were held up waiting at the buoys for a ship it was not unusual for the families of the crews to go on board. Swings for the children would be rigged under the paddle box bridge and music was provided by fiddlers to accompany the singing of hymns and songs on the after deck.

(d) Not all tugmen were hard drinkers as is often presumed. Many were teetotallers - one such skipper, nicknamed 'Hallelujah Harry', would not tolerate any drinkers in his crew.

(e) When tugs were moored for the night at their buoy berths the crew were sculled ashore in a skiff. One of the crew then took the scull home with him so that it would not be stolen, and it is suggested that he kept this under his bed so that it was handy if he was called out during the night.

(f) When crews were called on unexpectedly to work extra long hours the 'grub-boy' was sent around the crews' homes to collect their next day's dinner wrapped up in a handkerchief. It is said that this included the gravy! Naturally, to collect from, say, six houses took some time and one person is known to have taken his girl-friend with him on such errands because it was often the only chance he had of seeing her! I suspect that the grub collection must have taken quite some time.

(g) When out 'seeking' for jobs, sometimes as far as the English Channel, there was great rivalry between the individual tugs. Subterfuges were resorted to in an attempt to deceive the competition, even, it is said, by reversing the tug's port and starboard lights or running with none at all.

(h) Notwithstanding all the hardships and inconvenience to family life, the tugboatmen have a great spirit of comradeship and are very proud of their profession. One union member described tugboating as a disease - once acquired it was very difficult to be cured of it. Thus it was that men spent a lifetime on the tugs following their forebears and being followed by their sons and grandsons. They never felt the job was monotonous because the shipping scene and the river itself was different every day.

The present secretary of the Association is Bill Robson, engineer of *Holmsider,* who has held the post since May 1990. His predecessor was Harry J. Malcolm who had been secretary since 1947. Harry had worked on the Tyne Improvement Commission's tugs from 1923 when he became an apprentice member of the union following his father and grandfather who had both been tug skippers. He retired from the TIC in 1972 but continued as union secretary until 1990.

Other known office holders of the Association were:-

Presidents	John T. Boag	1935 - 1938	
	Ald. W.G. Spence	1939 -	
	James McRitchie	1944 - 1952	
	Norman Carman	1971 - 1986	(also chairman)
	John Burn (vice)	1971 - 1986	
		1986 -	(also chairman)
Chairmen	G.R. Strong	In 1934	
	J.J.P. Foster	In 1935	
	Henry S. Jobling	In 1954	
Secretaries	John W. Hudson	In 1866	
	R. Hedley	In 1878	
	Ambrose Legg	In 1890s	
	Counc. W. Steele	1896 - 1910	
	Walter Dunn	? - 1934	
	J. Lincoln	In 1934	
	H.A. Lowden	1934 - 1938	
	Miles M. Storer	1938 - 1947	

In 1934 the Association had 400 members, 140 in 1967 and only 36 today. Since 1986 the Association has been a member of the United Kingdom Tugboatmen's Association. This is not a trade union and is used as a vehicle for the exchange of ideas at its quarterly meetings. Needless to say the North East Coast Tugboatmen's Association is the only non-affiliated trade union member on that body.

At Sunderland there existed a separate union entitled The Wear Steam Packet Trade Protecting and Friendly Society whose chairman and secretary between 1935 and 1939 were respectively Thomas Scott and R. Metcalf. By 1951 this society had been disbanded and the tug crews on the River Wear were members of The National Union of General and Municipal Workers.

GENERAL APPENDIX 6

The Tyne Tug-Owners' Association

The Tyne Tugowners' Association was formed in 1899 to further the collective interests of the tugowners then operating on the River Tyne. The companies referred to in this history and in existence in 1889 were:-

- Lawson Steam Tug-boat Co.Ltd.
- J. Batey & Son
- Joseph Crosthwaite
- John and James Ridley
- Axel H. Andersen
- John Anderson and Robert Chater
- John Coulson
- John White
- George Brown
- John Dry
- Robert Stephenson Allen
- Robert Redhead & Son

It is not known how many of these companies were members of the Association but it is likely that most of them would have been.

Although 1899 is given as the date of the formation of The Tyne Tugowners' Association there is reference to a similar organisation entitled 'Tyne Steam Tug Towing Association' as far back as 1865. John Batey was chairman and Mr. Gibson secretary in 1872. Richard Stobbs is known to have been secretary from at least 1878 to 1898.

The purpose of the Association was to deal with matters of common interest to all the Tyne tugowner members. In particular this involved negotiations with the Tyne Improvement Commission and the Shipowners' Association regarding the setting of towage charges, and with the tugmen's union with respect to wages and conditions. This meant that the various companies would not be competing against each other for business, or for labour, on a financial basis. Nevertheless the companies did each have their contracts for the towage of ships of particular companies. These were referred to as the 'Block List'.

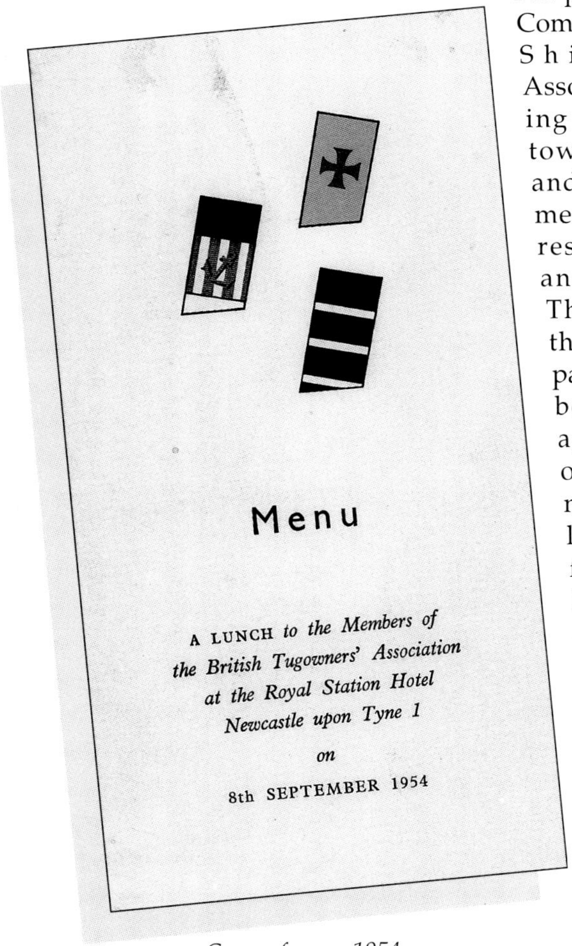

Cover of menu 1954.

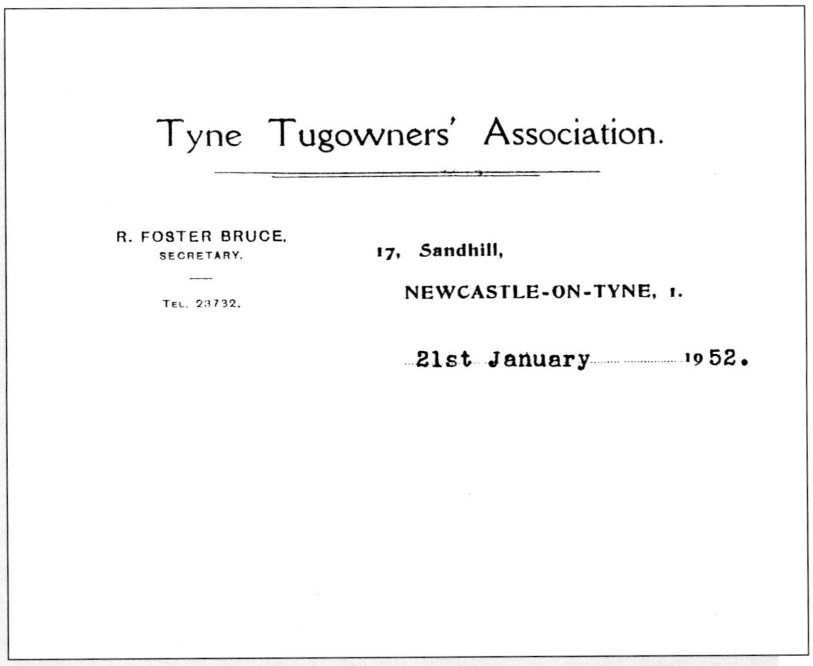

Tyne Tugowners' Association.

R. FOSTER BRUCE.
SECRETARY.

TEL. 23732.

17, Sandhill,
NEWCASTLE-ON-TYNE, 1.

21st January 1952.

Letterhead of 1952.

Each participating company nominated a representative to sit on the committee of the Association. From this committee a chairman was elected. Those known were:-

John Robert Lawson	1899 - 1901
Francis Batey	? - 1913
Joseph Lawson	1913 - 1927
Joseph Crosthwaite	1927 - 1930
Edwin Edwards	1934 - 1935
Frank Dawson (France Fenwick)	1943, 1947 and 1951
A.M.H. Wardlaw (Joseph Crosthwaite)	1944, 1948 and 1952
Frank Batey (Lawson-Batey)	1945
F.C. Batey (Ridley)	1946, 1949 and 1953
J.F. Batey (Lawson-Batey)	1950 and 1954

As will be seen, from at least 1943 the complete complement of the committee were each taking the chairmanship in rotation. Frank Batey, when he resigned in 1947, had served on the committee for about forty years.

The secretaries who are known to have served the Association are:-

J.A. McKenzie	1899 - c.1927
T.R.E. Abbott	1929 - 1935
R. Foster Bruce	1938 - 1954

With the amalgamation of the France Fenwick and Lawson-Batey fleets in 1977 the need for an Association of companies disappeared, since there was then only one company operating on the Tyne. The Tyne Tugowners' Association was, therefore, wound-up on 31st December 1977.

Until at least 1951 there was a similar organisation on the River Wear entitled The Sunderland Tugowners' Association. In 1935 Edwin Edwards was chairman and in 1951 A.M.H. Wardlaw held that position. The secretary was A. Hartley Cook in 1935 and G. Dryden from 1939 - 1951. Further information has not been discovered.

Rule Book of 1910. (Original size).

Tyne Tug-Owners' Association,
55, King Street, South Shields.

RATES OF TOWAGE

TO BE CHARGED BY

STEAM BOATS

ON

THE RIVER TYNE,

AS AGREED TO BY

THE RIVER TYNE COMMISSIONERS

AND

STEAM BOAT OWNERS,

MARCH, 1865.

J. A. McKenzie, Secretary.

North Shields:
PRINTED BY R. B. HENZELL, STEPHENSON STREET.

1910.

2

THE FOLLOWING

Scale of Twopence in the Shilling

To be added to and Charged in Addition
to the various

TOWAGE RATES

Set forth within this Book, commencing October 1st, 1872, as ordered by the River Tyne Commissioners at a meeting held September 12th, 1872.

Scale under £1.		Scale above £1.	
s.	£ s. d.	£	£ s. d.
1	0 0 2	1	0 3 4
2	0 0 4	2	0 6 8
3	0 0 6	3	0 10 0
4	0 0 8	4	0 13 4
5	0 0 10	5	0 16 8
6	0 1 0	6	1 0 0
7	0 1 2	7	1 3 4
8	0 1 4	8	1 6 8
9	0 1 6	9	1 10 0
10	0 1 8	10	1 13 4
11	0 1 10	11	1 16 8
12	0 2 0	12	2 0 0
13	0 2 2	13	2 3 4
14	0 2 4	14	2 6 8
15	0 2 6	15	2 10 0
16	0 2 8	16	2 13 4
17	0 2 10	17	2 16 8
18	0 3 0	18	3 0 0
19	0 3 2	19	3 3 4
20	0 3 4	20	3 6 8

3

TYNE TUG-OWNERS' ASSOCIATION.

RULES AND RATES
FOR TOWING VESSELS ON THE RIVER
TYNE WITHIN THE LIMITS.

1. The limits Seaward, Souter Point, South, Hartley Bates, North, on the East an imaginary line drawn three miles out to Sea from the shore, and on the West a line drawn southward across the river from the buoy on the Mussel Scarp, Low Lights, North Shields.

2. Shields Harbour to be deemed to extend from the Buoy on the Mussel Scarp to J. Young's West Dock Quay, South Shields.

3. The Rates hereinafter given to be charged on the British Registered Tonnage of all vessels of 150 Tons and upwards. *See* Pages 6 and 7. River Towage, Pages 8, 9, 10, 11, 12, 13.

In cases where a second tug is required £1 to be paid for second boat, and £2 for each additional boat, nett. These rates to apply below Swing Bridge only.

4. For the purposes of these Rates, a Ship one-third loaded to pay the loaded towage, below one-third to pay the light towage.

5. Vessels towing down the River and going into either Dock to pay the same as if going to Shields.

4

6. All vessels towing down the River and going to Sea direct, to pay the River towage and half the Sea towage.

7. Any towage services rendered not coming within the meaning of the regular rates, the amount to be paid for such services shall be fixed by agreement between the parties interested in each particular case; or in the event of disagreement, the dispute shall be decided by two Justices of the Peace, on the application of either party, in a summary way, and the decision of such Justices shall be final.

8. Ships towed from sea to either Dock, direct—Loaded, 1⅞d.; Light, 1¾d., with a single Boat, with Twopence to the Shilling added.

9. Ships towed out of the Docks into the Basin, or out of the Basin into the Dock, to pay the shifting towage. But ships towed out of the River into the Basin, or out of the Basin into the River, to pay for one Station.

10. An addition of Twopence to the Shilling to be charged upon the rates from Sea to Shields, or from Sea to either Dock direct.

STATIONS.

11. That the River be divided into Six Stations, making Shields one starting point, Newcastle the other:—

1st Station.—From Shields to either Dock, Jarrow, or Willington Gut.

5

2nd Station.—From Shields to Wallsend, or Gosforth.

3rd Station.—From Shields to Walker or Pelaw Main.

4th Station.—From Shields to Bill Quay or Felling.

5th Station.—From Shields to Dent's Hole, Tyne Main or St. Lawrence.

6th Station.—From Shields to Newcastle.

12. No Alterations to be made by the Steam Tug Association in any of the preceding Rules, except at a full Meeting of the Members, convened by 28 days' notice having been published in the Newspapers, and a like notice given in writing to the River Tyne Improvement Commissioners.

13. The Owners of the Steam Tugs will not be accountable for any damage that may occur while Towing.

SCREW STEAMERS' TOWAGE.

Screw Steamers, when light, whether steaming or not, same as other Ships.

Loaded, when Steaming, to be allowed one-fourth off the regular rate; not Steaming, same as other ships.

J. A. McKENZIE,
Secretary.

6

Towage Rates within the Limits at Sea into Shields Harbour, and to either Docks, and from either Docks to Sea, or a Single Station.
This Scale is for Vessels below 150 Tons.

LIGHT VESSELS.				LOADED VESSELS.			
	From Sea to Shields or Shields to Sea.	Sea to Dock or Dock to Sea.	Shields to Dock, Dock to Dock, or single station.		From sea to any part of Shields Harbour,	Shields to either Docks, or to Sea.	Sea to Dock or Dock to Sea direct.
Tons,	£ s. d.	£ s. d.	£ s. d.	Tons.	£ s. d.	£ s. d.	£ s. d.
40	0 10 0	0 14 0	0 6 6	40	0 10 0	0 9 0	0 14 0
50	0 10 6	0 15 0	0 7 0	50	0 11 0	0 9 6	0 15 0
60	0 11 0	0 16 0	0 7 6	60	0 12 0	0 10 0	0 16 0
70	0 11 6	0 16 6	0 8 0	70	0 13 0	0 10 6	0 17 0
80	0 12 0	0 17 0	0 8 6	80	0 14 0	0 11 0	0 18 0
90	0 12 6	0 17 6	0 9 0	90	0 15 0	0 11 6	0 19 0
100	0 13 0	0 18 6	0 9 0	100	0 16 0	0 12 0	1 0 0
110	0 13 6	0 19 6	0 9 0	110	0 17 0	0 13 0	1 1 0
120	0 14 0	1 0 6	0 9 0	120	0 17 6	0 14 0	1 2 0
130	0 14 6	1 1 0	0 9 0	130	0 18 0	0 14 6	1 2 6
140	0 15 0	1 1 6	0 9 0	140	0 18 6	0 15 0	1 3 0
149	0 15 6	1 2 0	0 9 6	149	0 19 0	0 15 6	1 3 6

7

A Towage Rate for all Vessels above 150 Tons British Register.

	Loaded Vessels per Register Tonnage.	Light Vessels per Register Tonnage.
	d.	d.
From Sea to any part of Shields Harbour	1½	1¼
,, Sea to Tyne, Northumberland, or Albert Edward Docks (*The same out to Sea*)	1⅞	1¾
,, Shields to either Dock, Jarrow, or Willington	1¼	0¾
,, Dock to Dock or a Single Station	1¼	0¾
,, Shields to Sea	1¼	1¼
Shifting Towage	Loaded Vessels per Register Tonnage.	Light Vessels per Register Tonnage.
	d.	d.
For Vessels above 150 *Tons* Register	0⅝	0½

~~Shifting Towage.~~

	Loaded Vessels Per Shift.	Light Vessels Per Shift.
For Vessels below 150 *Tons* Register	£ s. d.	£ s. d.
,, under 50 ,,	0 5 0	0 4 0
,, under 100 ,,	0 6 0	0 5 0
,, under 150 ,,	0 7 6	0 6 0

River Towage.—*A Towage Rate for Vessels above 200 Tons Register.*

From SHIELDS to the following Stations.

	Loaded Vessels per Register Tonnage.	Light Vessels per Register Tonnage.
	d.	d.
From Shields to either Dock, Jarrow, or Willington.... 1st Station	1¼	0¾
,, Shields to Wallsend, or Gosforth.............. 2nd Station	1½	1
,, Shields to Walker, or Pelaw Main.............. 3rd Station	1¾	1⅛
,, Shields to Bill Quay or Felling.............. 4th Station	1⅞	1¼
,, Shields to Dent's Hole, Tyne Main or St Lawrence 5th Station	2	1⅜
,, Shields to Newcastle .... 6th Station	2⅛	1½

8

River Towage.—*A Towage Rate for Vessels above 200 Tons Register.*

From NEWCASTLE to the following Stations.

	Loaded Vessels per Register Tonnage.	Light Vessels per Register Tonnage.
	d.	d.
From Newcastle to S. Lawrence, Tyne Main, Dent's Hole 1st Station	1¼	0¾
,, Newcastle to Felling or Bill Quay.. 2nd Station	1½	1
,, Newcastle to Pelaw Main, or Walker 3rd Station	1¾	1⅛
,, Newcastle to Gosforth, or Wallsend 4th Station	1⅞	1¼
,, Newcastle to Willington, or Jarrow 5th Station	2	1⅜
,, Newcastle to either Dock or Shields .6th Station	2⅛	1½

Newcastle Station to be deemed to extend from Tyne Bridge to Ouseburn.

9

River Towage.—*A Scale of Towage Rates for all Vessels below 200 tons Register.*
LIGHT VESSELS.—From Shields to the following Stations.

Tons.	1st Station, Docks, Jarrow, or Willington. £ s. d.	2nd Station, Wallsend or Gosforth. £ s. d.	3rd Station, Walker, or Pelaw Main. £ s. d.	4th Station, Bill Quay, or Felling. £ s. d.	5th Station. Dent's Hole, Tyne Main or S. Lawrence £ s. d.	6th Station. Newcastle. £ s. d.
40	0 6 6	0 8 0	0 9 6	0 11 0	0 12 6	0 14 0
50	0 7 0	0 9 0	0 10 6	0 12 0	0 13 6	0 15 0
60	0 7 6	0 9 6	0 11 6	0 13 0	0 14 6	0 16 0
70	0 8 0	0 10 0	0 12 6	0 14 0	0 15 6	0 17 0
80	0 8 6	0 10 6	0 13 0	0 15 0	0 16 6	0 18 0
90	0 9 0	0 11 0	0 13 6	0 15 6	0 17 6	0 19 0
100	0 9 0	0 11 6	0 14 0	0 16 0	0 18 0	1 0 0
110	0 9 0	0 12 0	0 14 6	0 16 6	0 18 6	1 0 6
120	0 9 0	0 12 6	0 15 0	0 17 0	0 19 0	1 1 0
130	0 9 0	0 13 0	0 15 6	0 17 6	0 19 6	1 1 6
140	0 9 0	0 13 6	0 16 0	0 18 0	1 0 0	1 2 0
150	0 9 6	0 14 0	0 16 6	0 18 6	1 0 6	1 2 6
160	0 10 0	0 14 6	0 17 0	0 19 0	1 1 0	1 3 0
170	0 11 0	0 15 0	0 17 6	0 19 6	1 1 6	1 3 6
180	0 11 6	0 15 6	0 18 0	1 0 0	1 2 0	1 4 0
190	0 12 0	0 16 0	0 18 6	1 0 6	1 2 6	1 4 6
200	0 12 6	0 17 0	0 19 0	1 1 0	1 3 0	1 5 0

10

River Towage.—*A Scale of Towage Rates for all Vessels below 200 Tons Register.*
LOADED VESSELS.—From Shields to the following Stations.

Tons.	1st Station. Docks, Jarrow, or Willington. £ s. d.	2nd Station. Wallsend, or Gosforth. £ s. d.	3rd Station, Walker, or Pelaw Main. £ s. d.	4th Station. Bill Quay, or Felling. £ s. d.	5th Station. Dent's Hole, Tyne Main or S Lawrence £ s. d.	6th Station. Newcastle. £ s d.
40	0 9 0	0 11 0	0 13 0	0 15 0	0 17 0	0 19 0
50	0 9 6	0 11 6	0 14 0	0 16 0	0 18 0	1 0 0
60	0 10 0	0 12 0	0 15 0	0 17 0	0 19 0	1 1 0
70	0 10 6	0 12 6	0 16 0	0 18 0	1 0 0	1 2 0
80	0 11 0	0 13 0	0 17 0	0 19 0	1 1 0	1 3 0
90	0 11 6	0 14 0	0 18 0	1 0 0	1 2 0	1 4 0
100	0 12 0	0 15 0	0 19 0	1 1 0	1 3 0	1 5 0
110	0 13 0	0 16 0	1 0 0	1 2 0	1 4 0	1 6 0
120	0 14 0	0 17 0	1 1 0	1 3 0	1 5 0	1 7 0
130	0 14 6	0 18 0	1 2 0	1 4 0	1 6 0	1 8 0
140	0 15 0	0 19 0	1 3 0	1 5 0	1 7 0	1 9 0
150	0 15 6	1 0 0	1 4 0	1 6 0	1 8 0	1 10 0
160	0 17 0	1 1 0	1 5 0	1 7 0	1 9 0	1 11 0
170	0 18 0	1 2 0	1 6 0	1 8 0	1 10 0	1 12 0
180	0 19 0	1 3 0	1 7 0	1 9 0	1 11 0	1 13 0
190	1 0 0	1 4 0	1 8 0	1 10 0	1 12 0	1 14 0
200	1 1 0	1 5 0	1 9 0	1 11 0	1 13 0	1 15 0

11

River Towage.—*A Scale of Towage Rates for all Vessels below 200 Tons Register.*
LIGHT VESSELS.—From Newcastle to the following Stations.

Tons.	1st Station. S. Lawrence, Tyne Main or Dent's Hole £ s. d.	2nd Station. Felling, or Bill Quay £ s. d.	3rd Station. Pelaw Main, or Walker. £ s. d.	4th Station. Gosforth, or Wallsend. £ s. d.	5th Station. Willington, or Jarrow. £ s. d.	6th Station. Docks, or Shields. £ s. d.
40	0 6 6	0 8 0	0 9 6	0 11 0	0 12 6	0 14 0
50	0 7 0	0 9 0	0 10 6	0 12 0	0 13 6	0 15 0
60	0 7 6	0 9 6	0 11 6	0 13 0	0 14 6	0 16 0
70	0 8 0	0 10 0	0 12 6	0 14 0	0 15 6	0 17 0
80	0 8 6	0 10 6	0 13 0	0 15 0	0 16 6	0 18 0
90	0 9 0	0 11 0	0 13 6	0 15 6	0 17 6	0 19 0
100	0 9 0	0 11 6	0 14 0	0 16 0	0 18 0	1 0 0
110	0 9 0	0 12 0	0 14 6	0 16 6	0 18 6	1 0 6
120	0 9 0	0 12 6	0 15 0	0 17 0	0 19 0	1 1 0
130	0 9 0	0 13 0	0 15 6	0 17 6	0 19 6	1 1 6
140	0 9 0	0 13 6	0 16 0	0 18 0	1 0 0	1 2 0
150	0 9 6	0 14 0	0 16 6	0 18 6	1 0 6	1 2 6
160	0 10 0	0 14 6	0 17 0	0 19 0	1 1 0	1 3 0
170	0 11 0	0 15 0	0 17 6	0 19 6	1 1 6	1 3 6
180	0 11 6	0 15 6	0 18 0	1 0 0	1 2 0	1 4 0
190	0 12 0	0 16 0	0 18 6	1 0 6	1 2 6	1 4 6
200	0 12 6	0 17 0	0 19 0	1 1 0	1 3 0	1 5 0

12

River Towage.—*A Scale of Towage Rates for all Vessels below 200 Tons Register.*
LOADED VESSELS.—From NEWCASTLE to the following Stations.

Tons.	1st Station. S. Lawrence. Tyne Main or Dent's Hole £ s. d.	2nd Station. Felling, or Bill Quay. £ s. d.	3rd Station. Pelaw Main, or Walker. £ s. d.	4th Station. Gosforth, or Wallsend. £ s. d.	5th Station. Willington, or Jarrow. £ s. d.	6th Station. Docks, or Shields. £ s. d.
40	0 9 0	0 11 0	0 13 0	0 15 0	0 17 0	0 19 0
50	0 9 6	0 11 6	0 14 0	0 16 0	0 18 0	1 0 0
60	0 10 0	0 12 0	0 15 0	0 17 0	0 19 0	1 1 0
70	0 10 6	0 12 6	0 16 0	0 18 0	1 0 0	1 2 0
80	0 11 0	0 13 0	0 17 0	0 19 0	1 1 0	1 3 0
90	0 11 6	0 14 0	0 18 0	1 0 0	1 2 0	1 4 0
100	0 12 0	0 15 0	0 19 0	1 1 0	1 3 0	1 5 0
110	0 13 0	0 16 0	1 0 0	1 2 0	1 4 0	1 6 0
120	0 14 0	0 17 0	1 1 0	1 3 0	1 5 0	1 7 0
130	0 14 6	0 18 0	1 2 0	1 4 0	1 6 0	1 8 0
140	0 15 0	0 19 0	1 3 0	1 5 0	1 7 0	1 9 0
160	0 15 6	1 0 0	1 4 0	1 6 0	1 8 0	1 10 0
160	0 17 0	1 1 0	1 5 0	1 7 0	1 9 0	1 11 0
170	0 18 0	1 2 0	1 6 0	1 8 0	1 10 0	1 12 0
180	0 19 0	1 3 0	1 7 0	1 9 0	1 11 0	1 13 0
190	1 0 0	1 4 0	1 8 0	1 10 0	1 12 0	1 14 0
200	1 1 0	1 5 0	1 9 0	1 11 0	1 13 0	1 15 0

13

INDEX
OF
Names of Tugs owned by companies included in this history

Active	CN1
Admiral	AST5, I6, L32
Africa/Nestor	L26, LB7
Agamemnon	I14
Albert Edward	CE4
Alnmouth	F49, LB42
Alnwick	F46, LB39
Amazon	B2
Annandale	S1
Anthony Irwing	I16
Appelsider	LB29
Ashbrooke	F45, LB38
Australia	S3
Balmoral	LH4
Bamburgh	F47, LB40
Battleaxe	R31, LB28 as *Hillsider* (1)
Beamish	F40
Belmont	CK1
Ben Ledi	CK6, RH15, RD8
Black Prince	RSA2
Black Watch	CN7
Blue Bell	CK3
Bob Chambers	B10
Bothal	BL4
Brothers (1)	RH6
Brothers (2)	D5
Brothers (3)	AST9, CK9 as *Rescue* (2)
Bull Dog	RH13
Bulldog	B18, LB15
Champion (1)	L2
Champion (2)	L19
Champion (3)	CE6
Charles Dickens	B16
Cinema Star	CK13
Clarence	CK16
Cleadon	F26
Collingwood	RH5
Comet	L22, LB4, BL15
Commodore	RH12
Conqueror (1)	I13
Conqueror (2)	RH19, RD1, F28
Contest	BR1
Cornhill	F43, LB36
Corsair (1)	W1
Corsair (2)	W2, BL8
Corsair (3)	W4, CN11, F20
Cragsider	LB33
Criccieth	F15
Criterion	L31, LB3
Cruiser (1)	L17, LB1
Cruiser (2)	CK7
Crusader	L9
Cullercoats	F23
Cumbrian	CN4
Daring	RH14
(Dawlish)	R24
Defence	AST6
Defiance	AST2
Dundee	RH11
Dunelm (1)	F17
Dunelm (2)	F50, LB43
Dunera/Dunera II	CN15, BL10
Earl of Beaconsfield (1)	BL1
Earl of Beaconsfield (2)	BL9
Earl of Glamorgan	CK8
Earsdon	CE8 as *Lewis Crosthwaite*, F39
Eastsider	LB24
Eddystone	LH5
(Ella)	R20
Empire Andrew	F/M3
Empire Jane	F/M4
Empire Mustang	RD/M1, F/M5
Empire Plane	F/M1
Empire Sara	F/M2
Empire Sybil	LB/M3
Enable	AHA1
Engineer	RH1, AST1
Eppleton Hall	LH10, F36
Exemplar	LB/M4
Expatriate	F/M6
Expert	L1
Favourite	R3
Fiery Cross	RSA1
Fiery Dragon	AST8
Florence	I4
Flying Arrow	I5, AST12
Flying Childers	AST11
Flying Foam	L24
Flying Scotchman	B11
Flying Scud	CN9, F21
Francis Batey	B19, LB18, BL21
Fulwell	F27
Ganges	L5
Gauntlet	D9, RD3
General Gordon	I8, BL3
George Dinsdale	BL6
George Peabody	BL5
George V (1)	RH17
George V (2)	RH18, RD7, F29, LB34
Gosforth	CN2
Grangetown	F41
Granville	I15, R13
Great Britain (1)	D6, BR4
Great Britain (2)	BR5
Great Emperor (1)	D4
Great Emperor (2)	D12, RD5, F32
Greatham	BL13
Guide	L14
Haakon	R8
Hardback	R10
Harmony	R2
Harwich	B15, S13, F7, CN14, BL11
Hector	W3, CN10, F19
Hendon (1)	S8, F3
Hendon (2)	F18
Hendon (3)	F52
Her Majesty	LH3
Hercules	L20, LB5
Hetton	LH6
Hillsider (1)	R31 as *Battleaxe*, LB28
Hillsider (2)	BL20
Holmsider	LB47
Homer	L29, LB11, BL18
Honor	CE1
Houghton	LH8, F34
Hutton	CK14
Imperial Prince	AST3, R4
Impetus	R29
Iron/Seaburn	F10
Ironsider	LB32
Isaac Wilson	R9
James Conley	D3
Joffre	L30, LB12

John Batey	B12, LB14
John Coulson	CN12
Joseph Crosthwaite	CE7, LB22 as *Northsider* (1), BL19
Jupiter	BL7
King Edward VII	I17, R19
Knight Templar	RSA3
Lambton	LH1
Langton	B17, LB16, BL16
Lewis Crosthwaite	CE8, F39 as *Earsdon*
Lion	L4
Little Western	RH4
Liverpool	RH10
Lizard	L10
(Loftus)	R18
Lord Derby	I10
Lord Warden	AST13, F11
Lumley	LH9, F35
Malta	RSA4, D13, RD6, F31
Marsden (1)	S7
Marsden (2)	S14, F8
Marsden (3)	F16
Marsden (4)	F48, LB41
Marty	R14
Maximus	R30, BL23, LB44
Mentor	L27, LB8
Mildred	R22
Mildred II	R26
Miss Roberts	CN5
Momentum	R28
Monarch	L13
Montrose	B3
Monty	R25
N.E.R. No.4	LB/M2
Nestor	See *Africa/Nestor*
Norfolk Hero	I9
North Star	CK4
Northsider (1)	CE7 as *Joseph Crosthwaite*, LB22, BL19
Northsider (2)	LB31
Ocean Bride	D2
Oscar	R7
Overgarth	AHA2, R21
Pactolus	I11
Panther	CN13, F22
Patriot	L6
Pearl	RH9
Pero Gomez	CK12
Pilot	I2
Plover	L28, LB6
Powerful	L7
Premier	R11
President	BR2, F25
Prestwick	F44, LB37
Prince	D10
Prince of Wales	CN8
Privateer	L33
Providence	B6
Quaysider	LB25
Rainbow (1)	CN3
Rainbow (2)	AST14, F14
Rapid (1)	B13, S6
Rapid (2)	R6
Reliant	R27
Renown	S2
Rescue (1)	CK5
Rescue (2)	AST9 as *Brothers* (3), CK9
Rescue (3)	AST16, F13
Robert L. Cook	CK11
Robert Redhead	RD9, F33
Roker	S4, F2
Rosamond	B4
Roughsider	BL17, LB27
Royal Briton	L25, LB10
Royal Charlie	RH7
Ruby	BL2
Ryhope	F42
St. Florence	LB/M1
Scalpa	R23
Scotia (1)	L8
Scotia (2)	L18
Scotland	L15
Seaburn	See *Iron/Seaburn*
Seasider (1)	LB20, BL14
Seasider (2)	CL1
Selina (1)	CE2
Selina (2)	CE3
Sensation	RH8
Silksworth	I7
Sir Colin Campbell	B7
Sir Walter Raleigh	L21
Snowdon	D11, RD4, CK17, F38
Snowdrop	LH2
Souter	S9, F5
Southsider	LB23
Spray	R5
Stag (1)	AST4, I3
Stag (2)	AST7
Stag (3)	AST10, CK10
Stag (4)	I12
Stag (5)	CK15, F37
(Staithes)	R17
Star	B9
Star of Hope	CN6
Steel	BL12
Storm King	B8
Sunderland	S11, F6
Symbol	S12
Taliesin	L23, LB9
Tartar	I1
Test	D7
Thames	RH16, RSA5
Thomas Nicholson	LH7
Tiger	RH2
Titan	B14, LB13
Transit	S15, F1
Triton	AST15, F12
Trover	R15
Tynesider (1)	LB19, BL22
Tynesider (2)	LB46
Ulysses	L16, LB2
United	S10, F4
Venus	B5
Vigilant	BR3
Vivid	L12
W.D.S.	B20, LB17
Wansbeck	CK2
Wards	RH3
Warrior	CE5
Washington	D8, RD2, F30
Waysider	LB21
Wearmouth	F24
Wearsider	LB45
Westsider (1)	LB26
Westsider (2)	LB30, BL24
Wexford	F9
Whitburn (1)	S5
Whitburn (2)	F51, LB35
Wild Rose	L11
William (1)	B1, D1
William (2)	L3
William Fallows	R12
Wonder (1)	R1
Wonder (2)	R16

INDEX
OF
Previous and/or later names of tugs owned by companies included in this history

For	Refer to
Abou Shanab	*Whitburn* (2)
America	*Robert L. Cook*
Archdale	*John Coulson*
Assistant (Built 1914)	*Sunderland*
Assistant (Built 1943)	*Empire Sybil*
Bacalan	*Wonder* (2)
Benachie	*Malta*
Betefdoe	*Robert L. Cook*
Birchrock	*Mildred II*
Brigadier	*Transit*
Brock	*Mentor*
C.111	*Cleadon*
Campaigner	*Battleaxe*
Caswell	*Empire Sybil*
Cervia	*Monty*
Champion	*Robert L. Cook*
Charing Cross	*Westsider* (1)
Coliemore	*Appelsider*
Colonel Thys	*Lewis Crosthwaite*
Conquérant	*Conqueror* (2)
Conqueror	*Criccieth*
Courier	*Albert Edward*
Cyclop	*Cullercoats*
D.P.C.74	*Ryhope*
D.P.C.95	*Grangetown*
D.P.C.99	*Cornhill*
D.P.C.100	*Whitburn* (2)
Diana	*Momentum*
Dundas Cross	*Empire Mustang*
Duneagle	*Empire Mustang*
Dzik	*Empire Andrew*
Elliott & Jeffery	*Panther*
Empire Cherub	*Tynesider* (1)
Empire Paul	*Beamish*
Eversand	*Expatriate*
Fergus Macivor	*Bothal*
(Fjellman)	*(Dawlish)*
Flying Squadron	*Bothal*
Frank Buckland	*Transit*
Fylde	*Plover*
Gaunt	*Gauntlet*
Gigas	*Ashbrooke*
Goole No.1	*Haakon*
Goole No.5	*Cinema Star*
Graythorpe	*Symbol*
Hannah Jolliffe	*Robert Redhead*
Hannah Spearing	*Cornhill*
Headman	*Hillsider* (2)
Hercule	*Champion* (3)
Heros	*Southsider*
Huskisson	*Joseph Crosthwaite*
Hussar	*Dundee*
J.C. Stevenson	*Dunelm* (1)
J.H. Lamey	*Sunderland*
Joachim	*Impetus*
Joseph Constantine	*Westsider* (1)
Jubilant	*Gui[illegible]*
Juno	*Mildred II*
Kastor	*Ryhope*
Knight Prendor	*Great Britain* (2)
Konigstein	*Trover*
Kronos	*Whitburn* (2)
Kyuquot	*St. Florence*
La Morinie	*Waysider*
Lady Elsie	*Monty*
Lady Hazel	*Monty*
Lizzie	*Scalpa*
Marie Lamey	*Momentum*
Mount Viso	*Great Britain* (2)
Mumbles	*Marsden* (2)
N.E.R.No.20	*Anthony Irwing*
(Natalyna)	*(Dawlish)*
Nelson	*Fulwell*
Nemea	*Hercules*
Neptun	*Thames*
Nervion	*Sir Walter Raleigh*
Nisos Syros	*Empire Mustang*
Nore Crest	*Alnwick*
Norman Prince	*Hetton*
Old Trafford	*Reliant*
Peewit	*Plover*
Pendennick	*Dunelm* (2)
Pilot	*Silksworth*
Pluto	*Eastsider*
Polaris	*Grangetown*
Presto	*Empire Sara*
Prince of Hesse	*Wonder* (1)
Queensgarth	*Beamish*
Rex Mellenger	*Beamish*
River Eskimo	*Empire Plane*
S.T.761	*Ryhope*
S.T.764	*Grangetown*
S.T.767	*Cornhill*
S.T.768	*Whitburn* (2)
Salt	*Earl of Beaconsfield* (2)
Salvage Steamer No.2	*George V* (1)
Samson	*Symbol*
Samurai Swordsman	*Wearsider*
(Sapphire)	*(Staithes)*
Sir Thomas	*Great Britain* (2)
Sostis	*Bamburgh*
Storesund	*Impetus*
Taioma	*Empire Jane*
Tamesa	*Monty*
Teazer	*Stag* (1)
Thiela	*Alnmouth*
Thunder Cape	*Beamish*
Towing Wizard	*Quaysider*
Toxteth	*Cleadon*
Transfer	*Transit*
Tuman	*Trover*
Tyne Dock No.1(1)	*Scotia* (2)
Tyne Dock No.2	*Champion* (2)
Tyne Dock No.4	*Enable*
Vegesack	*Exemplar*
Velmar	*Nestor*
Victor (Built 1870)	*Liverpool*
Victor (Built 1906)	*Mart[illegible]*
Vixen	*John Coulson*
Wakendam No.9	*Star of Hope*
Warrior	*Star*
Wellington	*Stag* (5)
West Hyde	*Seasider* (1)
William Gray	*Anthony Irwing*
William Wouldhave	*Stag* (4)

INDEX
OF
Tug Builders

Allsup, W., & Sons, Preston L21
Allsup & Co.Ltd., Preston CN12
American Shipbuilding Co., Buffalo, N.Y. F41, F43, F51
Ardrossan Drydock & Shipbuilding Co.Ltd. L29, L30
Armstrong, Mitchell, Sir W.G., & Co.Ltd., Low Walker L26
Armstrong, Whitworth, Sir W.G., & Co.Ltd., Low Walker RH17
Ash, J., & Co., Cubitt Town AST14

Bodewas, Gebr. G. & H., Scheepwerven N.V., Martenshoek LB24
Boele & Pot, Bolnes CE8
Boele, Wed.C., & Zonen, Slikkerveer LB21
Brodie, North Shields L8
Burdis, J.D., North Shields R5, AST4
Butchard, G., Milton, Nr.Gravesend R7

Chrichton, J., & Co.Ltd., Saltney LB/M1
Cochrane & Sons Ltd., Selby BL20, F24, F/M2, LB/M3
Cooper, J., & Co., North Shields RH11
Cox & Co., Falmouth AST15, F15
Crabtree & Co.Ltd., Great Yarmouth LB20
Cran, John, & Co., Leith CE7
Cumming & Ellis, Inverkeithing CK15, BR5

Decatur Iron & Steel Co., Decatur, Alabama F42
Deutsche Schiff und Maschinebau, A.G. Werk, Seebeck, Wesermünde LB/M4
Dixon, Sir Raylton, & Co., Middlesbrough R17
Dodgin, C.W., & Co., North Shields I11, RSA3, L9
Dowey, James., North Shields AST3
Driver & Hughes, Willington Quay L15
Dudgeon, J. & W., Cubitt Town RH16
Dunston, Richard, Ltd., Thorne F/M1, F/M3, BL17
Dunston, Richard (Hessle) Ltd., Hessle LB29, F49, LB30, F50, LB31, LB32, LB33, F52, LB45, LB46, LB47, CL1

Dyle & Bacalan, Soc.Anon.de Traveaux, Bordeaux R16

Elliott & Jeffery, Cardiff CN13
Eltringham, J.T., South Shields L13, I5, RSA4, AST5, BL7, S2, RH12, I10, L20, LH5, LH3, I4, S15, I16, F9, BL9, BL12

Eltringham, J.T. & Co., South Shields F10, R27, D12

Finch & Co., Chepstow L25

Gourlay Bros. & Co., Dundee B14
Green, R. & H., Blackwall BL4
Gunn, J. & M. & Co., Cardiff L22

Hall, A., & Co.Ltd., Aberdeen F/M4
Hall Russell & Co., Aberdeen CK6
Hall Russell & Co.Ltd., Aberdeen LB19
Harris, P.K. & Sons Ltd., Appledore F44, F45, F46, F47, F48
Harvey & Co., Hayle L14
Heckels Bros., North Shields CE2, CN5
Hepple & Co., North Shields F17, B12, RH19, R12, R9, CN15

Hepple & Co.Ltd., South Shields LH7, LH8, LH9, LH10, B19
Hepple, J., & Son, Low Walker L6
Hepple, T., Low Walker B4, CK4, RH8
Hepple, T., & Co., Low Walker B3, B5, B13, AST16, L10, CK12, B6, LH4

Hepple & Landells, Low Walker B7, L2, CK1
Hepple, W., & Son, Newcastle CK5

Jackson, James., South Shields S10
Jones, Neath CN9
Jones, John & Sons, Liverpool F26, F27

Lawson & Eltringham, South Shields D11
Lobnitz & Co., Renfrew R26
London Drydock Co., Deptford B18
Lübecker Maschinebau, Lubeck R28
Lytham Shipbuilding & Engineering Co.Ltd., Lytham AHA2

McNight, S. & Co., Ayr B17
Maxwell, J.R., South Shields W3, AST9, CN7, AST10, LH6
Murdoch & Murray, Port Glasgow L24
Mützelfeldwerft, GmbH, Cuxhaven R29, LB25, R30

Oliver, Edward., South Shields L18

Pile, Spence & Co., West Hartlepool I13
Pool, Skinner & Williams, Falmouth R22

Readhead, John & Co., South Shields. BR3, L16, CK11, RH14, BR2, CE6
Readhead & Softley, South Shields. BR1, B8, I7
Readhead, Softley & Co., South Shields. D4, BL5, AST13, D9, AST8, D8, CE5, B16, I14

Reid, J. & Co., Port Glasgow. L28
Rennoldson, J.P., South Shields. RH10, BL1, W2
Rennoldson, J.P. & Sons, South Shields. I8, S3, AHA1, W4, CK14, CK16, BL13, F23, RD9, R10, I17, S4, S5, S7, S8, S9, LB26
Rennoldson, J.P. & Sons Ltd., South Shields. S11, RH18, LB/M2, S14, F16, F18, R25

Robertson, A.W. & Co., Canning Town. R11
Robson, John., Gateshead. RH3

Scarr, Joseph & Son., Beverley R20
Schichau, F., Elbing R15
Schlesinger Davis & Co., Wallsend. B20
Scott, Thomas, Goole. CK13
Scott & Sons, Bowling. BL6
Seath, T.B. & Co., Rutherglen. L27
Seebeck, G., A.G., Wesermünde. F/M6
Softley Brothers, North Shields. L31
Softley, John & Co., South Shields. R8
Stephen, A. & Sons Ltd., Linthouse R14, R31
Stobbs, Richard., North Shields L19, L3, RH13, S1, RSA1, RSA2, L11

Tilkin, L. & Co., Liége. S12
Toward, T. & W., Newcastle. L33
Turnbull, John., North Shields. AST6
Tyneside Engine Works Co., Cardiff. L23

van Duijvendijks Sch. N.V., Lekkerkerk. LB23
Vaux, J.H., Harwich. B15
Vuijk, A., & Zonen, Capelle a/d Ysel. R24

Wallsend Slipway & Engineering Co.Ltd., Wallsend. R18
Watson, J.S. Ltd., Gainsborough. RD/M1, F40
Westwood Baillie & Co., Blackwall. I15
Winlo, J. & Co., South Shields. D6
Woodhouse, Andrew., South Shields. D7, AST2, AST11
Wouldhave, William., North Shields. I12
Wouldhave & Johnston, North Shields. CN6, CE3, AST7, CK7

INDEX
OF
Engine Builders

Ailsa Shipbuilding Co.Ltd., Troon — LB19
Aitcheson, Blair Ltd., Glasgow. — L29
Alblasserdam Maschinefabriek, Alblasserdam. — LB21
Allsup, W. & Sons., Preston. — L21, CN12
Almond, J.P., North Shields. — D7, L18, L19, BL2
Amos & Smith Ltd., Hull. — F/M2, LB/M3
Arthur, R., & Son., Newport. — R22

Baird Bros., North Shields. — D12
Baird & Barnsley, North Shields. — CN6, CE3, CK7, W3, AST9, CN7, AST10, D11, LH6
Butchard, G., Milton nr. Gravesend. — R7

Caterpillar Tractor Co., Peoria. — R24
Cochet, Les Hériteurs de G., Paris. — R16
Cox & Co., Falmouth. — AST15, F15
Crabtree & Co. Ltd., Great Yarmouth. — LB20
Cran, John & Co., Leith. — CE7, CK15

Deutsche Schiff und Maschinebau, A.G. Werk, Seebeck, Wesermünde — LB/M4

Earles Shipbuilding & Engineering Co.Ltd., Hull — BL20, F24
Elliott & Jeffery, Cardiff. — CN13

Fawcett, Preston & Co.Ltd., Liverpool. — LB/M1
Findley, R.B., & Son., North Shields. — B12, I12
Fishers Ltd., Paisley. — BL6
Fyenoord, Maats., Rotterdam. — CE8

General Motors Corporation, U.S.A. — F41, F42, F43, F44, F45, F46, F47, F48, F51, F40
Gourlay Bros. & Co., Dundee. — B14
Gunn, J. & M. & Co., Cardiff. — L22

Hall, A., & Co.Ltd., Aberdeen. — F/M4
Hall Russell & Co., Aberdeen. — CK6
Harvey & Co., Hayle. — L14
Heckels Bros., North Shields. — CE2, CN5
Hepple & Co., North Shields. — F17, RH19, R12, R9, CN15
Hepple & Co., South Shields. — L17, BL7, S2, I10, LH5, LH3
Hepple & Co.Ltd., South Shields. — LH7, LH8, R27, LH9, LH10, B19
Hepple, G.P., North Shields. — L33
Hepple, T., Low Walker. — L7, RH8
Hepple, T., & Co., Low Walker. — B13, L10, CK12, B6, AST16, LH4, R8, L17
Hepple & Landells, Low Walker. — B7, B4, CK1, CK4
Hepple & Landells, North Shields. — L2
Hepple, W., & Son., Newcastle. — CK5
Holmes, C.D. & Co.Ltd., Hull. — RD/M1. F40
Huntley, W., & Co., — I13
Hutson & Son, Glasgow. — BR5

Jones, John & Sons., Liverpool. — F27, F26

Kincaid, J.G. & Co.Ltd., Greenock. — L30
Klöckner-Humboldt-Deutz, A.G., Cologne. — R29, LB25, R30, BL17, LB29, LB30, LB31, LB32
Kromhaut Engineering Works, Amsterdam. — R18

Landells, Richard., North Shields. — CK3
Lister, R.A., Ltd., Dursley. — R24
Lobnitz & Co.Ltd., Renfrew. — R26
Lübecker Maschinenbau, Lubeck. — R28
Lytham Shipbuilding & Engineering Co.Ltd., Lytham. — AHA2

Manchester, Sheffield & Lincolnshire Railway Co., New Holland — L9
McKie & Baxter Ltd., Paisley. — F/M1
Marshall, J.D., South Shields. — AST2
Marshall, R.J., South Shields. — L31
Marshall Bros., South Shields. — AST6
Mirrlees Blackstone Ltd., Stamford. — LB45, LB46, LB47, CL1
Mirrlees Blackstone Ltd., Stockport. — LB33, F52
Motoren-Werke Mannheim A.G., Mannheim. — R24
Muir & Houston, Glasgow. — B17

North Eastern Marine Engineering Co.Ltd., Wallsend. — B20

Päglow, Gebr., Hamburg. — LB24

Rait & Lindsay, Glasgow. — L24
Rankin & Blackmore Ltd., Greenock. — L27, L28
Readhead, John & Co., South Shields. — BR3, L16, D6, CK11, RH14, BR2, CE6
Readhead & Softley, South Shields. — B8, RH13, I7
Readhead, Softley & Co., South Shields. — D4, BL5, AST13, AST8, D8, CE5, B16, I14
Rennoldson, J.P., South Shields. — AST3, R5, AST11, RH11, B9, RH10, L13, I5, RSA4, RH12, BL1, L20, W2
Rennoldson, J.P., & Sons., South Shields. — I4, S15, I8, S3, I16, F9, BL9, BL12, AHA1, F10, W4, CK16, BL13, CK14, F23, RD9, R10, I17, S4, S5, S7, S8, RH17, S9, LB26
Rennoldson, J.P., & Sons Ltd., South Shields. — S11, RH18, LB/M2, S14, F16, F18, R25, L31
Richardsons, Westgarth & Co.Ltd., Middlesbrough. — S13
Robertson, A.W., & Co., Canning Town. — R11
Robson, John., Gateshead. — RH3
Ross & Duncan, Glasgow. — L26
Ruston & Hornsby Ltd., Lincoln. — F49, F50

Schichau, F., Elbing. — R15
Scott, T.R., & Co., North Shields. — I11, RSA1, RSA2, L11, RSA3, AST5, L15, AST7
Scott, Thomas., Goole. — CK13
Scott, William., North Shields. — AST4, L3, L8, CN9
Seebeck, G., A.G., Wesermünde. — F/M6
Stephen, A., & Sons Ltd., Linthouse. — R14, R31
Stewart, J., London. — AST14, I15, B15

Tilkin, L. & Co., Liége. — S12
Tyneside Engineering Works Co., Cardiff. — L23, L25

Van Duijvendijks Sch.N.V., Lekkerkerk. — LB23
Wear Engine Works, Sunderland. — S10
Windus & Dunsmore, Millwall — B18
Worsley Mesnes Ironworks Ltd., Wigan. — F/M3

Young, J.A., Blackwall. — D10, RH16

NOTES

NOTES

HENDON